Analog Circuit Simulators for Integrated Circuit
Designers

Mikael Sahrling

# Analog Circuit Simulators for Integrated Circuit Designers

Numerical Recipes in Python

 Springer

Mikael Sahrling
IPG Photonics Inc.
San Jose, CA, USA

ISBN 978-3-030-64208-2      ISBN 978-3-030-64206-8   (eBook)
https://doi.org/10.1007/978-3-030-64206-8

This Springer imprint is published by the registered company Springer Nature Switzerland AG
The registered company address is: Gewerbestrasse 11, 6330 Cham, Switzerland

*For Nancy and Nicole*

# Preface

This book is intended as an aid in using modern simulators encountered in the design of both active and passive electronic circuits. These simulators have been developed over many years and have become quite complex. As a result, it can be hard to get a handle on all the sophisticated options one has at one's disposal as a circuit designer. Literatures describing these innovations are fairly extensive but they often target specialists, mathematicians, and computer scientists, and the level is often such that for a practical user it can be too time consuming to differentiate the advantages of different simulation options. This book goes through the simulation methods from the very basic algorithms and builds up understanding step by step by adding more sophisticated methods on top of what has already been discussed. The basic idea is to do this in such a way that the reader will better understand how simulators work without becoming a simulator design expert. This will be done by looking at fairly simple example codes that illustrate the methods without going into the details of what makes algorithms and code implementations production ready. Throughout the book, we will build a few simple simulators that demonstrate the basics of the lessons learned. The reader can then go out on their own to expand the code and learn more sophisticated techniques. The goal is not to build a full fledge modern simulator; however, the firm belief is that through actually doing numerical experiments, a deeper understanding of the difficulties and potential weaknesses behind modern implementations will develop. With increased understanding of the tools, a more efficient use of real simulators will follow. There are many references supplied for the interested reader wanting to dig deeper. A book such as this is by necessity heavy on examples and exercises to ensure the reader understands the sometimes-subtle workings of simulators and each chapter has plenty of both.

The use of simulators is intimately tied to modeling techniques, particularly of active devices. We therefore include a chapter where we discuss modeling techniques and various limitations that can impact the simulation results.

The final chapter describes good practices when using simulators gained from many years of active design work in the fast-paced semiconductor industry. Here, we address issues like how to handle foundry models including the supplied corner cases, and how to best simulate small circuits up to full chips in an efficient manner, among other topics.

We start with an overview of the basics in Chap. 2, how differential equations are translated into difference equations and common numerical solution algorithms. Nonlinear equations and their solution in terms of Newton-Raphson algorithms are highlighted for the simple one-dimensional case. It will also describe matrix equations and common solution techniques including iterative algorithms. In Chap. 3 we describe modeling techniques of active devices where we discuss mostly CMOS transistors, in particular the popular BSIM model and also the basics of the surface potential model. We further highlight a few common assumptions made by modern foundries when constructing models and what one should look out for as a circuit designer. We address bipolar devices briefly and mention some of the difficulties in their implementation. Chap. 4 discusses linear circuit simulators, and the lessons learned are applied to specific circuit examples as we go through the chapter. Chap. 5 extends the simulator discussion to the non-linear case. This chapter also includes steady-state simulators. The final chapter, Chap. 6, provides practical advice for using simulators. Limitations are highlighted and countermeasures are proposed. Chapter 7 contains a more detailed mathematical background to the operation of simulators, where some of the key theorems and properties of various integration methods are outlined. The appendix contains the complete python codes used in all the examples throughout the book. The python codes will also be available for download.

A book like this cannot be completed without the help and support of many people. First of all my family, Nancy and Nicole have both been unwavering in their loyalty and understanding of the sometimes-lonely task of putting a full manuscript together. My manager, Pirooz Hojabri, has been enthusiastic about the project from the very beginning. Also, my coworkers Dongwei Chen and Vincent Tso have read through the whole manuscript in detail and provided many useful tips for improvement. The same goes for Shahrzad Naraghi and Thomas Geers. Some of the professorial staff at UC Berkeley who I know indirectly have also provided great encouragement for the project. The code snippets have been reviewed by, in particular, Bart Hickman and others, and thanks to their comments, a much more readable code is the result.

San Jose, CA, USA                                                                            Mikael Sahrling

# List of Abbreviations

Common Symbols and Their Meaning

| Symbol | Value | Comment |
|--------|-------|---------|
| $C$ | | Capacitance |
| $C_{gs}$ | | Gate-Source Capacitance |
| $C_{ds}$ | | Drain-Source Capacitance |
| $C_{gd}$ | | Gate-Drain Capacitance |
| $C_{ox}$ | | Oxide Capacitance Per Unit Area |
| $D$ | $=\varepsilon E$ | Electric Flux Density Field |
| $E$ | | Electric Field |
| $\varepsilon$ | | Permittivity |
| $g_m$ | [mho] | Transistor Transconductance |
| $g_o$ | | Transistor Output Conductance |
| $\gamma$ | | Noise Density Correction Factor for CMOS |
| $J_{ij}$ | | Jacobian |
| $j$ | $j^2 = -1$ | Imaginary Factor |
| $L$ | | Inductance |
| $N_a$ | | Acceptor Impurity Density |
| $\omega$ | $=2\pi f$ | Angular Frequency |
| $q$ | $1.602 \cdot 10^{-19}\, C$ | Electronic Charge |
| $Q$ | | Electric Charge |
| $R$ | [ohm] | Resistance |
| $Y$ | [ohm] | Admittance |

# Contents

# About the Author

**Mikael Sahrling** is a principal electronics engineer working on developing high-speed electrical interfaces for the test and measurement and communication industries. He has about 25 years of experience developing integrated circuits. Mikael has worked for many of the leading analog chip companies like Semtech Corporation, Maxim Integrated, Tektronix Inc., and IPG Photonics where he is the lead analog design engineer. His special interest is in high-speed active and passive design with a bandwidth of several tens of gigahertz.

He is the author of "Fast Techniques for Integrated Circuit Design" published by Cambridge University Press in August 2019.

# Chapter 1
# Introduction

**Abstract** This chapter will introduce the general subject matter to the reader with no assumption regarding the reader's experience level. It will highlight the engineering environment where the tools are used and the importance of the subject itself both historically and in a modern engineering environment.

## 1.1  Background

The semiconductor industry is one of the marvels of modern society. We are constantly being presented with novel machines that utilize the extraordinary progress the industry has kept up over some seven decades of intense development. It was around 1965 that one of the Intel founders, Gordon Moore, coined what has become known as Moore's law stating that for every 2 years, the number of transistors in a given area is doubling. It is an exponential scaling law that has kept up until recently (Fig. 1.1).

This is an extraordinary development driven by consumer demand for higher and higher data processing. The advent of streaming where full movies can be viewed as they are downloaded to a particular device requires enormous data delivery rates. In the beginning of this industry epoch, a commercial integrated circuit had a few hundred transistors at most. Compare this to the latest tens of billions of devices in a modern central processing unit (CPU) integrated circuit.

One of the fundamental units one measures these devices with is the so-called transistor gate length. It is as presently known to the author at 3 nm scale with the latest so-called gate-all-around technology. A typical atom has perhaps a size scale of 0.1 nm, just 30 times smaller. We are at the realm of quantum physics for these devices. This is not new. Quantum effects like tunneling, where an electron can appear at the other side of a barrier with a certain probability, has been a source of the so-called leakage current for more than a decade.

Imagine now a modern integrated circuit or chip in industry parlance with some billion devices on it. The key development step of these products is the mask-making step. There might be some 50 masks needed for the latest technology, and on average such masks cost a few 100,000 US dollars each. This cost is a large portion of the design engineering cost, and the total cost of designing such chips can then be of the order tens of millions of US dollars. This is before mass production

© Springer Nature Switzerland AG 2021
M. Sahrling, *Analog Circuit Simulators for Integrated Circuit Designers*,
https://doi.org/10.1007/978-3-030-64206-8_1

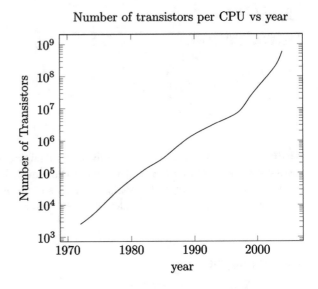

**Fig. 1.1** Moore's law exemplified by the number of transistors per CPU over the years. There are many similar scaling relationships, but this one shows that the transistor count in CPUs double every 2 years or so

starts. If something is wrong with the design, the masks need to be remade. How can one be reasonably sure that such enormous chips will be working when it comes back from the first fabrication run so a costly mask production step is avoided? The answer lies in the use of simulators, both digital, for the core data processing, and analog ones for the interfaces to the outside world among other things.

## 1.2 The Arrival of Simulators

Using simulators to prove out electronic circuitry is an old idea. The earliest attempts can be found in the 1960s where the US Department of Defense supported circuit simulation developments that were proprietary. The modern attempts to make simulators publicly available were started by researchers at the University of California at Berkeley, where the extraordinary vision by a handful of young professors and researchers has developed what became known as Simulation Program with Integrated Circuit Emphasis or SPICE. It was not without controversy in the beginning. A lot of contemporaries felt that simulators could not possibly capture the operations well and the effort was a waste of time. Instead the idea was to prototype the design using breadboards and discrete devices and then miniaturize on a chip. The Berkeley team persisted and it is now considered the original master code, and most simulators after this use many of the same features SPICE introduced to solve numerical problems. In fact the word spice has become a verb in that one often says of simulating a circuit as "spiceing" a circuit. Naturally many decades of innovation have produced a code that is quite a bit more complex than the first versions.

## 1.3   This Book

People often ask me why I would write a book about simulators: "You're a design engineer!" There are basically two answers to this question: first of all, I think it is important to have at least a basic understanding of the tools we use in our work. If one understands their strengths and weaknesses, one becomes a much better user. Second, the inner workings of these tools are really fascinating in their own right and a lot of fun to know more about.

This book concerns the inner workings of analog simulators and describes how various difficult problems that arose in their development were solved. The book shows the basics of the algorithms using Python as the code environment of choice. Python is readily available for virtually all operating systems, and the code examples can thus be run and examined pretty much everywhere. It is also unusually easy to both read and use the code itself. The codes are not intended to be the starting point for any full-fledged professional simulator. They are simple code test benches for the interested reader. All the algorithms behind the precise code implementation can be found in journals and books available to the public. This book will not present any secret algorithms or trade secret to the best of the authors' knowledge. Instead the intention is to show the basic working principles of the core algorithms.

Furthermore, the presentation is based on really simple examples rather than a detailed mathematical theory. This is so the basic ideas can be grasped quicker, and we leave the more detailed mathematical theory and its development to Chap. 7 that follows a more traditional and rigorous approach to the material.

It is the hope of the author that the provided gentle mathematics and plenty of code examples will inspire the reader to explore more on his or her own when it comes to simulators and what to watch out for and learn how to optimize their use. In addition to the provided code examples, the book also presents the results of professional simulators to ensure the reader we are on firm ground. For the curious mind, the basics of the formal mathematical theory behind the simulator algorithms can be found in one of the later chapters. They are there for the sake of completeness, and they contain nothing new for the sophisticated reader.

Toward the end, one of the chapters concerns a more subjective approach to designing analog parts of integrated circuits using analog simulators. It is based on the authors' many years in the semiconductor industry and incorporates lessons learned from one particular perspective. Other experienced engineers can have a somewhat different view on how to approach the design and production of mass-marketable chips. In either case, the approach the author describes is hardly controversial and has resulted in many successful market introductions of integrated circuits.

Chapter 2 presents a review of some of the basic numerical schemes frequently used in simulators. Chapter 3 follows with an overview of device modeling, specifically the complementary metal-oxide-semiconductor (CMOS) transistor. The idea is to point out that modeling can have a large impact on the simulation results and a basic knowledge of how this is done is essential to be an effective user of simulators. Chapters 4 and 5 constitute the core of the book with details on simulator implementation ranging from linear circuits all the way up to periodic steady-state algorithms simulating highly nonlinear phenomena. Chapter 6 highlights through some examples a good way to utilize simulators in a practical design environment and a good way to use corner device models all the way up to large-scale simulations. Finally, Chap. 7 presents more of the details of the mathematical theory behind commonly used simulator algorithms that were just sketched in the earlier chapters.

It is my hope that this book will inspire the reader to explore simulators further. For the interested reader, there are more codes and up-to-date errata and other information at http://www.fastictechniques.com.

# Chapter 2
# Overview of Numerical Methods

**Abstract** This chapter describes a number of topics that are full-fledged research subjects in and of themselves and to do them justice in a just a few pages is not possible. The hope of the author is to present the material in such a way as to wet the readers' appetite. The importance of numerical methods in solving various kinds of problems is paramount in modern product engineering and scientific research. Both the engineering and scientific communities are heavily involved in the development and use of these methodologies. In an effort to contain these vast subject matters, the focus will be on methods the reader is likely to encounter in an electrical engineering context and as such certain types of approximations to differential equations will be highlighted. The same thing goes for matrix equations where we here only use simple examples to highlight the fundamental ideas. The more advanced iterative methods that have been so successful in recent decades are mentioned briefly with an accompanying Python code example. Nonlinear equations and how to solve them efficiently is likewise another intense field of study, and over the years many methods have been developed that are in wide use in the scientific/engineering community today. Here we describe a method that is perhaps the most important to know due to its relative ease of implementation attributed to Isaac Newton, although other researchers such as Joseph Raphson were also involved over the years. Even though the presentation is held at a fundamental level, some basic familiarity with numerical methods, corresponding to an introductory class on the subject, will be helpful since we will be rather brief. We will start the chapter discussing differential equations and how one might implement them numerically. We will discuss implementations of what is called initial value problems where the state is known at a certain moment in time, and from then on, the system develops according to the governing equations. We will present implementations commonly used in circuit simulators. The chapter continues with nonlinear solution methods, and we wrap up the presentation with a description of matrix solvers. Rather than going through the mathematical theories behind these methods, we choose to present the basic ideas using examples, and for the interested reader a much more in-depth discussion of these issues can be found in Chap. 7 and the references at the end of the chapter. The importance of the subject matter presented in this chapter cannot be overstated, and it is the hope of the author the reader will explore the topic more deeply on his or her own.

© Springer Nature Switzerland AG 2021                                                    5
M. Sahrling, *Analog Circuit Simulators for Integrated Circuit Designers*,
https://doi.org/10.1007/978-3-030-64206-8_2

## 2.1  Differential Equations: Difference Equations

One difficulty when using numerical techniques to solve for a systems evolution in time comes when the solution at a particular point in time depends on the previous time points, when there is some kind of memory in the system. Most often this memory effect is expressed in terms of differential equations, and their numerical approximation is often a significant source of error. This section will briefly review such approximations, focusing on those that are common in circuit analysis. With some exceptions, we will discuss what is known as initial value problems in numerical analysis.

We will start with the basic idea behind approximating continuous time/space differential equations and look at simple examples with pseudocode suggestions. This we follow with differential equations common in circuit analysis that arise from typical circuit elements.

### *2.1.1  Initial Value Problems*

Let us start with a simple first-order equation:

$$-\frac{u(t)}{R} = C\frac{du(t)}{dt}, \quad u(0) = 1, \tag{2.1}$$

where $C, R$ are constants. This equation describes a parallel combination of a resistor with value $R$ and a capacitor with value $C$ (Fig. 2.1).

It has a well-known analytical solution:

$$u(t) = e^{-t/(RC)} \tag{2.2}$$

How can we approximate this equation numerically? We recall from basic calculus the derivative is defined as

**Fig. 2.1** RC parallel circuit with initial condition $u(0) = 1$

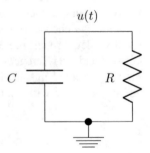

$$\frac{df(t)}{dt} = \lim_{\varepsilon \to 0} \frac{f(t+\varepsilon)-f(t)}{\varepsilon}$$

It is now natural to find a numerical approximation as

$$\frac{df}{dt} \approx \frac{\Delta f}{\Delta t} = \frac{f(t+\Delta t)-f(t)}{\Delta t}$$

With this approximation of the derivative, we find for the differential equation

$$-\frac{u(t)}{R} = C\frac{u(t+\Delta t)-u(t)}{\Delta t}, \quad u(0)=1,$$

or after rewrite

$$u(t+\Delta t) = u(t)\left(1-\frac{\Delta t}{RC}\right), \quad u(0)=1, \tag{2.3}$$

This formulation is knowns as Euler's forward method or sometimes Euler's explicit method. It is perhaps the most straightforward method to implement, but due to a notorious instability problem, where the solution blows up due to numerical errors, it is almost never used in practical situations. We will look at examples of this method in Chap. 4 where the stability issues will be clear.

There are many ways to formulate a differential equation in terms of difference equations, and so many choices are available. Great care must be taken when making the choice. Instability, accuracy, and other factors must be understood and accounted for. We will here discuss an overview of common ways to implement difference equations and mention various issues where appropriate. For a full analysis, we will refer to Chap. 7 and the ample literature, and the reader interested in digging deeper is encouraged to do so.

The problem we just discussed is part of a class of problems known as initial value problems. At some point in time, often chosen as $t = 0$, the solution is known, and the problem amounts to finding the solution at later times. This is a vast field of study, and many applications are found across science and engineering disciplines. One such application is circuit analysis, and in this section, we will discuss a handful of popular algorithms encountered in electrical engineering. Specifically, we will discuss Euler's methods, the trapezoidal method, and lastly the second-order Gear formulations. These last two methods, together with one of the Euler formulations, constitute the vast majority of numerical implementations in circuit simulators today. As we mentioned earlier, here we will just go through a couple of examples to give the reader a better practical understanding of how they work, and for the detailed analysis, we refer to the literature [1–17] and Chap. 7.

The next section discusses first two Euler's formulations and the trapezoidal implementation followed by the second-order Gear formulation. We will as an example consider numerical implementations of this simple equation:

$$i(t) = C \frac{du(t)}{dt} \qquad (2.4)$$

## 2.1.2  Euler's Methods

Perhaps the simplest numerical formulation is what is known as Euler's methods, the forward and backward Euler. We start with the most obvious implementation we just discussed in the previous section, the forward Euler's method:

$$i(t_n) = C \frac{u(t_{n+1}) - u(t_n)}{\Delta t} \rightarrow u(t_{n+1}) = u(t_n) + \Delta t \frac{i(t_n)}{C} \qquad (2.5)$$

This seems simple enough and it is certainly fair game to use. As we mentioned, one soon discovers it easily gets unstable and the solution can blow out of proportion quickly, and one is forced to sometimes take remarkably small timesteps to avoid the problem. We will not discuss the underlying reason here, it is deferred to Chap. 7, but it is fundamentally the reason why it is never used in practical implementations. Instead a really small reformulation fixes the problem:

$$i(t_{n+1}) = C \frac{u(t_{n+1}) - u(t_n)}{\Delta t} \rightarrow u(t_{n+1}) = u(t_n) + \Delta t \frac{i(t_{n+1})}{C} \qquad (2.6)$$

Please note the current is evaluated at the new timestep! We say this is an implicit formulation of the differential equation. The unknowns show up at both sides of the equal sign. This is the key, and it turns out this formulation does not suffer from the type of instability that plagues the forward Euler, and this is known as the backward, or implicit, Euler implementation. Most simulators provide this integration method, and it is fairly straightforward to implement as we shall see later.

## 2.1.3  Trapezoidal Method (Trap)

The trapezoidal method is based on integrating the differential equation over a short time using trapezoids as a way to approximate the functions, hence the name. We will not go into the details, but with this formulation, one finds the derivative is approximated as

$$\frac{df}{dt}(t+\Delta t) \approx 2\frac{f(t+\Delta t)-f(t)}{\Delta t}-\frac{df}{dt}(t) \tag{2.7}$$

It looks similar to the Euler formulation with the exception of the last term and the factor 2 in front of the first term. It has one well-known weakness we will discuss in Chap. 3, namely, an odd ringing which is very characteristic and fairly easy to recognize. Numerically, this is straightforward to implement. We have for the full formulation using the previous example

$$\frac{i(t+\Delta t)}{C}=2\frac{u(t+\Delta t)-u(t)}{\Delta t}-\dot{u}(t), \quad u(0)=1,$$

or after reformulation

$$u(t+\Delta t)=u(t)+\frac{\Delta t}{2}\left(\frac{i(t+\Delta t)}{C}+\frac{i(t)}{C}\right), \quad u(0)=1, \tag{2.8}$$

where we have replaced the voltage derivative at time $t$ with $i(t)/C$. Written this way, one can view it as a Crank-Nicolson scheme [4] where the derivative is evaluated at time $t+\Delta t/2$ and the current is the average of the current at time $t$ and $t+\Delta t$. In this way, both the derivative term of Eq. 2.4 and the left-hand side of Eq. 2.4 are evaluated at the same point in time. This results in better accuracy. Another interesting observation is that one can view the trapezoidal method as an average of the forward and backward Euler methods.

If we leave the current update to another subroutine, a pseudocode will look like

```
subroutine SolveDiffTrap
u(0)=1
deltaT=RC/100
for(i=1, i<N, i++) do
    u(i)=u(i-1)+deltaT*(i(i+1)/C+i(i)/C)/2
end for
end subroutine
```

## 2.1.4 Second-Order Gear Method (Gear2)

C. William Gear [5] published a book in 1971 that has since become one of the classics. He constructs a set of difference equations with varying truncation error that has some really nice properties. He shows how one can systematically formulate a difference equation of higher order and in the early days of SPICE one can incorporate several of them in the numerical solver. In recent decades, the second-order version has proven to be the most widely used, and in modern simulators, the Gear2 option is a standard integration method.

In the second-order Gear implementation of a derivative, we have

$$\frac{df}{dt}(t+\Delta t) \approx \frac{1}{\Delta t}\left(\frac{3}{2}f(t+\Delta t)-2f(t)+\frac{1}{2}f(t-\Delta t)\right).$$

It looks quite a bit different compared to the earlier implementations. This formulation is somewhat less precise than the trap method (we will see why later in Chap. 4) but does not suffer from its "ringing" weakness. Numerically, this is also straightforward to implement; we just need an initial derivative in addition to initial value and knowledge of the solution two timesteps back. We have for the full formulation using the previous example

$$\frac{i(t+\Delta t)}{C} = \frac{3u(t+\Delta t)/2-2u(t)+u(t-\Delta t)/2}{\Delta t}, \quad u(0)=1,$$

or after reformulation

$$u(t+\Delta t)=u(t)\frac{4}{3}-\frac{u(t-\Delta t)}{3}+\frac{2\Delta t}{3C}i(t+\Delta t), \quad u(0)=1, \qquad (2.9)$$

As before, ignoring the current update routine, a pseudocode will look like

```
subroutine SolveDiffGear2
u(0)=1
u(1)=1
deltaT=RC/100
for(i=2, i<N, i++) do
    u(i)=4u(i-1)/3-u(i-2)/(3C) deltaT -2I(i)/(3C)
end for
end subroutine
```

The second-order Gear method is an example of a multistep difference method where we need to know the solution two timesteps back.

## 2.1.5   Summary

These last three methods are overwhelmingly used in practical simulator implementations, and depending on the circuit to be studied, one usually prefers one over the other, and we will show examples of these situations as we go through the book.

## 2.1.6  Solution Methods: Accuracy and Stability

We have looked at four different ways to formulate a differential equation in terms of difference equations. Let us now try to quantify them each in terms of accuracy and stability.

### 2.1.6.1  Accuracy

The accuracy of the solution will depend on the circuit to be solved, the timestep, and which integration method was used. The accuracy of numerical approximations to differential equations can be estimated by their truncation error. One way to construct difference approximations is to use a Taylor series of a function around some point:

$$f(t+\Delta t) = f(t) + f'(t)\Delta t + \frac{1}{2}f''(t)\Delta t^2 + \ldots + \frac{1}{n!}f^n(t)\Delta t^n$$

A first-order accurate approximation will now be found by assuming all higher-order derivatives are negligible, and we are left with

$$f(t+\Delta t) = f(t) + f'(t)\Delta t + \frac{1}{2}f''(t)\Delta t^2 + \ldots$$

which after some rewrite gives

$$f'(t) = \frac{f(t+\Delta t) - f(t)}{\Delta t} + o(\Delta t) \tag{2.10}$$

where the $o$ symbol indicates behavior for small arguments.

This is the Euler forward approximation which is accurate to first order. It means the truncation error is $\sim\Delta t$. One can similarly show that the trapezoidal approximation and Gear2 approximation are of second-order accuracy where the error scales as $\Delta t^2$. What this means practically is that if the solution is a straight line, the first derivative is not changing with time, so the second-order derivative is zero, and all the three methods we have been discussing will have no truncation error in the evaluation of the derivative. If the solution is a second-order polynomial, the Euler method will start to have truncation errors, and one must reduce the timestep to reduce their impact. The trap and Gear2 method can trace out a second-order polynomial with no truncation error but will have errors when higher-order solutions are encountered. Generally the higher the order, the better when it comes to accuracy. The penalty is lengthy evaluations in time, so most simulators use up to third-order approximations in the derivative calculation.

The global accuracy is more difficult to predict since it is very dependent on the circuit being simulated. We will address these issues again in Chap. 4.

> If the solution is a polynomial of order $n$, a difference approximation that is accurate to this order will have no truncation errors.

### 2.1.6.2  Stability

Stability of numerical implementations of differential equations is obviously an important subject, and it has been well studied over many years (see, for example, Chap. 7 and [2–17]). We will here just show what is referred to as the regions of stability for the methods we just discussed in Sects. 2.1.2, 2.1.3, and 2.1.4.

**General Stability Theory**
What one means by a system being stable can be different from application to application. Here, as in most circuit theory books, we will adopt the bounded definition of stability, meaning the solution is considered stable if the signals will remain within a certain limited bound, $B$, at all times. We will follow the common thread in the literature and use simple linear systems to investigate the numerical stability.

To analyze stability, let us then consider a homogeneous linear differential system with constant coefficients

$$\frac{dx}{dt} = Ax$$

where we assume $|x(t)|_{max} \to \; \leq B$ as $t \to \infty$. This equation is referred to as a test equation. A set of linear equations like the test equation has the exact general solution

$$x(t) = \sum_{i=1}^{n} c_i e^{\lambda_i t} v_i$$

where $\lambda_i$, $v_i$ are the eigenvalues/vectors to the system obtained by solving

$$A v_i = \lambda_i v_i$$

$\lambda_i$ are in general complex numbers. If we assume the system is stable, it implies that $Re(\lambda_i) \leq 0 \; \forall \, i$. All eigenvalues are in the left half plane or on the imaginary axes. Let us illustrate in Fig. 2.2. It shows how the solution with positive real eigenvalues explodes beyond bounds due to an exponent that increases linearly with time on the right-hand plane, while eigenvalues with negative real values decreases with time on the left-hand side (Fig. 2.2).

**Fig. 2.2** Complex plane with indicated solution behavior as a function of eigenvalue location. For eigenvalues with no complex part, the behavior is strictly exponential with increasing response of the real value >0 and decreasing otherwise. For complex poles, the solution will be likewise oscillating with increasing or decreasing amplitude depending on the sign of the real part of the eigenvalue

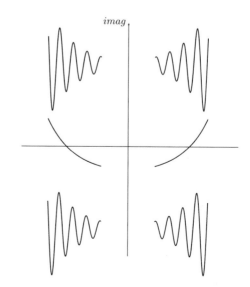

It turns out, [5, 6, 9, 17], that to investigate the stability of a numerical approximation, one simply applies it to the test equation and examines for which timesteps the numerical solution dies down, or stays within a certain bound with time.

**Region of Stability**

The region of stability is now the part of the complex plane defined by the product $\Delta t\, \lambda_i$ where the numerical implementation goes to zero with time. We will look at our four integration methods and simply quote the results; the details are given in, for example, [5, 6, 9, 17] and Chap. 7. The main point here is to get a sense of the region of the stability.

**Euler**

For Euler's implementation, the backward Euler is much more stable than the forward one. The stability region can be found to be in Fig. 2.3. What does this mean? If the system itself is stable, in other words all the eigenvalues have negative real components, the forward Euler implementation can still become unstable. In fact it is only within the small grayed-out circle, the product $\Delta t\, \lambda_i$ will result in stable behavior. One simply has to take smaller timesteps to achieve a stable behavior. We will quantify more in Chap. 4 what this means in practice.

Next we can look at backward Euler (Fig. 2.4). Here the behavior is totally different. The method is stable even if the system itself is not; note the grayed-out region extends into the right-hand plane. We will look at a system in Chap. 4 that behaves this way with Euler's method. The reader is certain to appreciate the peculiar property that with Euler's backward method, a system might simulate just fine but in reality it will become unstable!

**Fig. 2.3** Stability region
for forward Euler's method

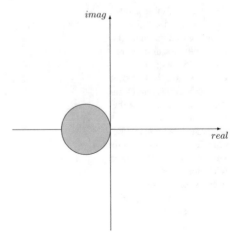

**Fig. 2.4** Stability region
for backward Euler's
method

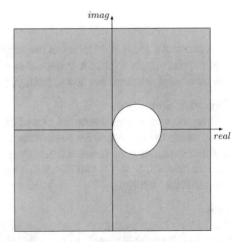

## Trapezoidal

The trapezoidal method suffers occasionally from the well-known trapezoidal ringing we will look at more in Chap. 4. Apart from this effect, the stability region can be found in Fig. 2.5. This region is much more reasonable than for the Euler methods. If the system is stable, the trap method will be stable. If the system is not stable, the trap method will show this also. This is fundamentally why the trapezoidal method is often the preferred way to go.

## Gear2

The second-order Gear method has the stability region shown in Fig. 2.6. The plot is similar to the backward Euler we just discussed, and if the system has some positive real eigenvalues, the method still remains stable.

**Fig. 2.5** Stability region for the trapezoidal method

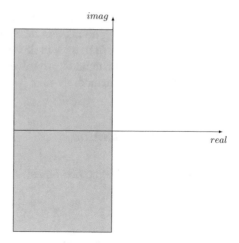

**Fig. 2.6** Stability region for the Gear2 method

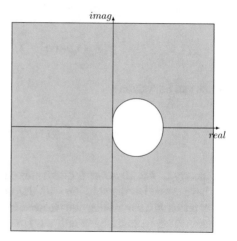

The second-order methods, trapezoidal and Gear2, are the most commonly used methods in circuit analysis. The forward Euler is virtually never used because of the small stability region.

## 2.2 Nonlinear Equations

Solving linear differential equations involves issues like stability and accuracy as we saw in the previous sections. The choice of method involves deciding on the needed accuracy avoiding regions of instability. A large class of equations is nonlinear, and the added difficulty this provides is hard to overstate. Issues like chaotic solutions where the system enters odd modes are not that uncommon. The study of nonlinear systems is important, and much work has been done as the reader is no doubt aware. In this section, we will describe perhaps the most common technique in the circuit simulator community to solve nonlinear initial value problems, namely, the Newton-Raphson method.

### 2.2.1  Newton-Raphson

Let us look at the following one-dimensional equation

$$f(x) = 0 \qquad (2.11)$$

where we need to solve for $x$. Let us do a Taylor expansion around a point $x_0$ which does not solve the equation but is close. We find

$$f(x = x_0 + \Delta x) = f(x_0) + \frac{df}{dx}(x_0)\Delta x = 0 \qquad (2.12)$$

which can be written as

$$\Delta x = -\frac{f(x_0)}{df / dx(x_0)} \qquad (2.13)$$

Clearly, if the higher-order derivatives are zero, we have a linear equation, and we have arrived at the solution with this calculation of $\Delta x$. In practice, the higher-order terms still contribute, and we need to iterate a few times to arrive at the correct solution. The method is generally easy to implement, and it is a standard work horse in almost all numerical implementations of nonlinear solvers. We will also use the multidimensional version later on in Chap. 5. Newton's method is guaranteed to converge if the starting point is close enough to the solution and the solution function is smooth (no discontinuities of the function itself or its derivatives) since then the higher-order derivatives are small. Clearly for functions that are not continuous or the derivative is not continuous, the method will easily get confused. This was historically one of the most notorious reasons transistor models had convergence issues. The continuity of the derivatives has been guaranteed in the latest transistor model implementations (Fig. 2.7).

## 2.3  Matrix Equations

We have until now only discussed a very simple system in terms of solution methods. In fact, the values of interest were a voltage and an associated current. The equation was exactly solvable, and we described a reformulation using difference

**Fig. 2.7** An illustration of the first two iterations of a Newton-Raphson algorithm. The slope of the curve f(x) predicts the next iteration. It is clear from the figure that if the curve is a straight line, the first iteration will find the correct solution. If not, as the iterations approach the correct answer, the curve will look more and more "straight," and eventually the correct answer is found

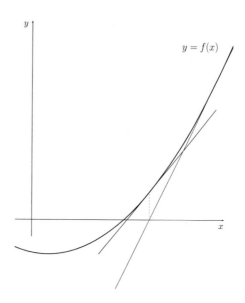

equations. In fact what we had was a one-dimensional matrix equation. If we have more nodes that are interdependent, it should come as no surprise we will end up with a higher-order matrix equation. In Chap. 4 we will show how to set up such a system of equations for circuit networks. The matrix equation occurs naturally in the circuit world since we have a finite number of node voltages and currents we need to solve for. But matrix equations show up almost everywhere when solving systems numerically, [12–15]. Fundamentally, the reason for all these cases of matrix formulations stem from the fact that most simulators assume some kind of grid within which the entities of interest are either constant or slowly changing following some low-order polynomial. There are a finite number of such grid points, so a matrix equation is quite natural. Another way to put it would be the solution space is quantized one way or other, so we end up with a finite number of unknowns.

It now comes as no surprise that matrix equations and their solutions are one of the most important and actively studied research fields. Today's matrix solvers are much superior to the ones that existed just a few decades ago, and they are often readily available for public and noncommercial implementations.

Different systems will have different matrix characteristics. For the circuit analysis systems, the matrix is often sparse in that the number of non-zero entries is of the same size as the number of rows and columns of the matrix. In the mathematical literature, one often says the matrix is of the same order (indicated by an $\mathcal{O}$ sign) as the number of rows/columns $\mathcal{O}(N)$ and not as one would expect $\mathcal{O}(N^2)$. This greatly simplifies the building up of the matrix and also the solution time. We will mention some of these methods in the next few sections and give ample references for the interested reader to study further. Other systems can have dense matrices, and then the set of solution methods will be different.

As we set out to do in the beginning, we will provide mere examples of various methods to give the reader some practical exposure to how they work. In the rest of the book chapters, we will use the matrix inverter routine built into the Python environment, and for the interested reader, we refer to the literature references so one can go and explore matrix inversions on one's own. The proofs and fundamental theorems we leave to Chap. 7 and [11, 18], for example.

### 2.3.1  Basic Matrix Formulation Based on N Unknowns

A simple matrix equation looks like

$$
\begin{cases}
a_{11}x + a_{12}y + a_{13}z = r_1 \\
a_{21}x + a_{22}y + a_{23}z = r_2 \\
a_{31}x + a_{32}y + a_{33}z = r_3
\end{cases}
\tag{2.14}
$$

This is written in matrix form as

$$
\begin{pmatrix}
a_{11} & a_{12} & a_{13} \\
a_{21} & a_{22} & a_{23} \\
a_{31} & a_{32} & a_{33}
\end{pmatrix}
\begin{pmatrix} x \\ y \\ z \end{pmatrix}
=
\begin{pmatrix} r_1 \\ r_2 \\ r_3 \end{pmatrix}
\tag{2.15}
$$

where one often assigns

$$
A =
\begin{pmatrix}
a_{11} & a_{12} & a_{13} \\
a_{21} & a_{22} & a_{23} \\
a_{31} & a_{32} & a_{33}
\end{pmatrix}
\tag{2.16}
$$

$$
x = \begin{pmatrix} x \\ y \\ z \end{pmatrix}
\tag{2.17}
$$

$$
y = \begin{pmatrix} r_1 \\ r_2 \\ r_3 \end{pmatrix}
\tag{2.18}
$$

We have

$$
Ax = y
\tag{2.19}
$$

The $x$-vector is referred to as the *unknown*, whereas $y$ is often called the *right-hand side* (*rhs* for short). The rhs is generally known and the matrix entries are also known. In this case, the matrix has size $3 \times 3$, and if the number of rows is equal to the number of columns, it is a square matrix and is referred to as a potentially solv-able system. If the number of rows is larger than the number of columns, it is an *overdetermined* system and a solution is unlikely; if the number of rows is less than the number of columns, it is an *underdetermined* system, and there is not enough information for a full solution. Here we will restrict ourselves to square matrices with size *nxn*, where *n* also refers to the number of unknowns. The unknowns are solved for with numerical techniques we will discuss briefly in the next few sections.

## 2.3.2   Matrix Solvers

Matrix solvers are an intense research field, and new extraordinarily efficient algo-rithms are invented at a furious pace. Inventing matrix inverters that are stable and accurate for arbitrary problems is a herculean task. Often various simplifying assumptions are made that restrict the problem and make it more manageable. In this section, we will discuss the traditional algorithms followed by a quick review of more recent ones.

For our purposes in this book, the precise matrix inverters we use is of less inter-est in principal. The reader is encouraged to search online for free implementations one can use for self-study. If you are interested in building a simulator for commer-cial purposes, you need to make sure you have the proper permission if you decide to use an external matrix solver.

### 2.3.2.1   Gauss Elimination

Gaussian elimination is a well-known method to solve matrix equations and is often part of elementary classes on linear algebra. It produces both the solution and the matrix inverse at the same time. The inverse matrix tends to suffer from round- off errors and using it to solve for other right-hand sides (rhs) can result in poor accu-racy. Its main weakness is it requires the right-hand side (rhs) to be known and manipulated along with the operations, and for the cases the inverse matrix is not needed, it takes up to three time longer to complete than other methods [11].

We will spend a bit of time on this method since it exemplifies some common issues. Let us consider a set of three equations:

$$\begin{cases} 3x + 2y + z = 7 \\ x + 3y + 2z = 5 \\ 2x + y + 3z = 12 \end{cases}$$

In matrix form, this becomes

$$A x = b$$

where

$$A = \begin{pmatrix} 3 & 2 & 1 \\ 1 & 3 & 2 \\ 2 & 1 & 3 \end{pmatrix} \qquad x = \begin{pmatrix} x \\ y \\ z \end{pmatrix} \qquad b = \begin{pmatrix} 7 \\ 5 \\ 12 \end{pmatrix}$$

It is straightforward to see the following properties are true:

- The rows in the matrix equation are interchangeable. It is just a matter of ordering the equations. The second equation can exchange places with the first, for example, with no change in the solution.
- Naturally we can add rows together, with a weight, at will as long as we also do the same operation on the rhs. For example, row1–3*(row2) will result in a new row $-7y - 5z = -8$ that does not contain any $x$. No information is added or destroyed when this new row is used in place of one of the original two rows.
- One can also interchange any two columns in $A$ but not without also interchanging the corresponding rows in $x$:

$$A = \begin{pmatrix} 3 & 1 & 2 \\ 1 & 2 & 3 \\ 2 & 3 & 1 \end{pmatrix} \rightarrow x = \begin{pmatrix} x \\ z \\ y \end{pmatrix}, \quad b = \begin{pmatrix} 7 \\ 5 \\ 12 \end{pmatrix}$$

Gaussian elimination uses one or more of the steps above to reduce the matrix $A$ to the identity matrix. When this is done, the rhs becomes the solution. Let us solve the matrix equation above to illustrate specifically:

$$\begin{cases} 3x + 2y + z = 7 \\ x + 3y + 2z = 5 \\ 2x + y + 3z = 12 \end{cases} \rightarrow \left\{ R_2 \rightarrow R_2 - \frac{1}{3}R_1 \right\} \rightarrow \begin{cases} 3x + 2y + z = 7 \\ \frac{7}{3}y + \frac{5}{3}z = \frac{8}{3} \\ 2x + y + 3z = 12 \end{cases} \rightarrow$$

$$\left\{ R_3 \rightarrow R_3 - \frac{2}{3}R_1 \right\} \rightarrow \begin{cases} 3x + 2y + z = 7 \\ \frac{7}{3}y + \frac{5}{3}z = \frac{8}{3} \\ -\frac{1}{3}y + \frac{7}{3}z = \frac{22}{3} \end{cases} \rightarrow \left\{ R_3 \rightarrow R_3 + \frac{1}{7}R_2 \right\} \rightarrow$$

$$\begin{cases} 3x + 2y + z = 7 \\ \dfrac{7}{3}y + \dfrac{5}{3}z = \dfrac{8}{3} \\ \dfrac{18}{7}z = \dfrac{54}{7} \end{cases} \rightarrow \left\{ R_3 \rightarrow \dfrac{R_3}{54/21} \right\} \rightarrow$$

The number, 18/7, is here called a pivot. For large numbers >1, this division is not a problem, but imagine if it is close to zero! In that case errors will be amplified and the inversion will fail. Here we find

$$\begin{cases} 3x + 2y + z = 7 \\ \dfrac{7}{3}y + \dfrac{5}{3}z = \dfrac{8}{3} \\ z = 3 \end{cases}$$

which is known as the echelon form of the matrix, or upper triangular form (Wikipedia). Proceeding along these lines and manipulating the equations so the unknowns become trivial to solve for, we find

$$\begin{cases} x = 2 \\ y = -1 \\ z = 3 \end{cases}$$

This whole scheme, incidentally called "Gaussian elimination with no pivoting" [11], works great as long as the factors we divide with, the pivots, to get to the unity matrix are not zero or too close to zero. It almost never works in practice for large matrices. Often the pivots are really small and something needs to be done. Usually, the rows *and* columns are interchanged to try to put a large number in front of the desired variable to avoid the divide-by-zero scenario. In simulators there are often options related to the pivot, like *pivrel* (this sets the biggest relative value of the pivot) and *pivabs* (the smallest acceptable size of a pivot element) that sets how the matrix inverter handles the pivoting. In almost no instances is there a need to adjust these parameters in modern simulators, but it is good to be aware they exist.

In modern simulators, there is rarely, if ever, any need to adjust pivoting-related parameters.

### 2.3.2.2   LU Decomposition

A popular type of matrix solvers is the LU decomposition method. Here one elimi-
nates the problem of the rhs by writing the matrix as a product of two other matrices
$L,U$ such that $A = LU$. The $L$ matrix has the lower-left triangle filled including the
diagonal, and $U$ has the upper-right triangle field with zeros in the diagonal. This
way of writing the equation results in another way of doing back substitution like
earlier, but it no longer depends on the rhs and as long as the matrix is not changing,
it is often a better method. In more detail

$$A x = (LU)x = L(Ux) = b$$

By annotating $y = Ux$, we have a new set of equations

$$Ly = b$$

and

$$Ux = y$$

The advantage here is that solving triangular equations is quite trivial; it is a matter of row by row direct substitution. For the details on how to perform the decomposition for the general case, we refer the interested reader to [11]. Here we can use
the previous example, and we note the Gaussian elimination steps produced a matrix
in echelon form or upper triangle form. This is $U$. We have

$$U = \begin{pmatrix} 3 & 2 & 1 \\ 0 & \dfrac{7}{3} & \dfrac{5}{3} \\ 0 & 0 & \dfrac{18}{7} \end{pmatrix}$$

To find $L$ in this case is straightforward since we know

$$A = LU = \begin{pmatrix} 1 & 0 & 0 \\ l_{21} & 1 & 0 \\ l_{31} & l_{32} & 1 \end{pmatrix} \begin{pmatrix} 3 & 2 & 1 \\ 0 & \dfrac{7}{3} & \dfrac{5}{3} \\ 0 & 0 & \dfrac{18}{7} \end{pmatrix} = \begin{pmatrix} 3 & 2 & 1 \\ 1 & 3 & 2 \\ 2 & 1 & 3 \end{pmatrix}$$

By doing specific matrix multiplications and identifying with the element in
$A$, we find

$$l_{21} = \frac{1}{3}, \; l_{31} = \frac{2}{3} \quad 2l_{31} + \frac{7}{3}l_{32} = 1 \rightarrow l_{32} = -\frac{1}{7}$$

$$L = \begin{pmatrix} 1 & 0 & 0 \\ 1/3 & 1 & 0 \\ 2/3 & -1/7 & 1 \end{pmatrix}$$

The reader will surely note the matrix coefficients are in fact the row operations we did for the Gaussian elimination scheme but with the sign reversed. We find from $L\,y = b$ that

$$y = \begin{pmatrix} 7 \\ 8/3 \\ 54/7 \end{pmatrix}$$

Finally, from $U\,x = y$, we have the solution

$$\begin{pmatrix} 3 & 2 & 1 \\ 0 & 7/3 & 5/3 \\ 0 & 0 & 18/7 \end{pmatrix} \begin{pmatrix} x \\ y \\ z \end{pmatrix} = \begin{pmatrix} 7 \\ 8/3 \\ 54/7 \end{pmatrix}$$

We find by back substitution

$$\begin{pmatrix} x \\ y \\ z \end{pmatrix} = \begin{pmatrix} 2 \\ -1 \\ 3 \end{pmatrix}$$

We can say the $LU$ decomposition is such the $L$ matrix keeps track of the row manipulations, so the rhs need to be adjusted accordingly with $L$. After those operations, it is just a matter of back substitution using $U$ to get the answer. As before, in the Gaussian elimination method, pivoting is a key step when factoring/decomposing the matrix. In modern simulators, there is rarely a need to adjust the parameters of the pivoting algorithm. The advantage here that we mentioned initially is that the decomposition (or factorization) is independent on the rhs. The pivoting in a real implementation is somewhat more subtle than what is apparent in this example (see [11]).

### 2.3.2.3   Iterative Methods

What really have impacted the matrix inversion speeds are iterative methods that can be very advantageous for large sparse systems. The basic idea is to start with a guess, $x_0$, to the solution of $A\,x = b$ and to come up with a way to minimize the residual $y = A(x - x_0)$ by somehow calculating a new solution from $x_1 = x_0 + \beta z_0$, where $z_0$ is some cleverly chosen direction. This goes on until desired accuracy (size of $|y|$) has been reached. There are a number of different ways of doing this that are called conjugate gradient, biconjugate gradient methods, generalized minimum residue methods, etc. (see [11, 18]). These methods are part of a larger class of algorithms that goes under the name Krylov subspace methods. They are generally quite easy to implement numerically that works well for sparse matrices, frequently encountered in circuit systems (see [18] for a good and detailed discussion of these techniques). How they work is somewhat more difficult to understand than the previously discussed techniques, and we provide a bit more details in Chap. 7. Instead we will show an example of such an iterative algorithm known as the GMRes (generalized minimal residual) method. Basically, this method minimizes the residual, $|Ax_m - b|$, using the least-mean-square (lms) method. There are a lot more details to this methodology, and the reader is highly encouraged to study the details more in [11, 18] and Chap. 7. A Python code implementation can be found in Sect. 2.6.1:

*Specific example*
To illustrate these methods, let us apply them to our example matrix equation from the previous section.

$$Ax = b$$

where

$$A = \begin{pmatrix} 3 & 2 & 1 \\ 1 & 3 & 2 \\ 2 & 1 & 3 \end{pmatrix} \qquad x = \begin{pmatrix} x \\ y \\ z \end{pmatrix} \qquad b = \begin{pmatrix} 7 \\ 5 \\ 12 \end{pmatrix}$$

Let us guess the first solution

$$x_0 = \begin{pmatrix} 1 \\ 0 \\ 0 \end{pmatrix}$$

and run this through our python code. We find the following result as a function of iteration (Table 2.1)

This was just a very simple example, and it hardly does the methods any justice. For large sparse matrices common in circuit analysis, the speedup can be significant compared to the direct methods. Please note the remarkable simplicity of the algorithm; it is just a handful of lines in this case.

**Table 2.1**  Residual error as a function of iteration for an implementation of GMRes. The error is less than 1% after 16 iterations

| Iteration | $x$ | $Y$ | $Z$ | $\text{Error} = \dfrac{\left\|(x - x_{exact})\right\|}{\left\|x_{exact}\right\|}$ |
|---|---|---|---|---|
| 1 | 1 | 0 | 0 | 0.886405 |
| 2 | 1.839753 | 1.007704 | 1.175655 | 0.726282 |
| 3 | 2.048373 | 1.266386 | 0.75507 | 0.852666 |
| 4 | 1.126685 | 0.953778 | 2.074173 | 0.623188 |
| 5 | 1.584919 | 0.547796 | 2.073904 | 0.494658 |
| 6 | 2.165718 | −1.17596 | 3.27605 | 0.098063 |
| 7 | 2.080217 | −0.97118 | 2.829113 | 0.051038 |
| 8 | 2.144762 | −1.02899 | 2.809989 | 0.06431 |
| 9 | 2.046818 | −0.93741 | 2.838481 | 0.047957 |
| 10 | 2.033982 | −0.88932 | 2.843 | 0.052136 |
| 11 | 2.043041 | −0.88589 | 2.836684 | 0.054476 |
| 12 | 1.980063 | −0.90503 | 2.891431 | 0.038918 |
| 13 | 1.970487 | −0.91966 | 2.923431 | 0.030693 |
| 14 | 1.974714 | −0.96654 | 2.969754 | 0.013819 |
| 15 | 1.966067 | −0.97667 | 2.983252 | 0.011881 |
| 16 | 1.979837 | −0.98308 | 2.985342 | 0.008052 |
| 17 | 2.001691 | −1.01337 | 3.011812 | 0.004789 |
| 18 | 1.994125 | −0.99197 | 3.003188 | 0.002792 |
| 19 | 1.996385 | −0.99334 | 3.001539 | 0.002067 |
| 20 | 2.004308 | −0.99362 | 2.99015 | 0.003341 |
| ⋮ | | | | |
| 50 | 2.000001 | −1 | 3 | 3.69E-07 |

#### 2.3.2.4    Summary

The main takeaway from this section is that some of these iterative projection algorithms are remarkably easy to implement and the reader is encouraged to do so for their own projects. Just bear in mind, becoming a professional matrix inverter algorithm developer is a huge task, and proper appreciation for the remarkable difficulties involved in the development is warranted. No doubt the reader will quite quickly learn to appreciate the sophisticated art that is matrix inversions.

The solution of matrix equations is a major research field, and a lot of progress is being made on a yearly basis. We have just highlighted some important algorithms and leave the reader to explore more on his or her own. Some simulators have options referring to entities like pivots, and this section has served as a quick reminder of the major steps one encounters when solving these types of equations. The techniques are really at the core of modern simulators, and it is well worth spending time keeping up to date on modern developments.

In the rest of the book, we will use standard built-in matrix inverters in numerical packages like Python, and we will not dig into this subject much further.

## 2.4   Simulator Options Considered

This chapter discussed the following simulator options specifically related to the matrix inversion routines:

- Pivrel
- Pivabs

## 2.5   Summary

This chapter has reviewed the basic of numerical implementations of differential equations. We presented integration methods that are common in electronic circuit analyses. We also gave a brief review of the important Newton-Raphson method to solve nonlinear equations. We presented these in one-dimensional context since that is often easier to build up an intuition as to how these work. In Chaps. 3 and 4, they will be applied to multidimensional systems.

## 2.6   Codes

### 2.6.1   Code 2.7.1

```
"""
Created on Sun Aug  4 17:26:20 2019

@author: msahr
"""

import numpy as np

Niter=50
h = np.zeros((Niter, Niter))
A = [[3,2,1],[1,3,2],[2,1,3]]
b = [7, 5, 12]
x0 = [1, 0, 0]

r = b - np.asarray(np.matmul(A, x0)).reshape(-1)
x = []
v = [0 for i in range(Niter)]

x.append(r)
v[0] = r / np.linalg.norm(r)

for i in range(Niter):
    w = np.asarray(np.matmul(A, v[i])).reshape(-1)
```

```
for j in range(i):
    h[j, i] = np.matmul(v[j], w)
    w = w - h[j, i] * v[j]
if i < Niter-1 :
    h[i + 1, i] = np.linalg.norm(w)
    if (h[i + 1, i] != 0 and i != Niter - 1):
        v[i + 1] = w / h[i + 1, i]

b = np.zeros(Niter)
b[0] = np.linalg.norm(r)

ym = np.linalg.lstsq(h, b,rcond=None)[0]
x.append(np.dot(np.transpose(v), ym) + x0)

print(x)
```

## 2.7  Exercises

1. Examine the GMRes code with different starting vectors for the example in Sect. 2.3.2.3.
2. Examine the forward Euler method, and discuss why it is unstable for such a large input space.

## References

1. Pedro, J., Root, D., Xu, J., & Nunes, L. (2018). *Nonlinear circuit simulation and modeling: fundamentals for microwave design* (The Cambridge RF and microwave engineering series). Cambridge: Cambridge University Press. https://doi.org/10.1017/9781316492963
2. Lapidus, L., & Pinder, G. F. (1999). *Numerical solution of partial differential equations in science and engineering.* New York: John Wiley.
3. Hinch, E. J. (2020). *Think before you compute.* Cambridge: Cambridge University Press.
4. Crank, J., & Nicolson, P. (1947). A practical method for numerical evaluation of solutions of partial differential equations of the heat conduction type. *Proceedings. Cambridge Philological Society, 43*(1), 50–67.
5. Gear, C. W. (1971). *Numerical initial value problems in ordinary differential equations.* Englewood Cliffs: Prentice-Hall.
6. Butcher, J. C. (2008). *Numerical Methods for Ordinary Differential Equations* (2nd ed.). Hobroken: John Wiley & Sons.
7. Kundert, K., White, J., & Sangiovanni-Vicentelli, A. (1990). *Steady-state methods for simulating analog and microwave circuits.* Norwell: Kluwer Academic Publications.
8. Kundert, K. (1995). *The designers guide to spice and spectre.* Norwell: Kluwer Academic Press.
9. Najm, F. N. (2010). *Circuit simulation.* Hobroken: John Wiley & Sons.
10. Bowers, R. L., & Wilson, J. R. (1991). *Numerical modeling in applied physics and astrophysics.* Boston: Jones and Bartlett Publishers.
11. Press, W. H., Teukolsky, S. A., Vetterling, W. T., & Flannery, B. P. (2007). *Numerical recipes.* Cambridge: Cambridge University Press.

12. Allen, M. P., & Tildesley, D. J. (1987). *Computer simulation of liquids*. Oxford: Oxford University Press.
13. Taflove, A., & Hagness, S. C. (2005). *Computational electrodynamics, the finite-difference time-domain method* (3rd ed.). Norwood: Artech House.
14. Gibson, W. C. (2014). *The method of moments in electromagnetics* (2nd ed.). New York: CRC Press.
15. Harrington, R. F. (1993). *Field computation by moment methods*. Piscataway: Wiley-IEEE.
16. Brayton, R. K., Gustavson, F. G., & Hachtel, G. D. (1972). A new efficient algorithm for solving differential-algebraic systems using implicit backward differentiation formulas. *Proceedings of the IEEE, 60*, 98–108.
17. Lambert, J. D. (1991). *Numerical methods for ordinary differential systems*. Chichester: Wiley & Sons.
18. Saad, Y. (2003). *Iterative method for sparse linear systems* (2nd ed.). Philadelphia: Society for Industrial and Applied Mathematics.

# Chapter 3
# Modeling Techniques

**Abstract** In order to fully utilize a simulator, one needs to have an understanding of the devices the simulator is modeling and the difficulties that might hide underneath the hood. This chapter will provide an overview of the CMOS transistor and briefly describe bipolar implementations with associated challenges. We will go into the basic physical description in some detail in order to connect such a model with the first computer models we encounter in Sect. 3.1.4. We will highlight the difficulties and what the designer needs to keep in mind when simulating circuits. We are making the assumption the reader has already encountered the basic physical properties of the transistors, and we present this as a review, and for the details, the reader should consult the references.

## 3.1 CMOS Transistor Model

The CMOS transistor is by far the most used active device in modern integrated circuits. It might seem deceptively simple in its construction, in particular if one is looking at popular text books. It is however quite difficult to model [1–15]. The BSIM (Berkeley short-channel IGFET Model) which exist in several different levels of sophistication can take advantage of hundreds of parameters when characterizing a transistor. In this section, we will motivate why this is so, and we will start with a basic review of the physics of these devices. This is followed by a section on the BSIM models where we follow the model development over the years. The BSIM starting point in the late 1980s is quite similar to the physical model we will discuss in Sect. 3.1.2 and and as such is a good transition from simple analytical modeling to more complex numerical ones. The BSIM development is also an interesting study in how improvements are done gradually as the need arises and how deeper understanding of the transistor devices and their construction develop with time.

© Springer Nature Switzerland AG 2021                                             29
M. Sahrling, *Analog Circuit Simulators for Integrated Circuit Designers*,
https://doi.org/10.1007/978-3-030-64206-8_3

### 3.1.1   CMOS Transistor Basics

Broadly speaking a CMOS transistor changes its behavior significantly as the terminal voltage changes. The device goes through several operating regions where different mechanisms are responsible for its response. This has traditionally caused some problems when modeling the devices which we will describe in Sect. 3.1.4. Here, we start with a broad description of the gate capacitance as we change the gate voltage (see Fig. 3.1).

> To improve accuracy of simulations, bias the transistor in such a way it stays in one region of operation.

The voltage across the structure is set by the parameter $V_g$. For negative voltage, the bottom plate (substrate) will attract the majority carriers, and the capacitance is simply

$$C_g = \frac{4\pi\varepsilon}{t_{ox}} \tag{3.1}$$

where $\varepsilon$ is the permittivity of the insulator material. This mode is referred to as accumulation mode. When the voltage $V_g$ is increasing, it will push away the positive charges and create a so-called space charge region. The result of this is the effective thickness of the capacitor dielectric is increasing, causing a reduction in capacitance. This is called the depletion region. As the voltage $V_g$ is increasing further passed a so-called threshold voltage $V_t$, an inversion layer is eventually created at the interface between the semiconductor material and the insulator. As a result, the effective capacitance is again increasing since the distance between the capaci-

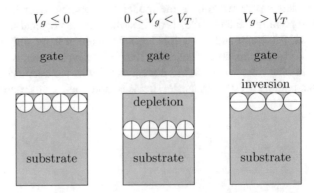

**Fig. 3.1** Simple semiconductor capacitor model as a function of the gate voltage $V_g$. The flatband voltage is assumed to be zero and $V_T$ is the threshold voltage

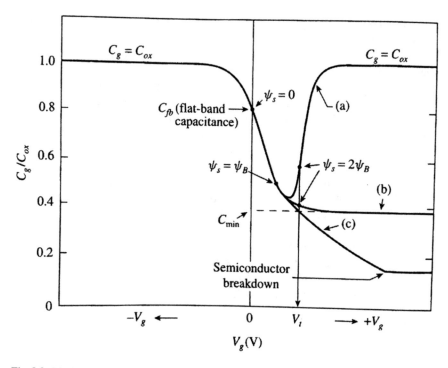

**Fig. 3.2** Idealized capacitance vs voltage. (© [2009] Cambridge University Press. Reprinted, with permission, from Cambridge University Press)

tor plates (where the charges are accumulating) is decreasing. We end up with a capacitance-voltage relationship as in Fig. 3.2.

This is all well-known, and here we use it as an example of the complex behavior of a MOSFET transistor as its terminal voltages are changing. Please note in Fig. 3.2 there is a frequency component to this simple description that changes the capacitance we just outlined.

## 3.1.2  CMOS Transistor Physics

In order to better follow the discussion in the modeling implementation in Chaps. 4 and 5, we will provide a brief overview of a more detailed model of a CMOS transistor where we make a basic two-dimensional approximation. The details we will leave to the references. We will along the way take a look at the important threshold voltage derivation among other things. These calculations will serve to motivate some of the model assumptions we will discuss in Sect. 3.1.4. We will follow what is known as the surface potential approximation. This is a convenient starting point for the transistor behavior in the various operating regions.

Surface potential models are based on the surface potential solution to Poisson's equation for a MOS transistor. The main difference to earlier models, like the first versions of BSIM, is instead of having the transistor operation divided into separate operating regions, the solution is continuous across all regions, and there is no need for various transition region modelings. The disadvantage is that the governing equation is inherently implicit and it can take time to find an acceptable solution. We will do a quick review of the basics in this section. We start with a basic two-dimensional approximation and show a way to solve it implicitly. We finally discuss the consequences for the design engineer. These are well-known calculations and we choose to follow [1] fairly closely.

Let us start the modeling by referring to Fig. 3.3. Here we will follow the estimation analysis flow [4] without spending too much time on the actual calculations since they have been done at other places in great detail.

We denote by $\psi(x, y)$ the intrinsic potential at $(x, y)$ with respect to the intrinsic potential of the bulk substrate (see Fig. 3.3). We define a voltage $V$ which is positive in the channel direction and is zero at the source terminal. There is a more precise technical definition of $V$, and the reader is encouraged to find out in for example [1]. For our review purposes, here it is sufficient to define it the way we have it.

**Simplify**  First let us simplify the problem so it becomes manageable for a simple model.

We will assume:

- The gradual channel approximation – the variation of the electric field in the $y$-direction – is much smaller than the variation in the $x$-direction [5]. We can then reduce the 2D Poisson equation to 1D slices. This approximation is valid for most of the channel regions except beyond pinch off.
- The hole current and generation and recombination currents are negligible. This means the current is the same along the direction of the channel ($y$-direction).
- We assume the voltage $V$ is independent of $x$ so $V = V(y)$. It is motivated by the fact that the current flows mostly in the source-drain or $y$-direction. At the source end $V(0) = 0$ and at the drain, we have $V(y = L) = V_{ds}$
- The inversion layer is really thin leading to what is known as the charge sheet model where the electric field changes abruptly across the inversion region. Denote that $V_{fb}$ is the flatband voltage, $\varepsilon_{si}$ the silicon permittivity, $q$ the electron charge, $N_a$ the doping concentration, and $\psi_s = \psi(0, y)$. The expression for the inversion charge becomes then

$$Q_i = -C_{ox}\left(V_{gs} - V_{fb} - \psi_s\right) - \sqrt{2\varepsilon_{si}qN_a\psi_s} \tag{3.2}$$

These assumptions lead to the following expression for the electron concentration at $(x, y)$.

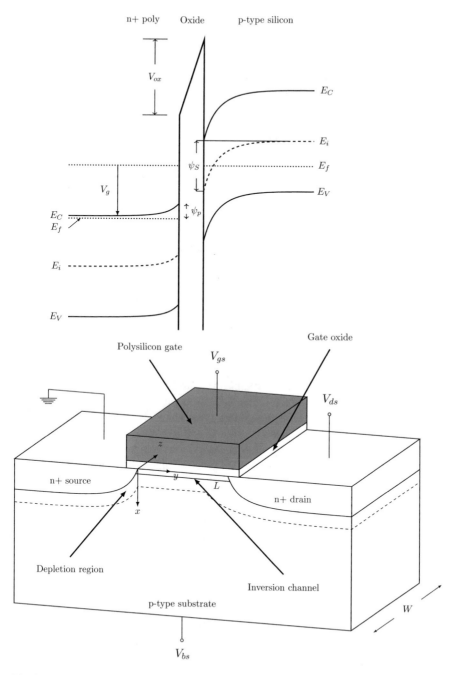

**Fig. 3.3** Basic NMOS approximation of transistor parameters. (© [2009] Cambridge University Press. Reprinted, with permission, from Cambridge University Press)

$$n(x,y) = \frac{n_i^2}{N_a} e^{q(\psi - V)/kT} \tag{3.3}$$

From Maxwell's equation, we know in the static approximation that $E = \nabla \psi$ and $\nabla \cdot D = en(x,y)$ we can find an expression for the electric field:

$$E^2 = \left(\frac{d\psi}{dx}\right)^2 = \frac{2kTN_a}{\varepsilon_{si}}\left(e^{-q\psi/kT} + \frac{q\psi}{kT} - 1\right) \\ + \frac{n_i^2}{N_a^2}\left(e^{-qV/kT}\left(e^{q\psi/kT} - 1\right) - \frac{q\psi}{kT}\right) \tag{3.4}$$

The surface inversion occurs when

$$\psi(0,y) = V(y) + 2\psi_B \tag{3.5}$$

where $2\psi_B = \psi_{s,s}$ is the surface potential at the source terminal.

We can now assert the electron current density at a point $(x,y)$ to be

$$J_n(x,y) = -q\mu_{eff} n(x,y) \frac{dV(y)}{dy} \tag{3.6}$$

where the generation and recombination currents have been neglected. $\mu_{eff}$ is the effective electron mobility in the channel based on averaging over the channel potentials. The total current at a point $y$ along the channel is obtained by multiplying the above equation with the channel width $W$ and integrating over the depth of the current carrying layer. The integration is carried out from $x = 0$ to $x_i$ where $x_i$ is a certain depth into the p-type substrate; the precise value is not important since the integrand is near zero in the bulk of the substrate. We have

$$I_{ds} = W\mu_{eff} \frac{dV(y)}{dy} \int_0^{x_i} qn(x,y)\,dx \tag{3.7}$$

**Solve** The expression inside the integral is simply a charge:

$$Q_i = -\int_0^{x_i} qn(x,y)\,dx \tag{3.8}$$

The expression for

$$I_{ds} = -W \mu_{eff} \frac{dV(y)}{dy} Q_i(V) \qquad (3.9)$$

In the last step, we have changed variables from $y$ to $V$ and can express $Q_i$ as a function of $V$ directly. We now take this idea one step further and express $V = V(\psi_s)$ such that $Q_i = Q_i(\psi_s)$.

One of our simplifications leads to the current in the channel and is independent of $y$. We can therefore integrate this expression and find

$$I_{ds} L = -W \mu_{eff} \int_{\psi_{s,s}}^{\psi_{s,d}} \frac{dV(\psi_s)}{d\psi_s} Q_i(\psi_s) d\psi_s \qquad (3.10)$$

The boundary value $\psi_s$ is determined by two coupled equations

$$V_{gs} - V_{fb} = \psi_s - \frac{Q_s}{C_{ox}} \qquad (3.11)$$

or the gate bias equation

$$Q_s = -\varepsilon_{si} E(\psi_s) \qquad (3.12)$$

or Gauss's law. One finds for the combined equation after some algebra and assuming $q\psi_s/kT \gg 1$

$$V_{gs} = V_{fb} + \psi_s + \frac{\sqrt{2\varepsilon_{si} kT N_a}}{C_{ox}} \sqrt{\frac{q\psi}{kT} + \frac{n_i^2}{N_a^2} e^{q(\psi_s - V)/kT}} \qquad (3.13)$$

This is an implicit equation for $\psi_s(V)$ given the voltages $V_g$, $V_s$. Their complexity requires numerical solution strategies in general. Please note the fact they are valid for any combination $V_{gs}$, $V_{ds}$ and as such do not require specific regions to be highlighted. We can rewrite the equation, so we express $V(\psi_s)$ and we find then

$$V = \psi_s - \frac{kT}{q} \ln \left[ \frac{N_a^2}{n_i^2} \frac{C_{ox}^2 (V_{gs} - V_{fb} - \psi_s)^2}{2\varepsilon_{si} kT N_a} - \frac{q\psi_s}{kT} \right] \qquad (3.14)$$

Its derivative is

$$\frac{dV}{d\psi_s} = 1 + 2\frac{kT}{q} \frac{C_{ox}^2 (V_{gs} - V_{fb} - \psi_s) + \varepsilon_{si} q N_a}{C_{ox}^2 (V_{gs} - V_{fb} - \psi_s)^2 - 2\varepsilon_{si} kT N_a} \qquad (3.15)$$

Substituting this derivative term and the equation for the inversion layer charge $Q_i$, Eq. (3.2) into the expression for $I_{ds}$ and assuming the term $kT/q$ is small, one can integrate the expression analytically and end up with, after some algebra,

$$I_{ds} = -\mu_{eff} \frac{W}{L} \left[ \begin{array}{c} C_{ox}\left(V_{gs} - V_{fb} + \dfrac{kT}{q}\right)\psi_s - \dfrac{1}{2}C_{ox}\psi_s^{\;2} \\[2mm] -\dfrac{2}{3}\sqrt{2\varepsilon_{si}qN_a\psi_s}\,\psi_s^{\;3/2} + \dfrac{kT\sqrt{2\varepsilon_{si}qN_a\psi_s}}{q} \end{array} \right]_{\psi_{s,s}}^{\psi_{s,d}} \tag{3.16}$$

**Verify** Similar expressions can be found in, for example [1],

**Evaluate** As we mentioned earlier, this equation covers all regions of MOSFET operation in a single continuous function. It has become the basis for all surface potential-based compact models for circuit simulations.

These expressions are quite cumbersome and need to be simplified to get specific expressions for various operating regions. We do this here just to show how this surface potential model relates to more common analytical calculations where the various operating regions are discussed. The early BSIM models have a similar division of the mode of operations, and this discussion will serve to show this relationship directly.

We will look at what happens when we break the charge sheet model into piecewise sections.

**Simplify** After the onset of inversion but before saturation, the linear region, we have from Eq. (3.5) that $dV/d\psi = 1$.

**Solve** Applying the intrinsic potential values at the source $\psi_{s,\,s} = 2\psi_B$ and drain $\psi_{s,\,d} = 2\psi_B + V_{ds}$, we obtain the drain current as a function of the gate and drain potentials.

$$I_{ds} = \mu_{eff}C_{ox}\frac{W}{L}\left(V_{gs} - V_{fb} - 2\psi_B - \frac{V_{ds}}{2}\right)V_{ds} - \frac{2\sqrt{2\varepsilon_{si}qN_A}}{3C_{ox}}\left((2\psi_B + V_{ds})^{3/2} - (2\psi_B)^{3/2}\right)$$

$$\tag{3.17}$$

**Verify** This is a well-known expression for the drain current [1].

**Evaluate** This expression can now be expanded in terms of $V_{ds}$, and we find three distinctive regions.

*Linear Region*

The linear region is characterized by $V_{ds}$ being small, and so we can expand the expression for $I_{ds}$ in terms of $V_{ds}$. We find to first order

$$I_{ds} = \mu_{eff} C_{ox} \frac{W}{L} \left( V_{gs} - V_{fb} - 2\psi_B - \frac{\sqrt{4\psi_B \epsilon_{si} q N_A}}{C_{ox}} \right) V_{ds} \qquad (3.18)$$

This can be formulated in terms of a threshold voltage

$$V_T = V_{fb} + 2\psi_B + \frac{\sqrt{4\psi_B \epsilon_{si} q N_A}}{C_{ox}} \qquad (3.19)$$

so that

$$I_{ds} = \mu_{eff} C_{ox} \frac{W}{L} \left( V_{gs} - V_T \right) V_{ds} \qquad (3.20)$$

This is the familiar expression that can be found in most circuit textbooks.

*Parabolic Region*

For larger $V_{ds}$, we need to include the second-order terms also, and then we find

$$I_{ds} = \mu_{eff} C_{ox} \frac{W}{L} \left( \left( V_{gs} - V_T \right) V_{ds} - \frac{m}{2} V_{ds}^2 \right) \qquad (3.21)$$

*Saturation Region*

The expression for $I_{ds}$ in the parabolic region indicates the current increases with $V_{ds}$ until it reaches a maximum where

$$I_{ds} = \mu_{eff} C_{ox} \frac{W}{L} \frac{\left( V_{gs} - V_T \right)^2}{2m} \qquad (3.22)$$

This expression is also familiar as the square law function of $I_{ds}$ vs $V_{gs}$ in the saturation region. These expressions represent an idealized linear to saturation region behavior, and they are a good starting point for constructing models. In fact, the earliest models used small varieties of these expressions, and we will present the BSIM models in the order in which they were constructed where these simple expressions where expanded upon to become much more sophisticated with time.

**Noise Model**

The most convenient noise model for hand calculations is to simply put a noise current source between the drain and source of the transistor as in Fig. 3.4.

**Fig. 3.4** MOS noise
model

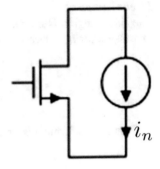

The noise current is modeled as

$$< i_{n,gm} >^2 = 4kT \gamma g_m \left[ \frac{A^2}{Hz} \right]$$

where $\gamma$ is a correction factor that will depend on the operating region of the transistor and the channel length of the transistor.

### 3.1.3   MOSFET Capacitance Modeling Details

Historically, the development of the FET capacitors models started in the early 1970s when the Meyer model was presented. This model included three nonlinear capacitors between the gate (G), source (S), drain (D), and body (B) terminals.

It was realized in the late 1970s that something was off with the model. It did not conserve charge in all situations. Instead an improved model by [13] addressed the issue. Here, more capacitances were added to the model, and the capacitors were assumed to be non-reciprocal, in other words $C_{ij} \neq C_{ji}$. This results in a total of 12 capacitors of which 9 are independent. To see why the capacitors are not necessarily reciprocal, we can use a simple argument. Let us imagine a MOS transistor in saturation. We ignore all side wall capacitances and just focus on the capacitance of the channel. Now, in saturation a small test voltage at the drain will not affect the charge at the gate terminal. However, the opposite is not true: a test voltage at the gate will certainly impact charge at the drain through the transistor action. We can then say that $C_{gd} \neq C_{dg}$, and this non-reciprocal nature of the transistor capacitances was a key to resolve the issue of charge conservation.

We know from basic physics that the rate of change with time of a charge $Q$ is equal to a current. For a FET, we find these four currents

$$i_g = \frac{dQ_g}{dt} \qquad i_b = \frac{dQ_b}{dt} \qquad i_d = \frac{dQ_d}{dt} \qquad i_s = \frac{dQ_s}{dt} \qquad (3.23)$$

We can expand these to get

$$i_g = \frac{\partial Q_g}{\partial v_{gb}} \frac{\partial v_{gb}}{\partial t} + \frac{\partial Q_g}{\partial v_{gd}} \frac{\partial v_{gd}}{\partial t} + \frac{\partial Q_g}{\partial v_{gs}} \frac{\partial v_{gs}}{\partial t}$$

$$i_b = \frac{\partial Q_b}{\partial v_{bg}} \frac{\partial v_{bg}}{\partial t} + \frac{\partial Q_b}{\partial v_{bd}} \frac{\partial v_{bd}}{\partial t} + \frac{\partial Q_b}{\partial v_{bs}} \frac{\partial v_{bs}}{\partial t}$$

$$i_g = \frac{\partial Q_d}{\partial v_{dg}} \frac{\partial v_{dg}}{\partial t} + \frac{\partial Q_d}{\partial v_{db}} \frac{\partial v_{db}}{\partial t} + \frac{\partial Q_d}{\partial v_{ds}} \frac{\partial v_{ds}}{\partial t}$$

$$i_s = \frac{\partial Q_s}{\partial v_{sg}} \frac{\partial v_{sg}}{\partial t} + \frac{\partial Q_s}{\partial v_{sb}} \frac{\partial v_{sb}}{\partial t} + \frac{\partial Q_g}{\partial v_{sd}} \frac{\partial v_{sd}}{\partial t}$$

These equations define 12 nonlinear non-reciprocal capacitances:

$$C_{ij} = \frac{\partial Q_i}{\partial v_{ij}} \quad \text{where} \quad i,j \in \{v,g,s,d\} \tag{3.24}$$

We require for charge conservation that

$$Q_g + Q_b + Q_d + Q_s = 0 \tag{3.25}$$

Which implies, by taking the time derivative, that

$$i_g + i_b + i_d + i_s = 0 \tag{3.26}$$

which should come as no surprise. One can easily show that this means that the capacitances are not all independent. One can in fact show that 3 of the 12 capacitances can be found from the other nine.

We conclude by saying that modern simulator implementations have equations to calculate the following nine capacitors from the model equations

$$\begin{pmatrix} C_{gb} & C_{gd} & C_{gs} \\ C_{bg} & C_{bd} & C_{bs} \\ C_{dg} & C_{db} & C_{ds} \end{pmatrix} \tag{3.27}$$

and derive the remaining three from these nine.

### 3.1.4   BSIM Models

We have spent the last few sections going through the basic physics of the MOS transistor. We are now all set for a review of various models. We will spend significant time on the Berkeley short-channel IGFET Model (BSIM) since it is by far the most used in the industry. We will discuss it from a historical perspective since it is easier to connect that flow with the simple physical model we just described. The earlier versions were quite similar to the simple physics, and as the demand grew, the models became much more sophisticated.

BSIM was developed in the late 1980s to account for what was at the time referred to as short-channel field effect transistors. It is physics based and continuous across the various regions and is perhaps the most used model family today. Over the years as the technology has progressed, so has the modeling effort and at present we are at BSIM6. The model has as of this writing split up into several models, namely, BSIM-BULK, BSIM-CMG (common multi-gate), BSIM-IMG (independent multi-gate), and BSIM-SOI (silicon on insulator). The name BSIM6 is no longer used. We will highlight some of the main features and difficulties with the models and alert the user that at times the default settings from the foundry might be inappropriate for the task at hand. The latest release contains hundreds of parameters, and it is outside the scope of this book to do justice to all the work that has gone into making these models. The reader should consult the references for an in-depth discussion.

This section will start with a basic introduction of the model and highlight some of the basic features. We will then show that for some cases, the default settings should be changed. The reader is encouraged to question all the model settings when doing simulations of in particular unusual sizes and/or operating regions; here we will just point out a couple of such examples.

#### 3.1.4.1   Basic Model

The BSIM model was developed more than 30 years ago by the device physics group at UC Berkeley. We will here describe the features of the model by following the historical development starting with one of the very first papers describing the model, and then we will follow development and improvements over the years. Many books and papers have been written about this model, and we have no room to do the model development full justice. Instead we will start with the initial model and follow the improvements by highlighting some of the new features as we go along. We believe this approach will make the model more accessible than start to describe the complex modern version from the beginning. The goal is to familiarize the reader enough with the model so he/she can explore the details on his/her own.

#### BSIM1 1987

The first paper describing the BSIM model came in 1987. Up until then, the difficulty with modeling CMOS transistors was the different physical effects that drove the currents depending on the bias of the transistor. BSIM was one of the first attempts to take everything into account in one implementation. We will here follow the presentation in [9]. The formulation is based on device physics of small geometry MOS transistors. Special effects included are:

(a) Vertical field dependence of carrier mobility
(b) Carrier velocity saturation
(c) Drain-induced barrier lowering
(d) Depletion charge sharing by the drain and the source
(e) Nonuniform doping for ion-implanted devices
(f) Channel length modulation
(g) Subthreshold conduction
(h) Geometric dependencies

The eight drain current parameters which directly appear in the threshold-voltage and drain-current expressions are as follows:

$V_{FB}$ flatband voltage
$\varphi_s$ surface-inversion potential
$K_1$ body-effect coefficient
$K_2$ source and drain depletion charge sharing coefficient
$\eta$ drain-induced barrier lowering coefficient
$U_0$ vertical field mobility degradation coefficient
$U_1$ velocity saturation coefficient
$\mu_0$ carrier mobility

With these parameters, the threshold voltage is modeled as

$$V_{th} = V_{FB} + \varphi_s + K_1 \sqrt{\varphi_s - V_{BS}} - K_2 (\varphi_s - V_{BS}) - \eta V_{Ds} \tag{3.28}$$

Notice the parameter $\eta$ that models the channel length modulation in addition to the drain-induced barrier lowering effect. This expression should now be compared to what we derived in Sect. 3.1.2. For $V_{BS} = 0$, we see they are very close to each other when we identify the parameter $\varphi_s$ with $2\psi_B$ in Eq. 3.19. The last two terms are not part of our simple model calculation.

Let us now look at the drain current model. In BSIM1 it is divided into separate physical regions depending on the terminal bias points:

1. Cutoff region, $V_{gs} \le V_{th}$

$$I_{DS} = 0$$

2. Triode region, $V_{gs} > V_{th}$ and $0 < V_{ds} < V_{ds, sat}$

$$I_{ds} = \frac{\mu_0}{1 + U_0 (V_{gs} - V_{th})} \frac{C_{ox} W / L}{1 + \dfrac{U_1}{L} V_{ds}} \left( (V_{gs} - V_{th}) V_{ds} - \frac{a}{2} V_{ds}^2 \right)$$

3.  Saturation region $V_{gs} > V_{th}$ and $V_{ds} \geq V_{ds,\,sat}$

$$I_{ds} = \frac{\mu_0}{1+U_0\left(V_{gs}-V_{th}\right)}\frac{C_{ox}W/L}{2aK}\left(V_{gs}-V_{th}\right)^2$$

The weak inversion region was modeled as

$$I_{ds,w} = \frac{I_{exp}I_{limit}}{I_{exp}+I_{limit}} \tag{3.29}$$

where

$$I_{exp} = \mu_0 C_{ox}\frac{W}{L}\left(\frac{kT}{q}\right)^2 e^{1.8}e^{\frac{q}{kT}\left(V_{gs}-V_{th}\right)/n}\left(1-e^{-V_{ds}q/kT}\right) \tag{3.30}$$

and

$$I_{limit} = \frac{\mu_0 C_{ox}}{2}\frac{W}{L}\left(3\frac{kT}{q}\right)^2 \tag{3.31}$$

This approach did not introduce any discontinuity in the derivative between the different regions, and so convergence was much improved. The Newton-Raphson method discussed in Chap. 2 needs the derivatives, and if there are discontinuities, the convergence is bound to have problems.

Comparing these expressions to the analytical model we just studied in Sect. 3.1.2, we see some real resemblance, so this first BSIM attempt was a way to code the analytical model based on real physical parameters.

Notice the output resistance model for the saturation case. The dependence on the current on $V_{ds}$ is only seen through the threshold voltage expression. The later versions of BSIM will bring a significant improvement to this feature.

**BSIM3 1993**
The BSIM3 version of the transistor model brought several improvements to the predictability, in particular concerning the output resistance. This version had several subversions we will not go into here. Instead we will list the major differences to the earlier version. But let us first look at the new threshold model:

$$V_{th} = V_{T0} + K_1\left(\sqrt{\varphi_s - V_{BS}} - \sqrt{\varphi_s}\right) - K_2 V_{BS} - \Delta V_{th} \tag{3.32}$$

Where $V_{T0}$ is the long-channel threshold voltage, the rest of the parameters are the same as in BSIM1. As the reader sees, the parameterization has improved, and the model is based on an analytical solution to a quasi-2d Poisson equation. Notice here the $\Delta V_{th}$ parameter. It is intended to capture short-channel effects and has an exponential dependence on the channel length and a length scale parameter, $l_t = \sqrt{3T_{ox}X_{dep}/\eta}$ . Here, $T_{ox}$ is the thickness of the thin-oxide region, and $X_{dep}$ is the depletion width near the transistor source. The expression for

$$\Delta V_{th} = D_{vt0}\left(e^{-L/2l_t} + 2e^{-L/l_t}\right)\left(2\left(V_{bi} - \varphi_s\right) + V_{ds}\right) \tag{3.33}$$

Notice here the dependence on $T_{ox}$. The thinner we make the oxide, the smaller the impact of short-channel effects! This is an important realization that drove a lot of the later years' development of SOI transistors and finFETs. Let us also look at the drain current model.

**Mobility Model**
Compared to BSIM1 where the mobility was modeled based on a simply physical analytical solution, the BSIM3 version introduced a more complex model to capture effects related to smaller geometries. We will refer to this mobility as $\mu_{eff}$ here and not discuss the details much further.

**Drain-Source Resistance**
The interconnect resistance to the drain and source through the metal contact and the diffusion region is modeled internally and is assumed to be symmetric for drain and source regions.

**Drain-Source Current**
As before in the strong inversion region, we have two subregions, the triode region and the saturation region.

1. Triode region, $V_{gs} > V_{th}$ and $0 < V_{ds} < V_{ds,\,sat}$

$$I_{ds} = \mu_{eff}\,\frac{C_{ox}W/L}{1 + \dfrac{1}{LE_{sat}}V_{ds}}\left(V_{gs} - V_{th} - V_{ds}/2\right)V_{ds}$$

2. Saturation region $V_{gs} > V_{th}$ and $V_{ds} \geq V_{ds,\,sat}$

$$I_{ds} = \mu_{eff}\,\frac{C_{ox}W/L}{2aK}\left(V_{gs} - V_{th}\right)^2\left(1 + \left(V_{ds} - V_{ds,sat}\right)/V_A\right)$$

where $V_A$ is the Early voltage which is introduced to model the output resistance. The weak inversion region is modeled as

$$I_{ds} = \mu_{eff}C_{ox}\frac{W}{L}\left(\frac{kT}{q}\right)^2 e^{\frac{q}{kT}(V_{gs} - V_{off})/n}\left(1 - e^{-V_{ds}q/kT}\right) \tag{3.34}$$

which is similar to BSIM1. However the transition region between weak inversion and strong inversion is modeled separately by introducing two cutoff points in the weak regime and strong inversion regime referred to as $V_{gslow}$, $I_{dslow}$ and $V_{gshigh}$, $I_{dshigh}$. Within these limits, the drain-source current and gate-source voltage are parameterized as

$$I_E = (1-t)^2 I_{dslow} + 2(1-t)tI_p + t^2 I_{dshigh} \tag{3.35}$$

$$V_{gs} = (1-t)^2 V_{gslow} + 2(1-t)tV_p + t^2 V_{gshigh} \tag{3.36}$$

With this simple parameterization, the transition region will be continuous and have continuous first derivatives.

As we just mentioned, the output resistance model is much more sophisticated in BSIM3, and it is modeled by the introduction of the Early voltage $V_A$. This voltage was modeled as being dominated by three separate and independent effects: channel-length modulation (CLM), drain-induced barrier lowering (DIBL), and substrate current induced body effect (SCBE). The effects add like their inverses

$$\frac{1}{V_A} = \frac{1}{V_{ACLM}} + \frac{1}{V_{ADIBL}} + \frac{1}{V_{ASCBE}} \tag{3.37}$$

where

$$V_{ACLM} = I_{dsat} \left( \frac{\partial I_{ds}}{\partial V_{dsat}} \frac{\partial V_{dsat}}{\partial L} \frac{\partial L}{\partial V_{ds}} \right)^{-1} \tag{3.38}$$

$$V_{ACLM} = I_{dsat} \left( \frac{\partial I_{ds}}{\partial V_{dsat}} \frac{\partial V_{dsat}}{\partial V_{th}} \frac{\partial V_{th}}{\partial V_{ds}} \right)^{-1} \tag{3.39}$$

$$V_{ASCBE} = I_{dsat} \left( \frac{\partial I_{ds}}{\partial V_{th}} \frac{\partial V_{th}}{\partial V_{bs}} \frac{\partial V_{bs}}{\partial I_{sub}} \frac{\partial I_{sub}}{\partial V_{ds}} \right)^{-1} \tag{3.40}$$

The BSIM3 model version brought significant improvements to the modeling. The introduction of an analytical solution to a quasi-2D Poisson equation and the much improved output resistance model brought for some 10 years with some sub-model improvements was the standard model for CMOS transistors. The total number of physics-based parameters needed was around 25 in the early versions. We will see shortly this was to change drastically with the release of BSIM4 which we will discuss next.

**Noise Model**

The noise model in BSIM3 can be chosen by the user to be one of three with the parameter *noiMod*. For *noimod = 1*, the noise is modeled as in (the hand calc model) with $\gamma = 2/3$ and the transconductance set by the sum $g_m + g_{mbs} + g_{ds}$. The choice $noiMod = 2$ picks a more sophisticated noise model governed by the inversion channel charge $Q_{inv}$.

$$\frac{4kT\mu_{eff}}{\mu_{eff}\left|Q_{inv}\right|R_{ds}\left(V\right)+L_{eff}{}^{2}}\left|Q_{inv}\right|$$

**BSIM4 2000**

BSIM4 was the fourth generation of models and brought with it hundreds of parameters to describe MOSFET operation. As the scaling down of feature size driven by the famous Moore's law kept decreasing the gate length and width to improve power consumption and increase processing power, the need for a much refined BSIM model became clear. As the previous generation, this generation has plenty of subversions that provide improvements.

**Overview**

We will not dig that deep into the models but just scratch the surface and highlight some of the improvements. There are quite a few improvements in the BSIM4 version compared to BSIM3, and to highlight just a few, we have:

- An improved substrate resistance model.
- The channel thermal noise is much more accurately modeled also featuring an improved model for the induced gate noise.
- A better $1/f$ noise model.
- An improved non-quasi-static (NQS) model that can properly account for NQS effects.
- The drain-source resistance model properly takes into effect asymmetrical and other bias-dependent effects.
- A more accurate gate tunneling model.
- A unified current saturation model that includes all mechanisms of current saturation-velocity saturation, velocity overshoot, and source-end velocity limit.
- The threshold voltage definition has changed to improve subthreshold response.
- The drain-induced barrier lowering (DIBL) effect and ROUT model are improved.
- Just to mention, a few of the many improvements this family of models brought to the field!

Overall there are a few hundred new parameters. We will now describe some of these effects in a bit more detail.

**Threshold Voltage**

First let us look at the threshold voltage:

$$V_{th} = V_{T0} + K_1\left(\sqrt{\varphi_s - V_{BS}} - \sqrt{\varphi_s}\right) - K_2 V_{BS} - \Delta V_{th} \tag{3.41}$$

It is modeled very similarly to earlier. One of the biggest changes is in the expression for short-channel adjustments:

$$\Delta V_{th} = D_{vt0} \frac{0.5}{\cosh \dfrac{L}{l_t} - 1} \left(2\left(V_{bi} - \varphi_s\right) + V_{ds}\right) \tag{3.42}$$

The earlier expression caused a phantom second $V_{th}$ roll-up when $L$ becomes really small. To account for a larger range of technologies, the parameters are further split into subparameters to increase the applicability of the models. We will not go into these details here.

Earlier in BSIM3, the thickness of the gate was such that the effective thickness of the induced charge layer next to the thin oxide layer was much smaller than the gate thickness. In BSIM4 the fact that the gate thickness is much closer to the charge layers, thickness causes the fundamental approximation of surface charge to break down. This introduced several new equations and accompanying parameters.

The physics of the various biasing regions is now much more complicated, and we will briefly touch on a few new effects that need to be modeled.

**Weak Inversion Region**

The drain current in this region is similar to earlier:

$$I_{ds} = \mu_0 \sqrt{\frac{q\epsilon_{si} NDEP}{2\varphi_s}} \frac{W}{L} \left(\frac{kT}{q}\right)^2 e^{\frac{q}{kT}\left(V_{gs} - V_{th} - V_{off}\right)/n} \left(1 - e^{-V_{ds}q/kT}\right) \tag{3.43}$$

**Gate Tunneling Effect**

As the gate thickness continued to shrink, a quantum mechanical effect known as tunneling started to become important. If the reader recalls the modern physics classes, an electron can be seen as either particle or a wave, and depending on the particular situation, one of these natures is apparent. In the wave case, one can show by solving Schroedinger's equation the electron wave function exist in the oxide region and even has a finite size in the transistor body. There is then a certain likelihood the electron will be able to exist beyond the gate; it "tunnels" through the thin-oxide region. This is a purely quantum mechanical effect, and it can cause significant leakage loss. This is taking into account in BSIM4.

**Unified Mobility Model**

The mobility is obviously a critical parameter since it directly affects the drain current. In BSIM4 a new universal mobility was introduced with the parameter MOBMOD = 2.

**Drain-Source Current**

BSIM4 introduced a uniform current model for the triode and saturation regions

$$I_{ds} = \frac{I_{ds0}}{\left(1+\dfrac{R_{ds}I_{ds0}}{V_{dseff}}\right)} NF\left(1+\frac{1}{C_{clm}}\ln\frac{V_A}{V_{Asat}}\right) f\left(V_{ADIBL},V_{ADITS},V_{ASCBE}\right) \qquad (3.44)$$

where

$$I_{ds0} = \mu_{eff} W Q_{ch0} \frac{V_{ds}}{L} \frac{\left(1-\dfrac{V_{ds}}{2V_b}\right)}{1+\dfrac{V_{ds}}{E_{sat}L}} \qquad (3.45)$$

## Noise Model
The noise model for a transistor consists of the thermal noise of the channel as well as flicker (1/$f$) noise.

### 1/$f$ noise
The flicker noise is modeled using two kinds of approximations: one where the expressions are simple and convenient to use for hand calculations and another referred to as the unified model takes more of the detailed physics behind the phenomenon into account including what operating region the device is in. The choice is up to the user, via the *fnoimod* parameter, and it is often instructive to take a look at the models provided by the foundry to see which one is used.

### Thermal Noise
The thermal noise modeling is also controlled by a user parameter *tnoimod* that chooses between three sets of approximations. The first, enabled by setting *tnoimod* = 0, is similar to what was already in BSIM3. The second, referred to as the *holistic* model, is accessed by setting *tnoimod* = 1. It is modeled as a noise current between source and drain, and in addition there is a noise voltage source in series with the transistor source terminal. The model is called holistic because all short-channel effects and velocity saturation effects are automatically included. Finally the *tnoimod* = 2 option uses two noise currents to model the noise where instead of a noise voltage source at the source terminal, a noise current is inserted between the gate and source terminals of the transistor. The published correlation with measurements [12, 15] shows good correlation with *tnoimod* set to 2.

### Other Noise Sources
The BSIM4 model also includes other noise sources stemming from drain-source resistances and shot noise due to gate tunneling currents.

## Model Parameter Tweaking
It is important to be aware of the modeling and its limitations since often the foundry will provide models with a standard set of parameters and parameter flags that work in most cases. However, oftentimes if an unusual size is needed for example, this

standard parameter set might not sufficient. The informed engineer will then go and adjust these flags/parameters on his/her own. A book like this cannot cover every parameter in the latest model generations. It would be a large volume and unnecessary since this type of information is already out there. Instead we will highlight some effects where the author and the team around him have encountered discrepancies with measurements. Hopefully this will encourage the reader to boldly investigate the appropriateness of the model setup for his/her particular application.

**Non-quasi-static Effect**

A FET channel can be modeled as a transmission line with a string of resistors connected together and from each interconnect point a capacitor to gnd. This can cause a delay in the propagation of the signal if the channel is wide. In the short channel region, this can often be simplified as a lumped model with one resistor and one capacitor. This effect is controlled with a switch one for transient simulations and one for AC simulations. Oftentimes a foundry model will have this NQS switch set to zero, meaning the distributed effect is ignored. This will help with the simulation time also. However, occasionally one might need a long channel device perhaps as a capacitor in a part of the circuit carrying high-speed signals, and then the NQS switch needs to be enabled so the distributed effect is taken into account. Failure to do so can result in overoptimistic delay of such a capacitor.

**Gate Resistance Model**

The gate can have significant resistance, in particular the small geometry CMOS transistors. BSIM4 includes a switch to take such a resistance into account; it is called rgatemod. Setting this parameter to 0 means no internal gate resistance is generated. In a modern PDK, the transistor is often instantiated in a passive network that includes a gate resistor, so oftentimes this internal modeling effect is disabled. For more information on how it is used, see [15].

**Substrate Resistance Network**

The substrate resistance can be modeled in BSIM4, and this is an important parameter for high-speed applications. The model switch is called rbodymod. When investigating a new technology, it is often worthwhile to see if this model switch is enabled or not. For more information, see [15].

**Charge Partitioning Model**

The charge partitioning tells how the inversion layer charge is divided into source and drain charge. There are typically three options: 50%/50%, 40%/60%, and 0%/100%. The 40%/60% model is considered the most physical [13]. The actual split will depend on the node voltages, something that is hard to solve analytically [13]. Depending on the application, the partitioning model used might give the wrong answer.

**Summary**

We have looked briefly at the modeling of BSIM4. We have touched on a few of the basic modeling assumptions and compared to the earlier BSIM versions. We see a significant increase in the sophistication has occurred, and many subtle effects that are important for small geometry CMOS can now be captured with reasonable accuracy. We have also highlighted some model switches that can be very important for correct behavior and correlation with silicon. The reader is highly encouraged to study the model files provided by the factory and make his/her own judgment on what the suitable parameter (switches) should be for the application under consideration.

### 3.1.4.2  BSIM6

As the technology has gone to more sophisticated physical implementations like system-on-insulator (SOI) and FinFET structures, the BSIM model team has been very busy following along. The latest generation of model under the BSIM6 umbrella contains many more implementations targeting these specific technologies and others. Due to space limitations, we will here only briefly mention what they do and leave the details to be explored by the reader. The reader is highly encouraged to study these implementations in detail along the lines we described in the previous sections.

BSIM-BULK: This is the new BULK BSIM model

BSIM-SOI: This is a compact model for SOI structures, based on BSIM3v3

BSIM-IMG: BSIM-IMG (independent multi-gate) model has been developed to model the electrical characteristics of the independent double-gate structures like ultra-thin body and BOX SOI transistors (UTBB). It allows different front- and back-gate voltages, work functions, dielectric thicknesses, and dielectric constants.

BSIM-CMG: (common multi-gate) is a compact model for the class of common multi-gate FETs. Physical surface potential-based formulations are derived for both intrinsic and extrinsic models with finite body doping. The surface potentials at the source and drain ends are solved analytically with poly-depletion and quantum mechanical effects. The effect of finite body doping is captured through a perturbation approach. The analytic surface potential solution agrees with 2D device simulation results well.

## 3.2  Bipolar Transistors

A bipolar transistor operation is quite different from a CMOS transistor. Traditionally the speed of a SiGe bipolar transistor was much faster than a CMOS transistor, but with the introduction of small geometry CMOS, the speeds of the two technologies are now more comparable. Perhaps the biggest advantage of CMOS is the fact that it is complementary in nature; there is a PMOS transistor that is just as fast as the NMOS. There are then many more architecture choices available.

One of the hardest aspects of implementing bipolar devices in a simulator is the strong exponential dependence of current vs voltage. This can easily cause blowup of solutions since large steps in the voltage cause an overflow in the updated current. We will see examples of this behavior later in Chap. 5. There are ways to work around this problem by, for example, limiting the voltage excursion to some number or construct a new exponential function that is not exponential beyond a certain input.

We start this section with a brief description of the general behavior and follow this with a discussion of a few popular models in Sects. 3.2.2, 3.2.3, 3.2.4, and 3.2.5.

### 3.2.1   General Behavior

The physics of bipolar transistors is quite complex. One advantage is that the physics is the same for many orders of magnitude of output current, and this gives rise to fewer discontinuities in model which helps with convergence. We will not have space to go through the details here but rather just quote the fundamental results. The reader can find a good discussion in [2].

A bipolar transistor has a common symbol exemplified in Fig. 3.5

The basic relationship between the output collector current and the base emitter voltage is the well-known formulae

$$I_C = I_s \left( e^{q V_{BE}/kT} - 1 \right)$$
$$I_B = I_C / \beta$$
$$I_E = I_C + I_B$$

The factor $\beta$ is the current amplification factor.

### 3.2.2   Ebers-Moll Model

A more complicated model and one of the first that was developed is called the Ebers-Moll model [14]. It was originally a static nonlinear model only but was refined over the years to include effects like charge storage, variation with current, and base-width modulation (causing finite output conductance). A large signal model describing the effect of capacitances and parasitic resistors such as it was implemented in spice in the early years can be found in Fig. 3.6.

The basic current relationship for the collector and emitter node is much more complicated than we indicated in Sect. 3.2.1 and is given by

**Fig. 3.5** A bipolar
transistor symbol. It
consists of three main
terminals, base, collector,
and emitter

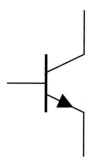

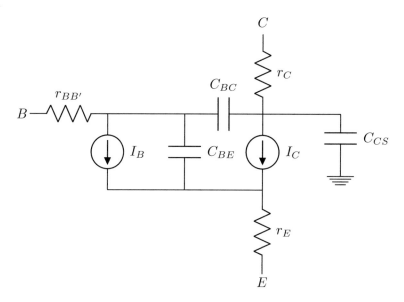

**Fig. 3.6** SPICE BJT large signal model. (© [1987] Republished with permission from McGraw
Hill "*Semiconductor device modeling with SPICE*", Massobrio G., Antognetti P.; permission conveyed through Copyright Clearance Center)

$$I_C = I_S \left[ \left( e^{qV_{BE}/kT} - e^{qV_{BC}/kT} \right) \left( 1 - \frac{V_{BC}}{V_A} \right) - \frac{1}{\beta_R} \left( e^{qV_{BC}/kT} - 1 \right) \right] + \left[ V_{BE} - \left( 1 + \frac{1}{\beta_R} \right) V_{BC} \right] G_{min}$$

$$I_E = I_S \left[ \frac{1}{\beta_F} \left( e^{qV_{BE}/kT} - 1 \right) + \frac{1}{\beta_R} \left( e^{qV_{BC}/kT} - 1 \right) \right] + \left[ \frac{V_{BE}}{\beta_F} + \frac{V_{BC}}{\beta_R} \right] G_{min}$$

### 3.2.3   Gummel-Poon Model

The Gummel-Poon model was the workhorse for many years and is now considered more or less obsolete. It expanded on the Ebers-Moll model by also taking into account effects of low currents and at high-level injection (Fig. 3.7) show the equivalent circuit of the *Gummel-Poon* model. The transistor operation is divided into four operating regions: normal active region, inverse region, saturated region, and off region.

*Normal Active Region*
Here, $V_{be} > -5n_f kT/q$ and $V_{bc} \leq -5\,n_f kT/q$. We then have

$$I_c = \frac{I_s}{q_b}\left(e^{\frac{qV_{be}}{n_f kT}} + \frac{q_b}{\beta_r}\right) + C_4 I_s + \left(\frac{V_{be}}{q_b} - \left(\frac{1}{q_b} + \frac{1}{\beta_r}\right)V_{bc}\right)G_{min}$$

$$I_c = I_s\left(\frac{e^{\frac{qV_{be}}{n_f kT}}}{\beta_f} - \frac{1}{\beta_f} - \frac{1}{\beta_r}\right) + C_2 I_s\left(e^{\frac{qV_{be}}{n_f kT}} - 1\right) - C_4 I_s + \left(\frac{V_{be}}{\beta_f} + \frac{V_{bc}}{\beta_r}\right)G_{min} \quad (3.46)$$

*Inverse Region*
Here, $V_{be} \leq -5n_f kT/q$ and $V_{bc} > -5\,n_f kT/q$. We then have

$$I_c = -\frac{I_s}{q_b}\left(e^{\frac{qV_{bc}}{n_f kT}} + \frac{q_b}{\beta_r}\left(e^{\frac{qV_{bc}}{n_f kT}} - 1\right)\right) + C_4 I_s\left(e^{\frac{qV_{bc}}{n_f kT}} - 1\right) + \left(\frac{V_{be}}{q_b} - \left(\frac{1}{q_b} + \frac{1}{\beta_r}\right)V_{bc}\right)G_{min}$$

$$I_c = I_s\left(\frac{1}{\beta_f} - \frac{1}{\beta_r}\left(e^{\frac{qV_{bc}}{n_f kT}} - 1\right)\right) - C_2 I_s + C_4 I_s\left(e^{\frac{qV_{bc}}{n_f kT}} - 1\right) + \left(\frac{V_{be}}{\beta_f} + \frac{V_{bc}}{\beta_r}\right)G_{min}$$

*Saturated Region*
Here, $V_{be} > -5n_f kT/q$ and $V_{bc} > -5\,n_f kT/q$. We then have

$$I_c = \frac{I_s}{q_b}\left(e^{\frac{qV_{be}}{n_f kT}} - e^{\frac{qV_{bc}}{n_f kT}} - \frac{q_b}{\beta_r}\left(e^{\frac{qV_{bc}}{n_f kT}} - 1\right)\right) - C_4 I_s\left(e^{\frac{qV_{bc}}{n_f kT}} - 1\right) + \left(\frac{V_{be}}{q_b} - \left(\frac{1}{q_b} + \frac{1}{\beta_r}\right)V_{bc}\right)G_{min}$$

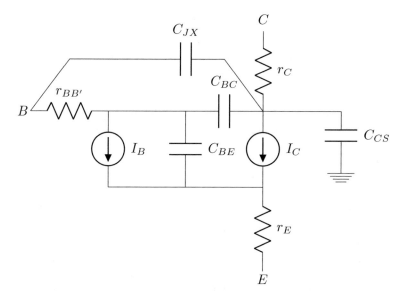

**Fig. 3.7** Gummel-Poon large signal model showing the distributed base-collector capacitance $C_{JX}$. (© [1987] Republished with permission from McGraw Hill *"Semiconductor device modeling with SPICE"*, Massobrio G, Antognetti P; permission conveyed through Copyright Clearance Center)

$$I_b = I_s \left( \frac{1}{\beta_f} \left( e^{\frac{qV_{be}}{n_f kT}} - 1 \right) + \frac{1}{\beta_r} \left( e^{\frac{qV_{bc}}{n_f kT}} - 1 \right) \right)$$

*Off Region*
Here, $V_{be} \leq -5n_f kT/q$ and $V_{bc} \leq -5\,n_f kT/q$. We then have

$$I_c = \frac{I_s}{\beta_r} + C_4 I_s + \left( \frac{V_{be}}{q_b} - \left( \frac{1}{q_b} + \frac{1}{\beta_r} \right) V_{bc} \right) G_{min}$$

$$I_b = -I_s \frac{\beta_f + \beta_r}{\beta_f \beta_r} - I_s \left( C_2 + C_4 \right) + \left( \frac{V_{be}}{\beta_f} + \frac{V_{bc}}{\beta_r} \right) G_{min}$$

In all these equations, $G_{min}$ represents the minimal conductance in parallel to each pn-junction.

### 3.2.4   Hi-Current Model (HiCUM)

The high-current model (HiCUM) is a modification to the basic Gummel-Poon model we just discussed. A few highlights of the improvements over Gummel-Poon models are:

- One of the problems with Gummel-Poon is the insensitivity to effects around the emitter periphery. These types of effects can play a significant role in modern high-speed transistors.
- Another issue is the distributed nature of the external base-collector region.
- Also the high-frequency small signal current crowding in the emitter needs to be addressed.
- Other convenient improvements involve the base-emitter isolation capacitance and the BC oxide capacitance.
- Compared to other models, the internal series resistance is taken into account by the model equations to some degree. This approach avoids the need for an internal resistor but also saves one node and makes the computational expense less.
- There is the possibility of a substrate parasitic transistor turning on at very low CE voltages, referred to as saturation region or even hard saturation region. Such a parasitic transistor is being taken into account by a simple transport model.

### 3.2.5   VBIC Model

VBIC is a bipolar junction transistor (BJT) model that was developed as a public domain replacement for the SPICE Gummel-Poon (SGP) model. VBIC is designed to be as similar as possible to the SGP model yet overcomes its major deficiencies. VBIC improvements on SGP:

- Improved Early effect modeling
- Quasi-saturation modeling
- Parasitic substrate transistor modeling
- Parasitic fixed (oxide) capacitance modeling
- Includes an avalanche multiplication model
- Improved temperature modeling
- Base current decoupled from collector current
- Electrothermal modeling
- Smooth, continuous model

## 3.3   Model Options Considered

In this chapter, we discussed the following BSIM model options specifically:

- Non-quasi-static (NQS) model
- Gate resistance model
- Substrate resistance model

## 3.4 Transistor Models Used

Implementing a full BSIM model is out of the scope for this book. Instead we will use simpler models to illustrate various aspects of analog simulators. We will use two different models for CMOS transistors and a simple exponential model for bipolar transistors.

### 3.4.1 CMOS Transistor Model 1

The new element is a VCCS that has a transfer function $g_m = Kv_g^2$. This is perhaps the simplest nonlinearity one can think of. The reader, no doubt, recognizes this as close to a CMOS transistor transfer function with a threshold voltage equal to zero. This is no coincidence. In order to proceed, we need to take a look at the definitions of the direction of currents as they apply to NMOS and PMOS transistors, respectively. Let us look at Fig. 3.8 where the direction of the source and drain currents is indicated. The convention is that a current going into a device (or subcircuit) is positive; outgoing currents are negative. For the NMOS transistor, we have then that the drain current

$$I_d = K\left(V_g - V_s\right)^2 = KV_{gs}^2 \tag{3.47}$$

For the PMOS, $I_d$ becomes negative:

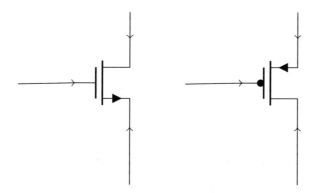

**Fig. 3.8** PMOS and NMOS with drain and source currents defined

$$I_d = -K\left(V_s - V_g\right)^2 = KV_{sg}^{\;2} \tag{3.48}$$

Note the subtle change of subscript for the PMOS. For this model, it has no significance, but it will play a role in the next model 2.

### 3.4.2   CMOS Transistor Model 2

The transistor model we have used so far was just a very basic one so we could illustrate how to implement nonlinear solvers. The model has infinite output impedance and no linear region, and what is worse it turns on reversely as the gate-source voltage goes below zero. It is not very useful. Let us now build a model that is a bit more realistic, still far from the BSIM quality. Let us use a drain current that varies with the port voltages like

$$I_d = \begin{cases} 0, V_{gs} - V_T < 0 \\ 2K\left(\left(V_{gs} - V_T\right)V_{ds} - \dfrac{1}{2}V_{ds}^{\;2}\right), & V_{ds} < V_{gs} - V_T \\ K\left(V_{gs} - V_T\right)^2\left(1 + \lambda V_{ds}\right) \end{cases} \tag{3.49}$$

where $V_T$ is the threshold voltage we here model as just a constant voltage, $K$ is the same constant we used earlier, and finally $\lambda$ is the parameter controlling the drain output impedance in saturation. The equation for a PMOS transistor will be very similar with some sign changes in line with what we saw for the simpler CMOS model:

$$I_d = \begin{cases} 0, V_{sg} - V_T < 0 \\ -2K\left(\left(V_{sg} - V_T\right)V_{sd} - \dfrac{1}{2}V_{sd}^{\;2}\right), V_{sd} < V_{sg} - V_T \\ -K\left(V_{sg} - V_T\right)^2\left(1 + \lambda V_{sd}\right) \end{cases} \tag{3.50}$$

### 3.4.3   Bipolar Transistor Model 3

The bipolar model we will use is given by the following expression

$$I_C = -I_E = \begin{cases} I_0 \exp\left(\dfrac{V_{BE}q}{kT}\right)\left(1 + \dfrac{V_{CE}}{V_A}\right), V_{BE} > 0 \\ 0, V_{BE} \le 0 \end{cases} \tag{3.51}$$

where we have used $V_A$ to denote the Early voltage. The model is similar in structure to its CMOS counterpart, but there are fewer regions. The exponential nature of this equation will cause some interesting problems in the implementation sections later.

## 3.5  Summary

We have in this chapter looked at basic device physics for key components of modern process technologies. Our premise is the designer needs to be aware of how the modeling is done in order to simulate the devices properly. After the basic physics review, we outlined how models are made and went through the historical development of BSIM from the UC Berkeley device group. Along the way, we highlighted various approximations and pointed out how nominal model files from modern foundries can sometimes have limitations, and we gave a few examples of things to watch out for. We also gave a brief overview of FinFET physics and showed how effects like DIBL are reduced, giving a much lower output conductance for transistors resulting in an overall higher basic gain, $g_m r_o$.

Finally we provided a set of nominal simulation sweeps that are helpful before starting to use a new set of models.

## 3.6  Exercises

1. Derive the quasi-2D physics for the finFET structure using the model in Fig. 3.9.
2. Discuss why finFER devices have superior transconductance and output conductance.

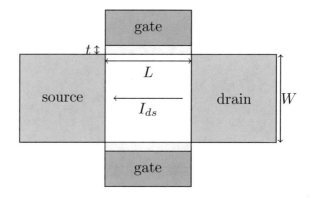

**Fig. 3.9** FinFET basic 2D approximation

# References

1. Taur, Y., & Ning, T. H. (2009). *Fundamentals of modern VLSI devices* (2nd ed.). Cambridge: Cambridge University Press.
2. Antognetti, P., & Massobrio, G. (1998). *Semiconductor device modeling with spice* (2nd ed.). India: McGraw Hill Education.
3. Hu, C. H. (2009). *Modern semiconductor devices for integrated circuits*. London: Pearson Publishing.
4. Sahrling, M. (2019). *Fast techniques for integrated circuit design*. Cambridge: Cambridge University Press.
5. Pao, H. C., & Sah, C. T. (1966). Effect of diffusion current on characteristics of metal-oxide (insulator)-semiconductor transistors. *Solid State Electronics, 9*(10), 927–937.
6. Sze, S. M. (2006). *Physics of semiconductor devices*. Hobroken: John Wiley.
7. Ashcroft, N. W., & Mermin, N. D. (1976). *Solid state physics*. New York: Brooks Cole.
8. Neamen, D. A., & Biswas, D. (2017). *Semiconductor physics and devices* (4th ed.). India: McGraw Hill Education.
9. Sheu, B. J., et al. (1987). BSIM: Berkeley short-channel IGFET model for MOS transistors. *IEEE Journal of Solid-State Circuits, 22*(4), 558–566.
10. Huang, J. H., et al. (1993). A robust physical and predictive model for depp-submicrometer MOS circuit simulation, Electronics Research Laboratory, UCB/ERL M93/57.
11. Chauhan, Y. S., et al. (2012). BSIM – Industry standard compact MOSFET models. *IEEE*, 30–33.
12. Xi, X., et al. (2004) The next generation BSIM for sub-100nm mixed-signal circuit simulation. *IEEE Custom Integrated Circuits Conference*, 13–16.
13. Ward, D. E., & Dutton, R. W. (1978). A charge-oriented model for MOS transistor capacitances. *IEEE Journal of Solid State Circuits, 13*, 703–708.
14. Ebers, J. J., & Moll, J. L. (1954). Large signal behavior of junction transistors. *Proceedings of IRE, 42*, 1761–1772.
15. Niknejad, A., et al. (2020). *BSIM4 manual*. http://bsim.berkeley.edu/models/bsim4/

# Chapter 4
# Circuit Simulators: Linear Case

**Abstract** This chapter discusses the basic of simulators where the focus is on linear systems. We start with a simple circuit network and use it as a demonstration of how an arbitrary circuit network can be built as a matrix equation. This is a natural environment for AC analysis, and we start with such a simulation and compare to a professional simulator. We then extend the analysis to include linear transient simulation where we include elements like capacitors and inductors so that one can demonstrate various integration methods.

## 4.1  Introduction

A circuit simulator is set up to solve Kirchhoff's current law (KCL) and voltage law (KVL) together with equations governing the electrical component's response to current and or voltage stimuli. This chapter will show the basic structure of simulators in the case of linear systems. We will start with a brief historical overview and then derive the basic matrix equation for a simple circuit as an example, and we will use this to motivate how to set up general matrix equations in a computer program. We will only hint at the formal derivations behind these general equations since this has been done thoroughly many times, and we will provide more of the details in Chap. 7 and also give plenty of references for the user to dig into the details if interested. We will first study the single-frequency tone transfer, and since we are looking at linear systems, this mean we will look at AC analysis first, which is a good starting point for understanding the general methodology. We follow this using a time-dependent or transient analysis. It follows naturally from the AC discussion. In Chap. 2, we mentioned four different ways to implement differential equations with the help of difference equations. These four implementations have different properties, and we will highlight those with specific example codes and netlists we will provide in the text.

© Springer Nature Switzerland AG 2021    59
M. Sahrling, *Analog Circuit Simulators for Integrated Circuit Designers*,
https://doi.org/10.1007/978-3-030-64206-8_4

## 4.2   Historical Development

The 1960s and 1970s saw many attempts in developing a way to build circuits using computer-aided design (CAD) techniques [1–8]. Many simulator codes in the 1960s were written under contract with the US Department of Defense and as such restricted in their public use. The design group at the University of California at Berkeley under Professor D.O. Pederson quickly became one of the leading teams proposing "Simulation Program with Integrated Circuit Emphasis" or SPICE as a way to implement circuit network equations in a computer and solve those using numerical techniques. At the time, it was often considered a waste of time. The thinking was that computer adaptations would be too inaccurate and not capable of producing meaningful results. A real transistor implementation being simply too difficult was some of the comments, but this did not discourage Pederson and his students. We all owe a debt of gratitude to Pederson and his group for their unwavering belief in computer implementations using SPICE.

The first version of SPICE was presented at a conference in Waterloo, Canada, in 1973 [7]. It was written by Laurence Nagel, a student of Professor Pederson, and is a derivative of an earlier proprietary program called Computer Analysis of Nonlinear Circuits, Excluding Radiation (CANCER). Pederson insisted the program to be rewritten such that the proprietary restrictions would be removed and the code could be placed in the public domain. The code was somewhat limited in what devices could be included, and it also used a fixed timestep to advance the equations. The popularity of the program increased with the release of SPICE2 in 1975 [6]. Now the timestep was variable, and there were more circuit elements included in the code. The first versions of SPICE were written in FORTRAN with the latest release SPICE2g.6 in 1983. The next generation of SPICE, SPICE3, was released in 1989 and was written in C where a graphical interface using the X Window System was added. The fact that the code was in the public domain and could be purchased for a nominal fee covering the cost of a magnetic tape contributed to the huge popularity of the code. The author took advantage of this offer from the development team as late as the mid-1990s. Simulating a circuit became synonymous to "spice" a circuit.

The development of SPICE was named a milestone in 2011 by IEEE, and L. Nagel was awarded the Donald O. Pederson Award in Solid-State Circuits in 2019 for the development of the initial code.

## 4.3   Matrix Equations

Matrix equations and their solutions were mentioned in Chap. 1. It is a key to many modern simulation methods, and a good knowledge on how one can set up and solve them is an essential piece of knowledge for the working engineer. This section will discuss the setup of such equations given an electronic circuit topology. Most students learn in elementary classes on circuit analysis how to analyze fairly simple circuit topologies, and this section will show how to systematize such methods when building simulator matrices. We will start with the fundamentals of *passive elements* and from a simple passive network build a full matrix equation. This should be familiar to most readers. We will then use that exercise to motivate a common way to set up simulation matrices. Much of the mathematical details and formal proofs we leave as references and outline in Chap. 7 for the interested reader to investigate further. Once such a methodology has been employed, we will use it to build a simple Python code that can be used to read in circuit netlists and set up matrix equations. Thereafter, we will use the built-in matrix solver in Python to solve the system of equations and compare to other SPICE implementations. The main motivation is to show that the methodology is fairly straightforward and dispels any mystery that might surround a circuit simulator. Naturally a fully professional simulator has many more optimization tricks, and details in the implementation can easily be differentiated between various products.

After showing a simple simulation setup with passive devices, we will then continue by defining *active devices* and show how they can be implemented in a similar fashion to the passive devices in building a full matrix.

### 4.3.1 Passive Element

A passive circuit element is a component that can only dissipate, absorb, or store energy in an electric or magnetic field. They do not need any form of electrical power to operate. The most obvious examples are resistors, capacitors, and inductors that we will be described in this section.

Let us first review the basic equation governing passive devices. Any device can be described by its current-voltage relationship:

$$i(t) = f\left(v(t), \frac{dv(t)}{dt}\right) \text{ or } v(t) = f\left(i(t), \frac{di(t)}{dt}\right) \qquad (4.1)$$

An element is called linear if $f$ is linear. A device is called a resistor if the relationship is set by

$$v(t) = i(t)R(t) \qquad (4.2)$$

where $R$ is the resistance we consider constant for now. Nonlinear resistors are an important circuit device, but we will not discuss them here. We also have inductors

$$v(t) = L(t)\frac{di(t)}{dt} \qquad (4.3)$$

and capacitors

$$i(t) = C(t)\frac{dv(t)}{dt} \qquad (4.4)$$

The last two are often referred to as dynamic elements, and we assume for now the variables $L$, $C$ are independent of current and voltage. $L$ is called the inductance and $C$ the capacitance. The fact that there is such a thing as nonlinear capacitors, for instance, in CMOS devices can significantly complicate the solution and accuracy and convergence of a simulator, and we will discuss a few examples of this later in Chap. 5. A network consisting of resistors only is referred to as a *resistive network*, and one that contains capacitors and inductors is often called a *dynamic network*. If the network only has passive devices, it is called a *linear network*.

Consider now a simple network as in Fig. 4.1.

We can write KCL at node $v_1$ as

$$i_1 = i_2 + i_3 \qquad (4.5)$$

At node $v_{in}$, KCL gives the simple

$$i_{in} = i_1 \qquad (4.6)$$

**Fig. 4.1** A simple linear network biased by a voltage $v_{in}$

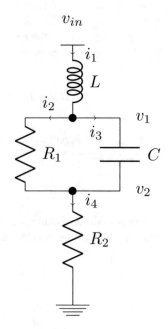

The currents for the devices themselves we can easily get from their element Eqs. 4.2, 4.3, and 4.4, and we end up with the following set of equations:

$$
\begin{cases}
i_1 = i_2 + i_3 \\[4pt]
i_4 = i_2 + i_3 \\[4pt]
v_{in} - v_1 = L\dfrac{di_1}{dt} \\[4pt]
i_2 = \dfrac{v_1 - v_2}{R_1} \\[4pt]
i_3 = C\dfrac{d(v_1 - v_2)}{dt} \\[4pt]
i_4 = \dfrac{v_2}{R_2}
\end{cases}
\tag{4.7}
$$

We have five equations and five unknowns, and we should be able to solve for them. We can also easily see there is a matrix relationship of sorts. We have four unknown currents and two unknown voltages. We assume $L, C$ are constants, so we have a set of linear equation relating these unknowns to each other. The derivative term might make things a bit confusing, but it will be clarified in Sects. 4.1 and 4.2. By rearranging a little bit, we can write these equations in matrix form as

$$
\left\{
\begin{pmatrix}
0 & 0 & 1 & -1 & -1 & 0 \\
0 & 0 & 0 & 1 & 1 & -1 \\
1 & 0 & L\dfrac{d}{dt} & 0 & 0 & 0 \\
-1 & 1 & 0 & R_1 & 0 & 0 \\
-C\dfrac{d}{dt} & C\dfrac{d}{dt} & 0 & 0 & 1 & 0 \\
0 & -1 & 0 & 0 & 0 & R_2
\end{pmatrix}
\begin{pmatrix}
v_1 \\ v_2 \\ i_1 \\ i_2 \\ i_3 \\ i_4
\end{pmatrix}
=
\begin{pmatrix}
0 \\ 0 \\ v_{in} \\ 0 \\ 0 \\ 0
\end{pmatrix}
\right.
\tag{4.8}
$$

where the derivatives are used as operators. We see the equations have somewhat different form. The first two comes from KCL at nodes $v_1$, $v_2$; it is often called *topological constraints*, since it involves a condition based on the circuit connectivity (with a more fancy word, topology). The remaining equations come from the elements themselves and how they relate current through them to the voltage across them; they are called element or *branch equations*. We will look at these equations first from a single-tone perspective in the next section, and then we will follow this with a time-dependent solution where the derivatives will be handled explicitly.

There are a couple of interesting things to note about this matrix. For one, there are a lot of zero elements. This is a common feature of circuit matrix equations. The matrix can most of the time be characterized as a sparse matrix where the number of non-zero elements are of $\mathcal{O}(N)$ ($N$ is the number of rows/columns). The solution of sparse matrices is a research area in itself since it is a common problem [9]. Another point of interest is the ground node does not appear in the circuit network equations. This is due to the fact that adding a ground node equation will overdetermine the system of equations; in fact, it will add no new information to the circuit analysis and can be left out (see Chap. 7 and [1–8, 10–15]).

### 4.3.2 AC Simulation

Traditionally, AC simulations are single-tone frequency analyses where a certain stimulus frequency is varied and the systems response is calculated. The circuit for these simulations is linear. If the underlying circuitry contains nonlinear elements, these have been simply linearized around some bias point. This section will look at such a simulation using the circuit from the previous section as a specific example. We will also describe how the analysis from this circuit suggests how to set up the matrix from an arbitrary circuit topology with some surprisingly simple rules. We will describe these rules for passive elements and also how they work for various sources (voltage/current) and multi-port controlled sources. We will then implement these rules in a Python code script and run it for a couple of circuit examples. The reader is then encouraged to explore more using the exercises at the end of this chapter.

If the system is linear and driven by a single frequency, $\sim e^{j\omega t}$, we can replace the derivatives $\dfrac{de^{j\omega t}}{dt} = j\omega e^{j\omega t}$ with a product $j\omega$ (the exponential term is the same everywhere and can be eliminated). We then have the three basic passive elements:

$$v(\omega) = i(\omega)R \qquad\qquad v(\omega) = j\omega L i(\omega) \qquad\qquad i(\omega) = j\omega C v(\omega) \quad (4.9)$$

Consider again the simple network in Fig. 4.1.
We can exchange the derivative operator in Eq. 4.8 and end up with

$$\begin{pmatrix} 0 & 0 & 1 & -1 & -1 & 0 \\ 0 & 0 & 0 & 1 & 1 & -1 \\ 1 & 0 & j\omega L & 0 & 0 & 0 \\ -1 & 1 & 0 & R_1 & 0 & 0 \\ -j\omega C & j\omega C & 0 & 0 & 1 & 0 \\ 0 & -1 & 0 & 0 & 0 & R_2 \end{pmatrix} \begin{pmatrix} v_1 \\ v_2 \\ i_1 \\ i_2 \\ i_3 \\ i_4 \end{pmatrix} = \begin{pmatrix} 0 \\ 0 \\ v_{in} \\ 0 \\ 0 \\ 0 \end{pmatrix} \qquad (4.10)$$

Before we get into the solution mode, let us take a step back and rewrite this matrix slightly. We see in Eq. 4.10 terms that look like an impedance, for example, $j\omega L$ or an admittance, $j\omega C$. This can in fact be shown to be quite general [4, 5, 10], and one can write the branch equations in either an impedance or admittance matrix form. To highlight this further, we can write the matrix in Eq. 4.10 as

$$\begin{pmatrix} 1 & -1 & -1 & 0 \\ j\omega L & 0 & 0 & 1 \\ 0 & 1 & 0 & -\dfrac{1}{R} \\ 0 & 0 & 1 & -j\omega C \end{pmatrix} = \begin{pmatrix} A & 0 \\ Z & Y \end{pmatrix} \tag{4.11}$$

where $Z$ is an impedance matrix

$$Z = \begin{pmatrix} j\omega L & 0 & 0 \\ 0 & 1 & 0 \\ 0 & 0 & 1 \end{pmatrix} \tag{4.12}$$

and

$$A = \begin{pmatrix} 1 & -1 & -1 \end{pmatrix} \text{ and } Y = \begin{pmatrix} \dfrac{1}{j\omega L} \\ -\dfrac{1}{R} \\ -j\omega C \end{pmatrix} \tag{4.13}$$

$A$ is often called the *reduced incidence* matrix in the literature [1–8, 10], and $Y$ is an *admittance* matrix. In short form, we have

$$\begin{pmatrix} A & 0 \\ Z & Y \end{pmatrix} \begin{pmatrix} i_1 \\ i_2 \\ i_3 \\ v_o \end{pmatrix} = \begin{pmatrix} 0 \\ v_{in} \\ 0 \\ 0 \end{pmatrix} \tag{4.14}$$

This is sometimes referred to as the *reduced form* of the *sparse tableau analysis* (STA) form in books. In this book, we will use this formulation throughout, since it keeps all the current and branch voltages as explicit unknowns and it then is easier to examine a particular variable. One can further reduce the systems of equations by rewriting most of the device currents as a function of the voltage across them. One then arrives at the *nodal analysis* form. It is rarely useful, and in simulators one most commonly use what is called the *modified nodal analysis* (MNA) form. Here some currents have been added back into the system of equations. The details on these various formulations we leave for Chap. 7 and [4, 5, 10–15]. These formulations are convenient since one can build up the matrix equations in linear time just upon reading the circuit netlist, following some simple rules. To see how this might be done, let us again look at Eq. 4.10:

$$\begin{cases} i_1 - i_2 - i_3 = 0 \\ -i_4 + i_2 + i_3 = 0 \\ i_1 j\omega L + v_1 = v_{in} \\ i_2 R_1 - v_1 + v_2 = 0 \\ i_3 - v_1 j\omega C = 0 \\ i_4 R_2 - v_2 = 0 \end{cases} \tag{4.15}$$

The first two equations are just KCL at nodes $v_1$, $v_2$ . The following four equations describe how our four unknown currents, $i_{1 \to 4}$, are determined. In general, we will have a certain number of nodes, $v_i$, $0 \le i < N$, and currents $i_j$, $0 \le j < M$, and each row in the matrix corresponds to KCL for the node if it is a voltage and the branch equations if it is a current:

$$\begin{pmatrix} v_0 \\ \vdots \\ v_{N-1} \\ i_0 \\ \vdots \\ i_{M-1} \end{pmatrix} \tag{4.16}$$

Let us look specifically at the resistor $R_1$ . It shows up in the first equation by sinking current out of node $v_1$. In the second equation, it shows up again by sourcing current into node $v_2$. It also shows up in the fourth equation as a branch equation variable:

$$\begin{matrix} v_1 \\ v_2 \\ \vdots \\ i_2 \end{matrix} \begin{pmatrix} & & \cdots & -1 \\ & & \cdots & +1 \\ \vdots & \vdots & \vdots & \vdots \\ -1 & 1 & \cdots & R_1 \end{pmatrix} \tag{4.17}$$

This form can be shown to be valid for any resistor in any netlist. We will here refer to this as a resistor's matrix *signature*. In general, we have for a resistor connecting nodes $a$, $b$ with a current through it annotated by $i_R$ motivated the following signature in the STA system:

$$\begin{array}{c} v_a \\ v_b \\ \vdots \\ i_R \end{array} \left( \begin{array}{cccc} & & & -1 \\ & & & +1 \\ & & & \vdots \\ -1 & +1 & \cdots & -R \end{array} \right) \left( \begin{array}{c} v_a \\ v_b \\ \vdots \\ i_R \end{array} \right) \tag{4.18}$$

In this discussion, we have kept the current $i_R$ as an unknown we need to solve for. This is not always necessary. At times we might not need to know this current explicitly. What happens then? Let us look at Eq. 4.10 one more time, but now we use the third equation to replace the current $i_2$ everywhere:

$$\begin{cases} i_1 - \dfrac{v_1 - v_2}{R_1} - i_3 = 0 \\[2mm] -i_4 + \dfrac{v_1 - v_2}{R_1} + i_3 = 0 \\[2mm] i_1 j\omega L + v_1 = v_{in} \\[2mm] i_3 - v_1 j\omega C = 0 \\[2mm] i_4 R_2 - v_2 = 0 \end{cases} \tag{4.19}$$

The resistor now only shows up in the KCL equation:

$$\begin{array}{c} v_1 \\ v_2 \\ i_2 \\ i_3 \end{array} \left( \begin{array}{cc} -1/R_1 & +1/R_1 \\ +1/R_1 & -1/R_1 \\ & \\ & \end{array} \right) \tag{4.20}$$

This is in fact the general signature for a resistor where the current through it is not of interest, and it is not necessary to keep it explicitly.

We have encountered two ways to look at a resistor in the matrix formulation: one where we keep the current through it and the other where we ignore that current. The latter is helpful if one wants to reduce the size of the matrix and so speed up the solution time.

It is always a good idea to limit the number of currents being saved since the resulting system of equations is smaller.

For the purposes of this book and to be able to access the device currents, we will as much as possible keep track of the current through any element more in line with the STA formulation, and we do this for illustrative purposes making the matrix larger, but the circuit examples will be small so it will not affect our simulation time much. A Python code snippet for a resistor could look like

```
if DeviceType = 'Resistor':
        Matrix[DeviceBranchIndex][DeviceNode1Index]=1
        Matrix[DeviceBranchIndex][DeviceNode2Index]=-1
   // This is the branch equation row, indicated by the variable
DeviceBranchRow,
      // The columns are set by the DeviceNode1Columns and
DeviceNode2Column variables
        Matrix[DeviceBranchIndex][DeviceBranchIndex]=-Resistance
        Matrix[DeviceNode1Index][DeviceBranchIndex]=1
        Matrix[DeviceNode2Index][DeviceBranchIndex]=-1
   // These two rows are the KCL equations
```

It is a simple matter to go through the inductor and capacitor the same way, and we find for the inductor due to its unique zero impedance at DC that it is only possible to implement when keeping the current.

For the inductor,

$$
\begin{array}{c}
\begin{array}{ccc} v_a & v_b & i_L \end{array} \\
\begin{array}{c} v_a \\ v_b \\ i_L \end{array}
\begin{pmatrix}
 & & -1 \\
 & & +1 \\
+1 & -1 & -j\omega L
\end{pmatrix}
\end{array}
\tag{4.21}
$$

And the capacitor when ignoring its current will look like

$$
\begin{array}{c}
\begin{array}{cc} v_a & v_b \end{array} \\
\begin{array}{c} v_a \\ v_b \end{array}
\begin{pmatrix}
+j\omega C & -j\omega C \\
-j\omega C & +j\omega C
\end{pmatrix}
\end{array}
\tag{4.22}
$$

And when keeping the current,

$$
\begin{array}{c}
\begin{array}{ccc} v_a & v_b & i_C \end{array} \\
\begin{array}{c} v_a \\ v_b \\ i_C \end{array}
\begin{pmatrix}
 & & -1 \\
 & & +1 \\
+j\omega C & -j\omega C & -1
\end{pmatrix}
\end{array}
\tag{4.23}
$$

The astute reader no doubt has a sense in reality that things are a bit more complicated, and yes when encountering real networks with various independent and dependent sources and active devices, the heat turns up a few notches. We will now continue by first describing how to incorporate various sources, which in the literature is often referred to as active elements.

### 4.3.3   Active Elements

An active circuit element is an electronic component that supplies energy to a circuit. This includes voltage and current sources of various flavors and of course transistors. In this book, we will model transistors using voltage sources, so we will limit the discussion of active elements to voltage and current sources.

An element is called active if one of the following conditions is met:

1. The voltage or current is a constant or a function of time, $v(t)$, $i(t) = f(t)$. It is then called an *independent* voltage or current source.
2. The voltage or current is a function of the current through an other network element or a function of the voltage across an other network element. It is then called a *dependent* voltage or current source.

There are two varieties of voltage-controlled devices, namely, voltage-controlled voltage source (VCVS) and voltage-controlled current source (VCCS). Likewise for current-controlled devices, we have current-controlled voltage source (CCVS) and current-controlled current source (CCCS). In this section, we will go through both the independent and dependent sources.

#### 4.3.3.1   Independent Sources

An independent source is either a voltage or a current source that outputs a voltage or current signal as a function of time or frequency independently of other devices or node parameters. A voltage source is most often considered a 0 ohms device, and the simulator is set up such that a voltage source can deliver whatever current the rest of the nodes require at any given point in time or frequency. A current source is analogously a device that has infinite impedance that delivers the desired current at whatever voltage the other devices on the connected nodes need to absorb the desired current. How this works in the simulator will soon become clear. We start here with discussing independent voltage sources and follow with independent current sources.

**Independent Voltage Source**
First, and this is the most simple, is the independent voltage source. In fact we have already looked at this in the example in the previous section. The independent voltage source simply shows up on the right-hand side of the matrix Eq. 4.10. We see

**Fig. 4.2** An independent
voltage source

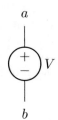

from that equation that the two nodes connected by the voltage source would be the
ground node and node $v_{in}$. Let us look at the more general case where a voltage
source sits between two nodes, $a$ and $b$ (see Fig. 4.2). We can write this mathematically as

$$v_a - v_b = V$$

This is the branch equation for an independent voltage source. Note in particular
the current of the voltage source does not enter the picture. This is a consequence of
the assumption that an independent voltage source can deliver any magnitude/phase
of current; it has 0 ohm impedance. The only requirement is the sum of all other
currents at nodes $a$ and $b$ should have the same magnitude but opposite sign. In
other words, at node $a$ the current going into the voltage source would be equal to
$i_v$, and the current going out of the voltage source at node $b$ is equal to $-i_v$. We follow the convention that current going into a device is positive and negative if it is
going out of a device.

If we treat this current as a separate matrix entry, we can formulate KCL at
node $a$ as

$$\cdots + i_V + \cdots = 0 \qquad \text{KCL at node } a$$

And for $b$

$$\cdots - i_V + \cdots = 0 \qquad \text{KCL at node } b$$

As a side note, what happens if there are two separate voltage sources connecting
two nodes in parallel as in Fig. 4.3? Since a current being supplied by a voltage
source is only defined by all other currents at the node, it should be clear that such
a system has no solution in general. Usually the simulator will issue a topology error
flag and ask the user to fix it.

It should now be clear that a voltage source will have the following signature
when reading in the netlist:

**Fig. 4.3** Two independent
voltage sources in parallel

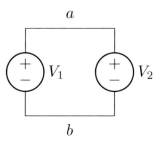

$$
\begin{array}{c}
\begin{array}{ccc} v_a & v_b & i_V \end{array} \\
\begin{array}{c} v_a \\ v_b \\ i_V \end{array}
\left(
\begin{array}{ccc}
& & -1 \\
& & +1 \\
+1 & -1 &
\end{array}
\right)
\end{array}
=
\left(
\begin{array}{c}
\\
\\
V
\end{array}
\right)
\tag{4.24}
$$

Note the voltage source value is now an entry in the rhs of the equation and the voltage $V$ is assumed known. In our example, the voltage source was connected between node $in$ and ground node (or zero). Therefore node $b$ is not present since the ground node equations do not add any new information to the system and is ignored.

A pseudocode snippet for an independent voltage source can look like

```
if DeviceType = 'Voltage Source':
     Matrix[DeviceBranchIndex][DeviceNode1Index] = 1
     Matrix[DeviceBranchIndex][DeviceNode2Index] = -1
     rhs[DeviceBranchIndex] = DeviceValue
// This is the branch equation row, indicated by the variable
DeviceBranchRow,
     // The columns are set by the DeviceNode1Columns and
DeviceNode2Column variables
     Matrix[Nodes.index[DeviceNode1Index][DeviceBranchIndex]=1
     Matrix[Nodes.index[DeviceNode2Index][DeviceBranchIndex]=-1
// These are the two KCL equation entries
```

**Independent Current Source**
An independent current source can be treated analogously. Let us assume we a have current source connecting node $a$ and $b$ like in Fig. 4.4. We get at node $a$ following a similar reasoning as for the independent voltage source

$$\cdots + i_A + \cdots = 0 \qquad \text{KCL at node } a$$

where we use a current $i_A = I$. Likewise for node $b$

**Fig. 4.4** An independent
current source

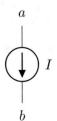

$$\cdots - i_A + \cdots = 0 \qquad \text{KCL at node } b$$

And for the current equation itself,

$$i_A = I$$

A matrix signature would look like

$$
\begin{array}{c}
\phantom{v_a} \\
\begin{array}{ccc} v_a & v_b & i_A \end{array} \\
\begin{array}{c} v_a \\ v_b \\ i_A \end{array}
\left(
\begin{array}{ccc}
 & -1 & \\
 & +1 & \\
 & +1 &
\end{array}
\right)
\end{array}
=
\left(
\begin{array}{c}
 \\
 \\
I
\end{array}
\right)
\qquad (4.25)
$$

A pseudocode snippet for an independent current source can look like

```
if DeviceType = 'Current Source':
     Matrix[DeviceBranchIndex][DeviceBranchIndex] = 1
     rhs[DeviceBranchIndex] = DeviceValue
// This is the branch equation row, indicated by the variable
DeviceBranchIndex,
     Matrix[Nodes.index[DeviceNode1Index][DeviceBranchIndex]=1
     Matrix[Nodes.index[DeviceNode2Index][DeviceBranchIndex]=-1
// These are the two KCL equation entries
```

Clearly we can directly eliminate the current row and we then find

$$\text{LHS} = I \qquad \text{KCL at node } a$$

And for $b$,

$$\text{LHS} = -I \qquad \text{KCL at node } b$$

We then find the following description when removing the explicit current equation:

$$v_a \begin{pmatrix} \\ \\ v_b \end{pmatrix} = \begin{pmatrix} -I \\ +I \end{pmatrix} \qquad (4.26)$$

These formulations are fairly intuitive. From the branch equation, we note the voltage between the output nodes can be anything; it does not enter the equation at all. We say that the current source has infinite impedance. The voltage between the output nodes will be completely determined by the other elements connecting to the nodes. Analogously to the independent voltage source case, we here cannot stack two independent current sources on top of each other, for such a construction to converge the currents needs to be exactly the same since the output impedance is infinite. A simulator will usually issue a complaint about the topology unless the two currents are identical.

### 4.3.3.2 Dependent Sources

The dependent sources are a little bit more complicated since there are sense terminals in addition to the drive terminals, and we will describe here how each of the four kinds of dependent sources operates and can be implemented in a matrix formulation. We will start with the voltage-controlled sources, VCVS and VCCS, and follow with the current-controlled sources, CCVS and CCCS.

**VCVS**

A simple picture of a voltage-controlled voltage source is shown in Fig. 4.5.

The sense ports, $s_1$, $s_2$, of the VCVS draw no current. The current is instead pulled from the output nodes $a$ and $b$ of the element. The voltage between the output nodes is amplified by a factor of $k$ compared to the sense nodes. We find for the new set of equations that KCL at nodes $a$ and $b$ is the same as for the independent voltage source in the previous section:

$$\cdots + i_v + \cdots = 0 \qquad \text{KCL at node } a$$

**Fig. 4.5** A simple example of a linear network with a VCVS

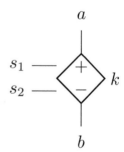

$$\cdots - i_v + \cdots = 0 \qquad \text{KCL at node } b$$

The branch equation for the VCVS element is now

$$v_a - v_b = k\left(v_{s1} - v_{s2}\right)$$

or equivalently

$$v_a - v_b - k\left(v_{s1} - v_{s2}\right) = 0$$

We see as with the independent voltage source the current through the output nodes can take any value.

It is now clear a VCVS has a signature as shown by the following matrix:

$$
\begin{array}{c}
\\
v_a \\
v_b \\
\vdots \\
i_v
\end{array}
\begin{array}{cccc}
v_a & v_b & v_{s1} & v_{s2}
\end{array}
\left(
\begin{array}{cccc}
 & & & +1 \\
 & & & -1 \\
 & & & \\
+1 & -1 & -k & +k
\end{array}
\right)
\qquad (4.27)
$$

Compared to the independent voltage source, there is no rhs entry. The behavior is controlled by the matrix itself:

```
if DeviceType = 'VCVS':
        Matrix[DeviceBranchIndex][DeviceNode1Index]=1
        Matrix[DeviceBranchIndex][DeviceNode2Index]=-1
    // This is the branch equation row, indicated by the variable
DeviceBranchRow,
        // The columns are set by the DeviceNode1Columns and
DeviceNode2Column variables
        Matrix[DeviceBranchIndex][DeviceNode3Index]=DeviceValue
        Matrix[DeviceBranchIndex][DeviceNode4Index]=-DeviceValue
        Matrix[DeviceNode1Index][DeviceBranchIndex]=1
        Matrix[DeviceNode1Index][DeviceBranchIndex]=-1
    // These two rows are the KCL equations
```

## VCCS

A VCCS is a device that outputs a current for a given input voltage. It can be viewed as an ideal transconductor. Most transistor elements in electronics can be viewed as a transconductor, so this element can be used as an ideal transistor. We will use it as such later on in this chapter, and later in Chap. 5, we will use it as a nonlinear trans-conductor. In this section, we will examine the equation assuming a linear gain

**Fig. 4.6** A simple linear
network with a VCCS

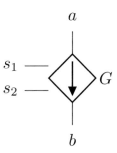

between the input sense terminals and the output terminals. Similarly for a VCVS, we find a simple implementation in Fig. 4.6.

This is similar to the VCVS section we just looked at, but here the result is a current at the output terminals. To capture this, we need to add two equations:

$$\cdots + i_G + \cdots = 0 \qquad \text{KCL at node } a$$

$$\cdots - i_G + \cdots = 0 \qquad \text{KCL at node } b$$

The branch equation for the element is now

$$i_G = G\left(v_{s1} - v_{s2}\right) \tag{4.28}$$

or equivalently

$$i_G - G\left(v_{s1} - v_{s2}\right) = 0 \tag{4.29}$$

As before, this is straightforward to incorporate into our matrix description. We end up with this signature:

$$
\begin{array}{c}
\begin{array}{ccccccc} v_a & v_b & v_{s1} & v_{s2} & \cdots & i_G \end{array} \\
\begin{array}{c} v_a \\ v_b \\ \vdots \\ i_G \end{array}
\left(
\begin{array}{cccccc}
 & & & +1 & \\
 & & & -1 & \\
 & & & & \\
 & -G & +G & -1 &
\end{array}
\right)
\end{array}
\tag{4.30}
$$

We see as before for the VCVS there is no entry on the rhs matrix. This is a reflection of the fact that these are dependent sources.

We will use several versions of the ideal transconductor, VCCS, in the rest of this chapter due to its similarity to the transistor transfer functions.

A code snippet for a VCCS could look like

```
    if DeviceType = 'VCCS':
        Matrix[DeviceBranchIndex][DeviceNode3Index]=-DeviceValue
        Matrix[DeviceBranchIndex][DeviceNode4Index]=DeviceValue
        Matrix[DeviceBranchIndex][DeviceBranchIndex]=-1
   // This is the branch equation row, indicated by the variable
DeviceBranchIndex,
   //
        Matrix[DeviceNode1Index][DeviceBranchIndex]=1
        Matrix[DeviceNode1Index][DeviceBranchIndex]=-1
   // These two rows are the KCL equations for Node1 and Node2
```

## CCVS

A CCVS is exemplified in Fig. 4.7. It senses a current through some element, most often a voltage source, and produces a voltage at the output terminals. In terms of KCL, it looks just like the voltage sources we have already examined:

$$\cdots + i_E + \cdots = 0 \qquad \text{KCL at node } a$$

$$\cdots - i_E + \cdots = 0 \qquad \text{KCL at node } b$$

The branch equation that relates current to voltage can be described as

$$v_a - v_b = E\, i_s$$

We see again this equation is not dependent on $i_E$, so the voltage source can absorb any current.

These equations we can easily incorporate into the matrix equations resulting in the following signature:

**Fig. 4.7**  A simple linear
network with a CCVS

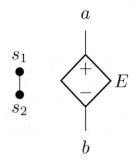

$$
\begin{array}{cccc}
& v_a \quad v_b & i_S & i_E \\
\begin{array}{c} v_a \\ v_b \\ i_S \\ i_E \end{array} &
\left(\begin{array}{cccc}
& & +1 & \\
& & -1 & \\
& & & \\
+1 \quad -1 & & -E &
\end{array}\right)
\end{array}
\tag{4.31}
$$

It has a similar structure to the dependent sources we presented earlier.
A code snippet for a CCVS could look like

```
if DeviceType = 'CCVS':
    Matrix[DeviceBranchIndex][DeviceNode1Index]=1
    Matrix[DeviceBranchIndex][DeviceNode2Index]=-1
    Matrix[DeviceBranchIndex][DeviceSenseIndex]=-DeviceValue
// This is the branch equation row, indicated by the variable
DeviceBranchRow,
// The columns are set by the two Device Nodes and the Sense cur-
rent, iS
//
    Matrix[DeviceNode1Index][DeviceBranchIndex]=1
    Matrix[DeviceNode2Index][DeviceBranchIndex]=-1
// These two rows are the KCL equations
//
```

## CCCS
The last dependent source is the CCCS. It senses a current through some element
and outputs a current between two output nodes. We will follow the same steps we
did for the other dependent sources and come up with a simple expression to use.
Let us look at the implementation of a CCCS in Fig. 4.8. As before, we can write the
branch equation as

$$A \cdot i_s = i_A$$

**Fig. 4.8** A simple linear
network with a CCCS

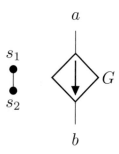

The currents at the output nodes follow the now recognizable pattern:

$$\cdots + i_A + \cdots = 0 \qquad \text{KCL at node } a$$

$$\cdots - i_A + \cdots = 0 \qquad \text{KCL at node } b$$

To incorporate this into our admittance description, we simply write the matrix signature as

$$
\begin{array}{c c}
\begin{array}{cccc} v_a & v_b & \quad i_S & \quad i_A \end{array} & \\
\begin{array}{c} v_a \\ v_b \\ i_S \\ i_A \end{array}
\left(
\begin{array}{cccc}
 & & +1 & \\
 & & -1 & \\
 & & & \\
 & -A & 1 &
\end{array}
\right) & \qquad (4.32)
\end{array}
$$

We see here that no voltage appears anywhere. The output impedance is infinite, and the element just senses a current which is a 0 ohm input.

A code snippet for a CCCS could look like

```
if DeviceType = 'CCCS':
    Matrix[DeviceBranchIndex][DeviceSenseIndex]=-DeviceValue
    Matrix[DeviceBranchIndex][DeviceBranchIndex]=1
// This is the branch equation row, indicated by the variable
DeviceBranchIndex,
// The columns are set by the two Device Nodes and the Sense cur-
rent, iS
//
    Matrix[DeviceNode1Index][DeviceBranchIndex]=1
    Matrix[DeviceNode2Index][DeviceBranchIndex]=-1
// These two rows are the KCL equations
//
```

### 4.3.4  Summary

The passive and active elements are incorporated into the basic admittance equation through really simple branch element equations between various voltages and current in addition to the KCL equations, where the current generated by the device is added and subtracted according to the connectivity. Independent voltage and current sources also have entries on the right-hand side of the matrix equation.

## 4.4  **Matrix Building: AC Simulation**

Let us know look at how such a code to build an STA matrix system can be put together. We will need a code that reads a netlist, where we will use SPICE3 netlist format for simplicity. Let us first define the environment for our codes. We use Python 3 for the code base. Python is a readily available code base, and it is perhaps the most popular analysis framework around both for its ease of use and available codes. We use it to demonstrate the various algorithms we discuss. For basic code variable definitions and netlist syntax, see Appendix A.

In the following code segment, we have a Python implementation of such a code:

```
Python code
#!/usr/bin/env python3
# -*- coding: utf-8 -*-
"""
Created on Thu Feb 28 22:33:04 2019

@author: mikael
"""
import numpy as np
import matplotlib.pyplot as plt
import analogdef as ana
#
# Initialize Variables
#
MaxNumberOfDevices=100
DevType=[0*i for i in range(MaxNumberOfDevices)]
DevLabel=[0*i for i in range(MaxNumberOfDevices)]
DevNode1=[0*i for i in range(MaxNumberOfDevices)]
DevNode2=[0*i for i in range(MaxNumberOfDevices)]
DevNode3=[0*i for i in range(MaxNumberOfDevices)]
DevModel=[0*i for i in range(MaxNumberOfDevices)]
DevValue=[0*i for i in range(MaxNumberOfDevices)]
Nodes=[]
#
# Read modelfile
#
modeldict=ana.readmodelfile('models.txt')
ICdict={}
Plotdict={}
Printdict={}
Optdict={}
#
```

```
# Read the netlist
#
DeviceCount=ana.readnetlist('netlist_4p1.txt',modeldict,ICdict,
Plotdict,Printdict,Optdict,DevType,DevValue,DevLabel,DevNode1,Dev
Node2,DevNode3,DevModel,Nodes,MaxNumberOfDevices)
#
# Create Matrix based on circuit size. We do not implement strict
Modified Nodal Analysis. We keep instead all currents
# but keep referring to the voltages as absolute voltages. We
believe this will make the operation clearer to the user.
#
MatrixSize=DeviceCount+len(Nodes)
#
# The number of branch equations are given by the number of
devices
# The number of KCL equations are given by the number of nodes
in the netlist.
# Hence the matrix size if set by the sum of DeviceCount and
len(Nodes)
#
    STA_matrix=[[0  for  i  in  range(MatrixSize)]  for  j  in
range(MatrixSize)]
    STA_rhs=[0 for i in range(MatrixSize)]
#
# Loop through all devices and create matrix/rhs entries accord-
ing to signature
#
NumberOfNodes=len(Nodes)
for i in range(DeviceCount):
    if DevType[i]=='capacitor' or DevType[i]=='inductor':
        DevValue[i]*=(0+1j)
    STA_matrix[NumberOfNodes+i][NumberOfNodes+i]=-DevValue[i]
    if DevNode1[i] != '0' :
        STA_matrix[NumberOfNodes+i][Nodes.index(DevNode1[i])]=1
        STA_matrix[Nodes.index(DevNode1[i])][NumberOfNodes+i]=1
    if DevNode2[i] != '0' :
        STA_matrix[NumberOfNodes+i][Nodes.index(DevNode2[i])]=-1
        STA_matrix[Nodes.index(DevNode2[i])][NumberOfNodes+i]=-1
    if DevType[i]=='capacitor':
        STA_matrix[NumberOfNodes+i][NumberOfNodes+i]=1
        if DevNode1[i] != '0' : STA_matrix[NumberOfNodes+i][Nodes.
index(DevNode1[i])]=-DevValue[i]
        if DevNode2[i] != '0' : STA_matrix[NumberOfNodes+i][Nodes.
index(DevNode2[i])]=+DevValue[i]
    if DevType[i]=='VoltSource':
```

```
            STA_matrix[NumberOfNodes+i][NumberOfNodes+i]=0
            STA_rhs[NumberOfNodes+i]=DevValue[i]
    #
    #Loop through frequency points
    #
    val=[0 for i in range(100)]
    for iter in range(100):
        omega=iter*1e9*2*3.14159265
        for i in range(DeviceCount):
            if DevType[i]=='capacitor':
                if DevNode1[i] != '0' :
                    STA_matrix[NumberOfNodes+i][Nodes.index(DevNode
1[i])]=DevValue[i]*omega
                if DevNode2[i] != '0' :
                    STA_matrix[NumberOfNodes+i][Nodes.index(DevNode
2[i])]=-DevValue[i]*omega
            if DevType[i]=='inductor':
                    STA_matrix[NumberOfNodes+i][NumberOfNodes+i]=DevVal
ue[i]*omega
        STA_inv=np.linalg.inv(STA_matrix)
        sol=np.matmul(STA_inv,STA_rhs)
        val[iter]=abs(sol[2])
    plt.plot(val)
    End Python
```

**Code 4-1**

```
    *
    v1 in 0 1
    vs vs 0 0
    l1 in a 1e-9
    r1 a b 100
    c1 a b 1.2e-12
    r2 b vs 230
    netlist 4.1
```

There are only two subcircuit calls in the code (readnetlist and the matrix inverter numpy.linalg.inv), and the rest is made up of a matrix/rhs build segment and a loop through the frequency points.

*Verify*

We can now verify the code works by comparing the simulation output to an ngspice simulation of the same circuit (Fig. 4.9) using netlist 4.1. We find the agreement is such the max error is less than 0.001%.

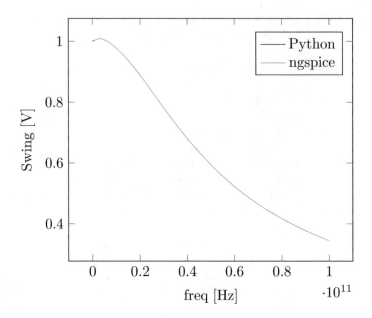

**Fig. 4.9** Comparison of SPICE vs code simulation of the network in . The two curves are on top of each other

This is in a nutshell how a simulator works. This particular code segment handles AC analysis and is perhaps the easiest implementation to understand since it is linear by definition, and the operating point is not important for the building of the matrix itself. We will in the following sections expand on this experience and show how a linear transient simulator can be set up and go from there to the more complex cases of nonlinear elements in the next chapter. The reader will hopefully appreciate how simple this code implementation is. No doubt the reader suspects a real implementation is a bit more complex, and sure enough it is. For example, the issue of convergence has not entered since everything is linear and no difficult derivative operators appear. We will discuss how to deal with such situations later on in this chapter and also in Chap. 5.

The heart of this type of solvers is the matrix inverter, and here we are using the built-in matrix inverter in Python. This is a general matrix inverter designed to handle most practical situations. A circuit matrix inverter is much more specialized and as such much faster. Also, more efficient ways to fill the matrix are needed to make a "real" simulator, and the example we have here is just to showcase the overall structure of such simulators. There is a lot of ongoing effort to make these numerical systems more efficient, and the interested reader should consult the references [12–17], for example, for more of the details. We will now move forward discussing linear transient simulations. The nonlinear discussion will follow in Chap. 5, where we also will discuss steady-state implementations.

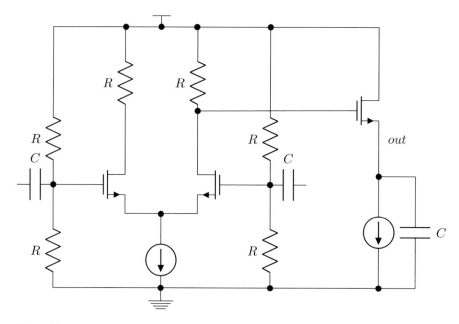

**Fig. 4.10** Simple transconductor with resistive/capacitive load and a source follower output stage

### Amplifier with Linear Transistors

Armed with the analysis we have just done, we can also look at circuits containing linear "transistors." Such devices can be described using a VCCS we investigated in Sect. 3.3. Let us look at a simple amplifier given by Fig. 4.10.

We can use the code snippets for VCCS we developed earlier, and the AC simulation is then extended by just a few lines (see Code 4.1 in Sect. 4.9.1). Let us simulate the circuit with the adjusted code, Code 4.2 in Sect. 4.9.1. See Fig. 4.11 for the result.

```
netlist
vss vss 0 0
vdd vdd 0 0.9
vinp in1 0 1
vinn in2 0 1
r1 vdd inp 100
r2 inp vss 100
r3 vdd inn 100
r4 inn vss 100
r5 vdd outp 100
r6 vdd outn 100
c1 in1 inp 1e-12
c2 in2 inn 1e-12
i1 vs vss 1e-3
m1 outn inp vs nch1
```

```
m2 outp inn vs nch1
m3 vdd outp op nch
i2 op vss 1e-3
c3 op vss 1e-13
```
**netlist 4.2**

Clearly we can see the response of the zero and the pole in the system. One can also compare to an exact calculation and convince oneself this is a correct result.

The input voltage is set by the high-pass filter at the input:

$$v_i = v_{in} \frac{R/2}{R/2 + 1/(j\omega C)} = v_{in} \frac{j\omega CR}{j\omega CR + 2}$$

This voltage is amplified at the output to be

$$v_o = v_i g_m R$$

The final transfer function to the output is now an easy algebraic calculation:

$$v_{out} = v_o \frac{g_{m,2}}{g_{m,2} + j\omega C}$$

And finally we get

$$v_{out} = v_{in} \frac{j\omega CR^2 g_m}{j\omega CR + 2} \frac{g_{m,2}}{g_{m,2} + j\omega C}$$

with a magnitude

$$|v_{out}| = v_{in} \frac{\omega CR^2}{\sqrt{(\omega CR)^2 + 4}} \frac{g_m}{\sqrt{(\omega C/g_{m,2})^2 + 1}}$$

This response is also included in Fig. 4.11. The two solutions coincide with a maximum relative error of 0.0004%.

We will in the rest of Sect. 4.4 go through a set of applications of AC analysis. Once the linear response of a circuit is known, a lot can be learned about its small signal behavior that is very useful for the circuit designer. These additional simulation techniques all use the AC formulation we just described. The linear response of circuits is very important which is why all these additional characterization methods have been developed.

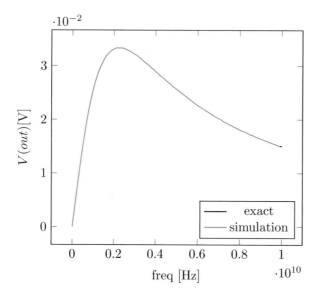

**Fig. 4.11** AC simulation of two-stage amplifier with linear transistors

## 4.4.1  Noise Analysis

Noise analysis is traditionally an AC-like analysis since noise voltages are usually small (some exceptions exist in, e.g., oscillators). The user needs to define the output node, and the simulator adds noise sources where the devices are and simulates the gain from each component to the output as a function of frequency. The noise model for various devices varies in sophistication depending on the complexity of the device. For example, the noise from resistors is usually modeled as a current source shunting the resistor with a value:

$$< i_{n,res} >^2 = \frac{4kT}{R} \left[ \frac{A^2}{Hz} \right] \tag{4.33}$$

where <> indicates root-mean-square (rms) value over time. One can of course also model it as noise voltage in series with the resistor. The noise model for active components like FETs, BJTs, and diodes is nowadays significantly more complicated compared to earlier efforts, and we mentioned these briefly in the modeling Chap. 3. One can summarize a noise simulation with the following pseudocode:

```
Subroutine noise_sim(Devices, OutputNode)
NoisePower=0
Foreach Device
Pwr=Integrate(Simulate_AC(Device,OutputNode)^2,MinFreq,MaxFreq)
    NoisePower+= Pwr
End Subroutine noise_sim
```

Let us implement this in Python and apply on a simple test case first where we know the result and then run it on a larger circuit to get an idea on how it works (see Sect. 4.9.2).

We will run this code on the following simple network found in Fig. 4.12.

The noise powers will add from the two resistors since their noise is uncorrelated. There are plenty of ways to do the calculation, but we will here assume there is a noise current across the resistor and calculate the noise transfer from this noise current to the output. We get for the noise contribution of $R_1$

$$v_{o1} = i_{n1} \frac{j\omega C R_2 R_1}{j\omega C (R_1 + R_2) - \omega^2 C^2 R_1 R_2}$$

Likewise for $R_2$,

$$v_{o2} = i_{n2} \frac{j\omega C R_2 R_1}{j\omega C (R_1 + R_2) - \omega^2 C^2 R_1 R_2}$$

The sum of the two noise voltages squared is

$$v_o^2 = \left(i_{n1}^2 + i_{n2}^2\right) \frac{\left(\omega C R_2 R_1\right)^2}{\left[\omega C (R_1 + R_2)\right]^2 + \left[\omega^2 C^2 R_1 R_2\right]^2} \tag{4.34}$$

We can simulate the same circuit with the netlist:

```
*
Vin in vss 1
R1 out in  50
R2 out vss 50
C1 out vss 50
vss vss 0 0
Netlist 4p3
```

We see the result in Fig. 4.13.

The simulation is of a maximum error of 0.17%.

Let us then apply this to a larger network and compare with an ngspice simulation in Fig. 4.14.

**Fig. 4.12** A simple noise
transfer circuit

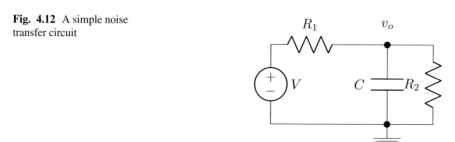

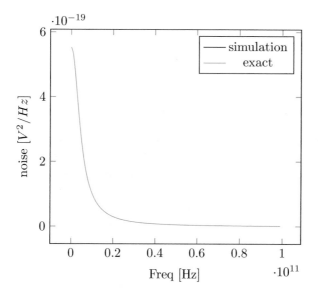

**Fig. 4.13** Comparison of Python and exact solution for network in Fig. 4.12. The two solutions overlap

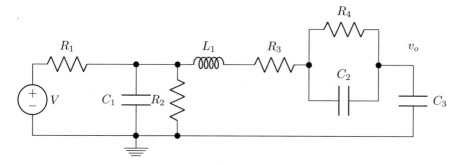

**Fig. 4.14** Larger test network for noise simulation

The netlist is

```
*
vdd in 0 1
vss vss 0 0
r1 in a 100
r2 a vss 50
c1 a vss 1e-12
l1 a b 1e-9
r3 b c 10
r4 c out 100
c2 c out 1e-13
c3 out vss 1e-11
netlist 4p4
```

and the resulting comparison is displayed in Fig. 4.15.

We see here a similar good correlation with a SPICE simulator. Max error is around 0.07%.

The limitations of simulating noise analysis as an AC simulation is that it is performed on a circuit linearized around the DC operating point. The noise performance of mixers, switched capacitors, sample and hold, ADC, DAC, and oscillators cannot be accurately predicted using this type of noise analysis. It completely ignores mixing effects where noise sources can be multiplied up in frequency due to nonlinear multiplication. Consider Fig. 4.16, where we model a VCO as a cross-coupled pair with an LC load. The output currents from the active stage multiply the

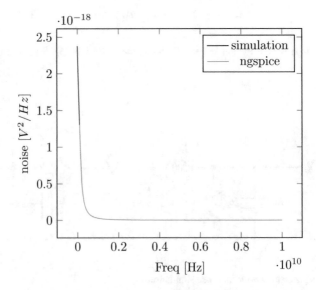

**Fig. 4.15** Comparison of Python and SPICE for noise simulation

**Fig. 4.16** Simple VCO
model with noisy current
sink

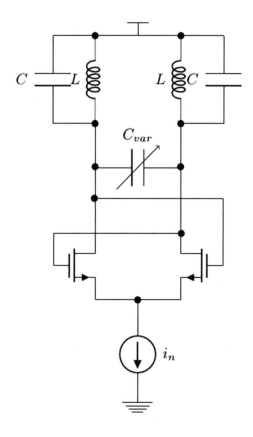

noise of the current sink with the oscillation frequency. Let us consider the current
sink having a noise source at $f_n$, and the oscillator has a oscillation frequency of $f$.
We then have the noise current from the current sink:

$$i_n = A \sin 2\pi f_n t \qquad (4.35)$$

The oscillator can be modeled as

$$i_{osc} = B \sin 2\pi f t \qquad (4.36)$$

The output current is now a product of the two:

$$i_{out} = i_{osc}\, i_n = \frac{AB}{2}\left(\cos 2\pi\left(f - f_n\right)t - \cos 2\pi\left(f + f_n\right)t\right) \qquad (4.37)$$

The noise current has created two side bands around the oscillator frequency.
Imagine now the noise source has a 1/f noise component. This DC noise will appear
around the oscillator tone (it turns out it will often look like 1/f^3 due to the oscilla-

tor tank response). This phenomenon, called phase noise, cannot be captured by traditional noise analysis. Instead periodic noise is needed, and we will discuss this in Sect. 5.5.4.

Even with simpler systems like amplifiers, the assumption behind noise analysis is not correct. Imagine a system where there are harmonics due to nonlinearity. Similar calculations will show that the same kind of mixing that occurs in oscillators also occurs in amplifiers.

### 4.4.2   Stability Analysis

Stability analysis is usually applied to circuit loops where questions like phase and gain margin needs to be answered. Stability theory is discussed at depths elsewhere, and here we will just provide a short summary as a reminder to the reader.

**Brief Stability Theory Review**
Basically a loop will oscillate if Barkhausen's criterion is met:

> **Barkhausen's criterion** A system will oscillate if the gain of the loop is greater than one and the phase shift is 360 degrees.

This criterion is certainly helpful, but it says nothing of margins or how close to oscillation a system might get. In a sense, one needs to be "far away" from the point of oscillation, and this needs to be quantified. The most general stability theory is the one discussed in, for example, [18–20] and commonly called Nyquist's criteria for stability. In most cases, this theory can be simplified to the concepts of gain margin and phase margin. The key to this discussion is the system loop needs to be opened up, and the analysis is done on that open loop where the open-loop gain and phase shift are simulated (in an AC sim commonly) and plotted as a function of frequency.

Mathematically, the concept of phase margin can be studied in terms of complex transfer functions (see [20, 23]).

**Numerical Algorithms for Stability Simulations**
Any modern circuit we try to build with the sophisticated models available now has huge numbers of poles and zeros, and trying to do a complete mathematical analysis is impossible, hence the need for a stability analysis tool. There is a well-known paper by Professor Middlebrook from the California Institute of Technology in 1975 [21] where he describes a general algorithm to simulate stability. It involves biasing the circuit, breaking the loop, and performing two AC simulations. It works really well in most cases and we discuss it first below. A further generalization that is nowadays more often used can be found in [22]. We will provide a short review of the theory by discussing Middlebrook's analysis first and then generalize it.

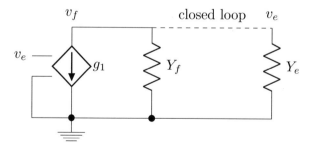

**Fig. 4.17** A loop modeled as a one-port system

**Middlebrook Stability Analysis**

For loop stability, let us first discuss it as described in [21]. This was one of the first theoretical discussions on linear loop gain and has been used as a basis for stability simulations
for many years. The basic idea is to use injections to be able to break the loop at any point.

Let us look at Fig. 4.17. The loop is broken, and in one end, there is a driving
source with a certain admittance, and at the other end, there is a load admittance. We
find with KCL following the notation in [22]

$$g_1 v_e + v_f \left( Y_e + Y_f \right) = 0$$

or, defining the loop gain

$$T = -\frac{v_f}{v_e} = \frac{g_1}{\left( Y_e + Y_f \right)}$$

The voltage and current injections are now performed in such a way as to null the
forward current. We then find

$$g_1 v_e + v_f Y_f = 0$$

with the null current ratio

$$T_n^i = -\frac{v_f}{v_e} = \frac{g_1}{Y_f}$$

In the next simulation, we provide voltage and current injections such that the
forward voltage drop is zero:

$$-i_f Y_e = g_1 i_e$$

or

$$-\frac{i_f}{i_e} = \frac{g_1}{Y_e}$$

We have the so-called null voltage gain:

$$T_n^v = -\frac{i_f}{i_e} = \frac{g_1}{Y_e}$$

From these we now find

$$T = \frac{T_n^v T_n^i}{T_n^v + T_n^i}$$

This technique has been widely used, and the name Middlebrook is sometimes used as a verb "Middlebrook the loop."

**Generalized Middlebrook Stability Analysis**
The Middlebrook analysis in the previous section suffers from shortcomings in terms of the signal flow direction. There is only one current source, and the assumption is the signal flows in only one direction. This approximation is often adequate for low-frequency applications. However, for higher frequencies, there is the possibility that there is a reverse loop flow [22]. We can then generalize this concept to the model in Fig. 4.18. Here, using the same notation as [22], we find the loop gain is given by

$$T = \frac{g_1 + g_2}{Y_e + Y_f}$$

Imagine now that we inject both current and voltage at the loop break point. The resulting parameters, $v_f$, $i_e$, can then be derived using the network ABCD approach:

$$\begin{pmatrix} i_f \\ v_e \end{pmatrix} = \begin{pmatrix} A & B \\ C & D \end{pmatrix} \begin{pmatrix} i_{inj} \\ v_{inj} \end{pmatrix}$$

where trivially

$$B = i_f \big|_{i_{inj}=0, v_{inj}=1} \quad D = v_e \big|_{i_{inj}=0, v_{inj}=1}$$

$$A = i_f \big|_{i_{inj}=1, v_{inj}=0} \quad C = v_e \big|_{i_{inj}=1, v_{inj}=0}$$

We now simply need to relate the ABCD entities with $g_1$, $g_2$, $Y_e$, $Y_f$, and we can calculate the loop gain. We find from the loop equations

$$-Y_e v_e + i_{inj} + i_f - g_2 \left( v_e - v_{inj} \right) = 0$$

$$g_1 v_e + i_f + Y_f \left( v_e - v_{inj} \right) = 0$$

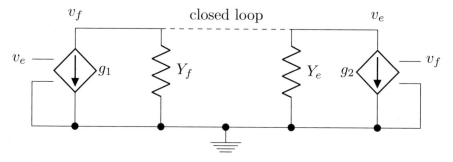

**Fig. 4.18** Two-port model of a loop

We can now identify the ABCD variables from these two relationships and we find

$$A = \frac{-g_1 - Y_f}{g_1 + g_2 + Y_f + Y_e} \qquad B = \frac{-g_1 g_2 + Y_f Y_e}{g_1 + g_2 + Y_f + Y_e}$$

$$C = \frac{1}{g_1 + g_2 + Y_f + Y_e} \qquad D = \frac{g_2 + Y_f}{g_1 + g_2 + Y_f + Y_e}$$

We then find for

$$g_1 = \frac{AD - BC - A}{C} \qquad g_2 = \frac{AD - BC + D}{C}$$

$$Y_f = \frac{BC - AD}{C} \qquad Y_e = \frac{1 - AD + BC + A - D}{C}$$

And for the loop gain,

$$T = \frac{2(AD - BC) - A + D}{2(BC - AD) + A - D + 1}$$

We can recover the original Middlebrook result by setting $g_2 = 0$. We then find

$$T_{forward} = \frac{AD - BC - A}{2(BC - AD) + A - D + 1}$$

The procedure is now to:

- DC bias the circuit.
- Break the loop somewhere.
- Insert a series voltage source.

- Run an AC sim.
- Notice $v_f$, $i_e$.
- Insert a shunt current source.
- Run an AC sim.
- Notice $v_f$, $i_e$.
- Calculate the open-loop gain.

To get a better feel for how this would work, let us use the simple example in Fig. 4.19, where we can do this calculation by hand and then we can compare to a Python implementation. The injected current is directed as indicated in the figure and the voltage likewise. For simplicity let us also assume the output capacitor is zero so the calculations go a little smoother. The input impedance to the transistor gain pair is infinite, so there is no current going in that way. We find

$$B = i_f\big|_{v=1,i=0} = 0$$

Now for the voltage response, $D = v_e|_{v=1,\,i=0}$, we need to calculate it. We find at the input to the gate of the transistor we have $v_e = v_f + v$, and since the follower is biased with an ideal current source, we find the following loop equation:

$$\frac{(v_f+v)g_m}{2}Z_L = v_f \rightarrow v_f\left(1 - \frac{g_m Z_L}{2}\right) = v\frac{g_m Z_L}{2}$$

We then see

$$D = v_e = v_f + v = v\left(\frac{g_m Z_L}{2 - g_m Z_L} + 1\right) = v\left(\frac{2}{2 - g_m Z_L}\right)$$

The remaining parameters are now

$$A = i_f\big|_{v=0,i=1} = -1$$

$$C = v_e\big|_{v=0,i=1}$$

Actually since $B = 0$, $C$ is not a player. We find for

$$T = \frac{2(AD-BC)-A+D}{2(BC-AD)+A-D+1} = \frac{-D+1}{D} = \frac{-\left(\dfrac{2}{2-g_m Z_L}\right)+1}{\left(\dfrac{2}{2-g_m Z_L}\right)} = \frac{-2+2-g_m Z_L}{2} = -\frac{g_m Z_L}{2}$$

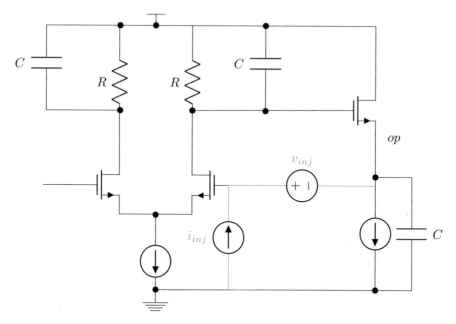

**Fig. 4.19** Circuit loop for stability check

which is the open-loop gain as we had expected. The load impedance is

$$Z_L = \frac{R_L}{R_L j\omega C_L + 1}$$

It is then straightforward to show the gain and phase margin.

We can implement the basic algorithm in a Python code:

**Code 4p4**

We can apply this code to the circuit in Fig. 4.19 and see if our estimation analysis is correct. The netlist is

```
*

  *
  vdd vdd 0 0
  vss vss 0 0
  vinp inp 0 1
  * stab
  * probes
  istab vss inn 0
  vstab inn op 0
  r5 vdd outp 1000
  r6 vdd outn 1000
```

```
c1 vdd outp 1e-12
c2 vdd outn 1e-12
i1 vs vss 0
m1 outn inp vs nch
m2 outp inn vs nch
m3 vdd outp op nch
i2 op vss 0
c3 op vss 1.5e-12
*
```

**Netlist 4p5**

We find the gain and phase response as in Fig. 4.20.

This code is easy to implement in principle, but keep attention to the signs of the currents and voltages. The directions of those entities are critical in getting the correct response.

Note that this is a linear response analysis which has limitations. Often in practice, one should, in addition, excite the circuit with a step function when the loop is closed to see if there is any ringing. Nonlinear effects causing instability can thereby be investigated.

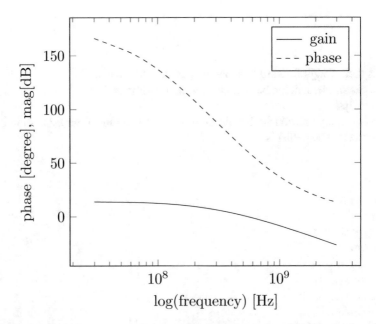

**Fig. 4.20** Gain and phase response as a result of simulation and compared to exact result. The second pole is intentionally set where the unity gain frequency of the first pole is. This gives roughly 45-degree phase margin at the unity gain frequency

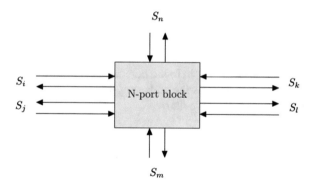

**Fig. 4.21** General system with multiple input/output ports © [2019] Cambridge University Press. Reprinted, with permission, from Cambridge University Press

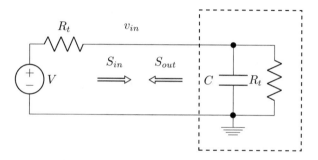

**Fig. 4.22** A schematic test setup for S11

### 4.4.3  S-Parameter Analysis

Another modern adaptation of AC analysis is in S-parameter simulations. An S-parameter is a linear response of a given circuit, and so AC simulation is appropriate. The basic idea behind the S-parameter analysis is that all sources and sinks have a certain termination impedance (in many cases, this is just 50 ohms real impedance). The S-parameter response is then the gain between ports (Fig. 4.21).

The gain between ports $i, j$ is annotated $S_{ji}$ where the $j$ index refers to the output port and the $i$ index is the input port. $S_{21}$ is then the gain between port $1 - > 2$. This gain is then the same thing as AC gain of a circuit that is terminated at all ports. The only bit of confusion can arise regarding same port gain like $S_{11}$ (Fig. 4.22). One can show, see [23, 24], that

$$S11 = \frac{S_{out}}{S_{in}} = \frac{Z - R_t}{Z + R_t} \tag{4.38}$$

where $Z$ is the impedance looking in to the port and $R_t$ is the termination resistance. The input impedance is defined as the voltage at the input node, $v_{in}$, divided by the current going into the input, $i_{in}$. We find

$$v_{in} = \frac{Z}{R_t + Z} V \qquad v_{R_t} = \frac{R_t}{R_t + Z} V \tag{4.39}$$

This gives

$$S11 = \frac{v_{in} - v_{R_t}}{V} = \frac{v_{in} - (V - v_{in})}{V} = \frac{2v_{in} - V}{V} = \frac{v_{in} - V/2}{V/2} \tag{4.40}$$

By convention it is common to use $V = 2 + j * 0$, a real voltage at the voltage source.

Simulating $S_{11}$ as an AC simulation is then a matter of having a resistor, say 50 ohms, in series with the input signal voltage source. Measure the AC response *after* the termination resistor, remove 1, and take the complex magnitude, and we are done:

$$S_{11} = 20 \log \left( |v_{in} - 1| \right) \tag{4.41}$$

S21 is easier to understand (see Fig. 4.23). It is simply an AC simulation with all ports terminated 50 ohms to ground. Still using the convention the driving voltage source has amplitude $V = 2$, we find the voltage at the input is $\approx 1$. The AC magnitude at the output port then directly gives the forward gain:

$$S_{21} = 20 \log |v_{out}| \tag{4.42}$$

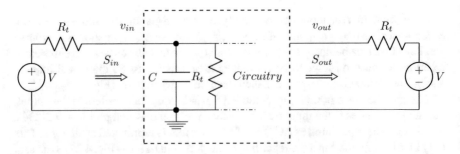

**Fig. 4.23**  S21 schematic setup

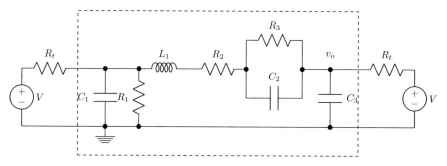

**Fig. 4.24**  Schematic network for S-parameter analysis

This is straightforward to implement in Python:
**Code 4p5**
Let us do an S-parameter analysis of the circuit in Fig. 4.24. We have the netlist

```
*
vp1 p1 0 0
vss vss 0 0
r1 p1 in 50
r2 in vss 50
c1 in vss 1e-12
l1 in b 1e-9
r3 b c 10
r4 c out 100
c2 c out 1e-13
c3 out vss 1e-11
r5 out p2 50
vp2 p2 vss 0
```
**netlist 4p6**

The result is shown in Fig. 4.25.

## 4.4.4  Transfer Function Analysis

A transfer function analysis is similar to an AC simulation in that it simulates the
gain (or transfer function) between inputs and outputs. The advantage with transfer
function analysis is that it computes the gain from inputs like independent sources
to a single output. In this way, properties like gain, PSRR, etc. can be accessed
through one simulation. In the early days of SPICE, this analysis performed a DC

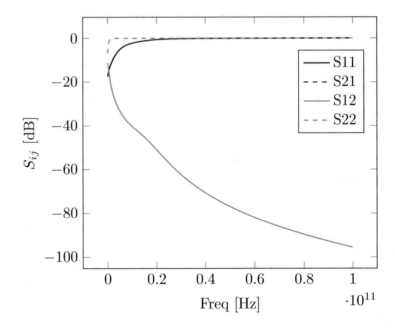

**Fig. 4.25** S-parameter result of network in Fig. 4.24

linearized transfer between a specified input source and an output node. It also frequently provided the input and output resistance. In modern implementations, one can do a broader range of investigations. This is in contrast to AC analysis where you have one (or a few) stimulus in the circuit and you want to find out its effect on various parts of the circuit; the output points are larger than the number of input or stimulus points. With a transfer function analysis, you can calculate the response from several starting points to a single (or a few) output; the number of inputs is larger than the number of outputs. This is straightforward to implement in Python, and we leave it to the reader to pursue.

### 4.4.5  Sensitivity Analysis

Sensitivity analysis is another AC analysis offshoot that is very convenient. The analysis lists the sensitivity of a user-defined output to a bunch of inputs that are also user-defined. Modern implementations also use DC simulations like operating point analysis to simulate sensitivity as shown in Fig. 4.26. We will bypass the explicit Python implementation here and leave as an exercise to the reader.

**Fig. 4.26** Figure
indicating how various
parameters can affect the
"output" in different ways.
The sensitivity analysis
option can quantify this via
AC analysis simulations

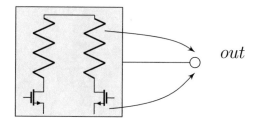

## 4.4.6   Odd Cases: Things to Watch out for

We indicated the impossibility of implementing an inductor eliminating its current
equation with the motivation that its impedance is zero, resulting in an infinite current going through it for a finite, non-zero, voltage across it. The same can be said
about a resistor with 0 ohm. Such a device must be done as an explicit current
device. If not, it is sort of obvious the matrix elements will become infinite and
poorly defined. A sophisticated simulator will detect this problem and take precau-
tions, but readers beware. Another more subtle problem occurs when we have an LC
resonator with ideal LC. A series combination of such will cause a 0 ohm impedance at the resonance. Let us investigate what happens.

**Ideal LC Resonator**
It should be fairly obvious that unless we hit the resonance exactly, where the
impedance is zero for a series combination, the algorithm will work. And in the
unlikely event, we do hit the resonance spot on the matrix will become singular as
expected. The guarantee to prevent this is a non-zero ohm resistor in series with the
inductor.

**Floating Elements**
It is often the case in practice that all elements are not connected to other elements;
they are floating. The matrix for such a case will be non-singular since the voltage
on the terminals of such an element is completely undeterministic; any value will
work. How should a simulator handle this problem? The traditional solution is to
add a resistor and capacitor to ground from all nodes with a resistance value of
$1/g_{min}$, where $g_{min}$ is a simulator variable that can be changed by the user. Also a
minimum capacitance, $c_{min}$, is usually defined at each node to reduce the nodes'
bandwidth. See a more detailed discussion of this in Sect. 5.2.

## 4.5   DC Linear Simulation

A DC solution for a linear system is just as straightforward as the AC solution. All
inductors are shorted and capacitors are removed. In effect we are only left with
linear resistors and DC sources. We can take the previous example and see how it

**Fig. 4.27** A linear
network from Fig. 4.1 with
the inductors shorted and
capacitors removed

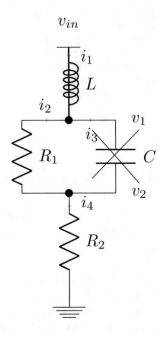

works. We first look again at Fig. 4.27 and remove the capacitors and short the
inductors.

The set of equations are now simplified like the following:

$$\begin{cases} -i_4 + i_2 = 0 \\ v_1 = v_{in} \\ i_2 R_1 - v_1 + v_2 = 0 \\ i_4 R_2 - v_2 = 0 \end{cases} \tag{4.43}$$

The solution is here straightforward. We will bypass a specific code example and
leave it as an exercise to the reader. The difficulties with DC simulations are usually
related to nonlinear aspects of the branch equations, and we will leave that for
Sect. 5.2.

## 4.6   Transient Linear Simulation

Once we have a DC solution, we can start a transient or time-dependent simulation.
Such a simulation is mathematically known as an initial value problem. The initial
values are known at a certain time, time zero is often a convenient choice, in

transient analysis, and from there the simulation progresses to later times. This section will describe how such a simulation can be implemented and will show how different implementations of differential equations lead to different characteristics of the solution. As before we will use the Python environment to build the simulation, but the reader is encouraged to use his or her favorite simulator in addition to verify the various concepts. We will stick to linear elements in this section. As an example, we will again use Fig. 4.1, but this time we will look at a solution in the time domain.

The key in transient simulations is implementing a convenient difference equation to approximate the differential equations. The following sections discuss the most common difference equation approximations using the circuit in Fig. 4.1 as a basis. One of the main difficulties with these approximations is the resulting so-called local truncation error (LTE), and we will examine this effect in Sect. 4.6.6 and how circuit simulators usually handle it. Finally, we end the section on transient linear simulations with a look at an ideal LC resonator and see how the different integration methods behave for this case.

**Timestep Adjustments**

It is well-known how to implement a variable timestep, and we will show such an implementation in Chap. 5, but in this chapter, we will stick to a fixed timestep. The nonlinear circuits we will discuss in Chap. 5 will benefit much more from an adjustable timestep, something we will demonstrate there. The flow here will be to first pick a timestep and run the simulation and observe the behavior. After discussing the LTE estimate, we will simply run the simulation, and if that LTE error condition is not met at some point in time, we will simply adjust the timestep and rerun the whole simulation. The benefit with the fixed timestep is that the error in all the difference equations is under better control. It is not unusual to employ this strategy in practical circuit situations where one can set the max timestep variable to some small value to ensure the timestep taken by the simulator is uniform. This will often improve accuracy as we will also see in Chap. 5.

## 4.6.1   Forward Euler

The forward Euler is the simplest implementation and the easiest to start a discussion. It is almost never used in circuit simulators due to poor stability, but it serves well as an initial point of discussion. Let us look at a capacitor from a differential equation view point:

$$i(t) = C \frac{du(t)}{dt} \tag{4.44}$$

where $u(t)$ is the voltage across the capacitor and $i(t)$ is the current through it at time $t$. To find a numerical implementation, we assume we know the solution at time $t = t_n$, and we want to know the solution at $t = t_{n+1}$. Using the forward Euler implementation, we find as in Chap. 1

$$i(t_n) = C \frac{u(t_{n+1}) - u(t_n)}{\Delta t} \rightarrow u(t_{n+1}) = u(t_n) + \Delta t \frac{i(t_n)}{C} \qquad (4.46)$$

Note that all entities on the right-hand side are known. The equation states that the new voltage across the capacitor is determined by the old voltage across it plus another known term. It looks like an independent voltage source. We already know the matrix equation signature of such a voltage source from Sect. 3.3. The equation we have is a bit simplified in the voltage $u(t)$ across the capacitor, and we need to spell it out. We assume the capacitor sits between two nodes $a$ and $b$, and we find $u(t) = v_a(t) - v_b(t)$.

$$
\begin{array}{c}
\begin{array}{ccc} v_a & v_b & i_V \end{array} \\
\begin{array}{c} v_a \\ v_b \\ i_V \end{array}
\left(
\begin{array}{ccc}
 & -1 & \\
 & +1 & \\
+1 & -1 &
\end{array}
\right)
=
\left(
\begin{array}{c}
\\
\\
v_a(t_n) - v_b(t_n) + \Delta t \dfrac{i_V(t_n)}{C}
\end{array}
\right)
\end{array}
\qquad (4.47)
$$

A pseudocode could look like

```
if DeviceType = 'Capacitor':
    Matrix[DeviceBranchIndex][DeviceNode1Index] = 1
    Matrix[DeviceBranchIndex][DeviceNode2Index] = -1
    rhs[DeviceBranchIndex] = V[DeviceNode1Index]-V[DeviceNode2I
ndex]+deltaT*I[DeviceBranchIndex]/DeviceValue
    // This is the branch equation row, indicated by the variable
DeviceBranchRow,
    // The columns are set by the DeviceNode1Columns and
DeviceNode2Column variables
        Matrix[Nodes.index[DeviceNode1Index][DeviceBranchIndex] = 1
        Matrix[Nodes.index[DeviceNode2Index][DeviceBranchIndex] = -1
    // These are the two KCL equation entries
```

Likewise looking at an inductor, we get

$$u(t) = L\frac{di(t)}{dt} \qquad (4.48)$$

With the forward Euler implementation, we find

$$u(t_n) = L\frac{i(t_{n+1}) - i(t_n)}{\Delta t} \rightarrow i(t_{n+1}) = i(t_n) + \Delta t \frac{u(t_n)}{L} \qquad (4.49)$$

This looks like an independent current source, and an inductor implementation will have a matrix equation signature like

$$
\begin{array}{c}
\begin{array}{ccc} v_a & v_b & i_L \end{array} \\
\begin{array}{c} v_a \\ v_b \\ i_L \end{array}
\begin{pmatrix} & & -1 \\ & & +1 \\ & & 1 \end{pmatrix}
\end{array}
=
\begin{pmatrix} \\ \\ i_L(t_n) + \Delta t \dfrac{v_a(t_n) - v_b(t_n)}{L} \end{pmatrix}
\qquad (4.50)
$$

where we have expanded the voltage across the inductor $u(t) = v_a(t) - v_b(t)$.

```
if DeviceType = 'Inductor':
    Matrix[DeviceBranchIndex][DeviceBranchIndex] = 1
    rhs[DeviceBranchIndex] = I[DeviceBransIndex]+deltaT*(V[Devi
ceNode1]-V[DeviceNode2])/DeviceValue
    // This is the branch equation row, indicated by the variable
DeviceBranchIndex,
    Matrix[Nodes.index[DeviceNode1Index][DeviceBranchIndex]=1
    Matrix[Nodes.index[DeviceNode2Index][DeviceBranchIndex]=-1
    // These are the two KCL equation entries
```

## Simulation

We have constructed a way to implement a capacitor and an inductor using Euler's forward method. The resistor implementation remains the same as we had in Sect. 3.1. We can now build a simple simulator code where the timestep is a constant, and we simulate to a stop time (see Sect. 4.9.5).

```
*
v1 in 0 sin(1G)
vs vs 0 0
l1 in a 1e-9
```

```
r1 a b 100
c1 a b 1.2e-12
r2 b vs 230
netlist 4.7
```

Let us run this code with netlist 4.7 with a few choices of timestep. Let us try
deltaT = 1 ps, 10 ps, 100 ps. The results are shown in Fig. 4.28

What is happening? It seems that to get anything reasonable, we need to take
really small timesteps. As the reader has probably guessed already, this is the
infamous instability of the forward Euler method. The timestep required is set by
smallest capacitor, and you can imagine in a real circuit implementation, where
there are plenty of very small capacitors, how difficult it is to use in practice.
Fundamentally, this is the reason why this method is never used. It will take forever to simulate basic circuits. We need to come up with something better and the
next section will do so.

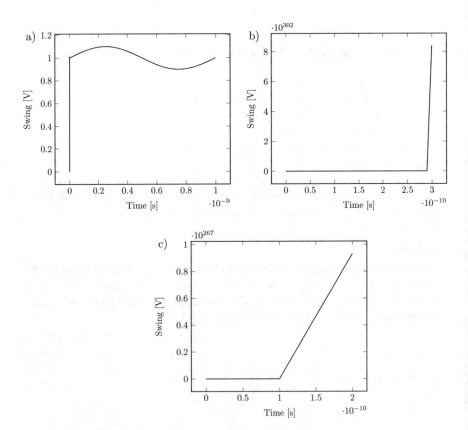

**Fig. 4.28** Forward Euler simulation with timestep (**a**) 1 ps, (**b**) 10 ps, (**c**) 100 ps

## 4.6.2   Backward Euler

The forward Euler implementation as we saw is fairly straightforward. The drawback using that method is its poor stability. We saw in the previous section we needed to take small timesteps to get anywhere. The backward Euler is a tad more complex as we will now show, but it is much more stable.

Let us start with the difference equation implementation of a capacitor using backward Euler:

$$i\left(t_{n+1}\right) = C\frac{u\left(t_{n+1}\right)-u\left(t_{n}\right)}{\Delta t} \rightarrow u\left(t_{n+1}\right) = u\left(t_{n}\right)+\Delta t\frac{i\left(t_{n+1}\right)}{C} \tag{4.51}$$

The right-hand side of this equation depends on the new timestep, so a matrix signature will be more complicated, but let us express it in terms of a current source so we can minimize the size of the matrix like we discovered in the discussion on inductors:

$$i\left(t_{n+1}\right) = \frac{C}{\Delta t}\left(v_{a}\left(t_{n+1}\right)-v_{b}\left(t_{n+1}\right)\right)-\frac{C}{\Delta t}\left(v_{a}\left(t_{n}\right)-v_{b}\left(t_{n}\right)\right) \tag{4.52}$$

The matrix equation now becomes

$$\begin{array}{c} \\ v_{a} \\ v_{b} \\ i_{C} \end{array} \begin{pmatrix} \begin{array}{ccc} v_{a} & v_{b} & i_{C} \\ & & 1 \\ & & -1 \\ -\dfrac{C}{\Delta t} & \dfrac{C}{\Delta t} & 1 \end{array} \end{pmatrix} = \begin{pmatrix} \\ \\ -\dfrac{C}{\Delta t}\left(v_{a}\left(t_{n}\right)-v_{b}\left(t_{n}\right)\right) \end{pmatrix} \tag{4.53}$$

A pseudocode will look like

```
if DeviceType = 'Capacitor':
    Matrix[DeviceBranchIndex][DeviceNode1Index] = 1
    Matrix[DeviceBranchIndex][DeviceNode2Index] = -1
    rhs[DeviceBranchIndex] = V[DeviceNode1Index]-V[DeviceNode2I
ndex]+deltaT*I[DeviceBranchIndex]/DeviceValue
    // This is the branch equation row, indicated by the variable
DeviceBranchRow,
    // The columns are set by the DeviceNode1Columns and
DeviceNode2Column variables
    Matrix[Nodes.index[DeviceNode1Index][DeviceBranchIndex] = 1
    Matrix[Nodes.index[DeviceNode2Index][DeviceBranchIndex] = -1
    // These are the two KCL equation entries
```

Likewise for an inductor, we find with backward Euler

$$u\left(t_{n+1}\right) = L\frac{i\left(t_{n+1}\right) - i\left(t_{n}\right)}{\Delta t} \rightarrow i\left(t_{n+1}\right) = i\left(t_{n}\right) + \Delta t\frac{u\left(t_{n+1}\right)}{L} \tag{4.54}$$

A matrix implementation becomes

$$
\begin{array}{c}
\begin{array}{cc} v_a & v_b \end{array} \\
\begin{array}{c} v_a \\ v_b \\ i_L \end{array}
\left(\begin{array}{ccc}
 & & 1 \\
 & & -1 \\
-\dfrac{\Delta t}{L} & \dfrac{\Delta t}{L} & 1
\end{array}\right)
\end{array}
=
\left(\begin{array}{c}
 \\
 \\
i_L\left(t_n\right)
\end{array}\right)
\tag{4.55}
$$

Example code

```
if DeviceType = 'Inductor':
     Matrix[DeviceBranchIndex][DeviceBranchIndex] = 1
     rhs[DeviceBranchIndex] = I[DeviceBransIndex]+deltaT*(V[Dev
iceNode1]-V[DeviceNode2])/DeviceValue
     // This is the branch equation row, indicated by the variable
DeviceBranchIndex,
     Matrix[Nodes.index[DeviceNode1Index][DeviceBranchIndex]=1
     Matrix[Nodes.index[DeviceNode2Index][DeviceBranchIndex]=-1
     // These are the two KCL equation entries
```

**Simulation**

We can change the system build routine straightforwardly (Sect. 4.9.6) and look at the result of simulating netlist 4.7 in Fig. 4.29.

The major difference from earlier is the stability of the algorithm. We can take much larger timesteps and get accurate results. The backward Euler is often implemented even in a normal simulation setup, in particular close to what is known as break points which we will go over later in Sect. 5.4.5 in much more detail.

### 4.6.3   Trapezoidal Method

The Euler methods we discussed in the previous section are of first-order accuracy like we described in Chap. 2. The more common integration methods are of second-order accuracy, and in this section, we will discuss how one can implement the trapezoidal method. We discussed this also in Chap. 1, and with a capacitor, the method will look like

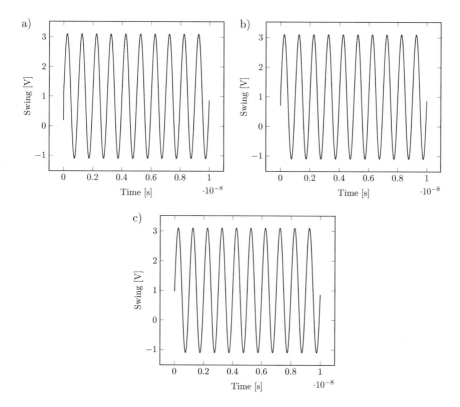

**Fig. 4.29** Backward Euler simulation with timesteps (**a**) 1 ps, (**b**) 10 ps, (**c**) 100 ps

$$i\left(t_{n+1}\right) = C\left(2\frac{u\left(t_{n+1}\right)-u\left(t_n\right)}{\Delta t}-u'\left(t_n\right)\right) \tag{4.56}$$

The last term we can approximate is

$$u'\left(t_n\right) \approx \frac{i\left(t_n\right)}{C} \tag{4.57}$$

and we end up with

$$i\left(t_{n+1}\right) \approx C\left(2\frac{u\left(t_{n+1}\right)-u\left(t_n\right)}{\Delta t}-\frac{i\left(t_n\right)}{C}\right) = C\left(2\frac{u\left(t_{n+1}\right)-u\left(t_n\right)}{\Delta t}\right)-i\left(t_n\right) \tag{4.58}$$

Similar to the backward Euler, we have now for the matrix implementation signature

$$
\begin{array}{c}
\phantom{v_a}\begin{array}{cc} v_a & v_b \end{array}\\
\begin{array}{c} v_a\\ v_b\\ i_C \end{array}
\left(
\begin{array}{ccc}
 & & 1\\
 & & -1\\
-\dfrac{2C}{\Delta t} & \dfrac{2C}{\Delta t} & 1
\end{array}
\right)
\end{array}
=
\left(
\begin{array}{c}
\\
\\
-2C\dfrac{v_a(t_n)-v_b(t_n)}{\Delta t}-i(t_n)
\end{array}
\right)
\tag{4.59}
$$

where all the known entities, evaluated at the previous timestep, can be found on the rhs. Also the voltage across the capacitor has been spelled out, $u(t) = v_a(t) - v_b(t)$.

A pseudocode can look like

```
if DeviceType = 'Capacitor':
    Matrix[DeviceBranchIndex][DeviceNode1Index] = 2*DeviceValue
    Matrix[DeviceBranchIndex][DeviceNode2Index] = -2*DeviceValue
    Matrix[DeviceBranchIndex][DeviceBranchIndex] = 1
    rhs[DeviceBranchIndex] =2*(V[DeviceNode1Index]-V[DeviceNode
2Index])+deltaT*I[DeviceBranchIndex]/DeviceValue
    // This is the branch equation row, indicated by the variable
DeviceBranchRow,
    //  The columns are set by the DeviceNode1Columns and
DeviceNode2Column variables
    Matrix[Nodes.index[DeviceNode1Index][DeviceBranchIndex] = 1
    Matrix[Nodes.index[DeviceNode2Index][DeviceBranchIndex] = -1
    // These are the two KCL equation entries
```

Doing similar calculation with the inductor equations, we see

$$
u(t_{n+1}) = 2L\frac{i(t_{n+1})-i(t_n)}{\Delta t} - Li^{\cdot}(t_n) \rightarrow i(t_{n+1}) = i(t_n) + \Delta t\frac{u(t_n)}{2L} + \Delta t\frac{u(t_{n+1})}{2L}
$$
$$\tag{4.60}$$

where we have used $i^{\cdot}(t_n) \approx u(t_n)/L$. A matrix implementation will then look like

$$
\begin{array}{c}
\phantom{v_a}\begin{array}{cc} v_a & v_b \end{array}\\
\begin{array}{c} v_a\\ v_b\\ i_L \end{array}
\left(
\begin{array}{ccc}
 & & 1\\
 & & -1\\
-\dfrac{\Delta t}{2L} & \dfrac{\Delta t}{2L} & 1
\end{array}
\right)
\end{array}
=
\left(
\begin{array}{c}
\\
\\
i_L(t_n)+\dfrac{\Delta t}{2L}(v_a(t_n)-v_b(t_n))
\end{array}
\right)
\tag{4.61}
$$

A pseudocode can look like

```
if DeviceType = 'Inductor':
    Matrix[DeviceBranchIndex][DeviceBranchIndex] = 1
    Matrix[DeviceBranchIndex][DeviceNode1Index]    =  -deltaT/
(2*DeviceValue)
    Matrix[DeviceBranchIndex][DeviceNode2Index]   =   deltaT/
(2*DeviceValue)
    rhs[DeviceBranchIndex] = I[DeviceBransIndex]+deltaT*(V[Dev
iceNode1]-V[DeviceNode2])/(2*DeviceValue)
// This is the branch equation row, indicated by the variable
DeviceBranchIndex,
    Matrix[Nodes.index[DeviceNode1Index][DeviceBranchIndex]=1
    Matrix[Nodes.index[DeviceNode2Index][DeviceBranchIndex]=-1
// These are the two KCL equation entries
// These Matrix entries need not be updated any further as time
moves along
// Lastly the RHS depends on the solution at the previous time
point so it needs to be updated continuously
    STA_rhs[NumberOfNodes+i] = sol[NumberOfNodes+i] + deltaT*(
sol[ Nodes.index( DevNode1[i]) ] - sol[Nodes.index(DevNode2[i])])/
(2*DevValue[i])
```

## Simulation

We can change the system build routine straightforwardly (Sect. 4.9.7). Let us look at simulations of the same netlist we used for the previous two integration methods in Fig. 4.30.

The major difference from earlier is the accuracy of the algorithm since it is of second order.

## 4.6.4   Second-Order Gear Method

Finally, the last method we will discuss is the second-order Gear method. It is together with trapezoidal method the most commonly used in circuit simulators. We have from Chap. 1 the basic approximation:

$$\frac{df}{dt}(t+\Delta t) \approx \frac{1}{\Delta t}\left(\frac{3}{2}f(t+\Delta t)-2f(t)+\frac{1}{2}f(t-\Delta t)\right) \qquad (4.62)$$

If we were to implement this with our capacitor model, we would get

$$i(t_{n+1}) = \frac{C}{\Delta t}\left(\frac{3}{2}u(t_{n+1})-2u(t_n)+\frac{1}{2}u(t_{n-1})\right) \qquad (4.63)$$

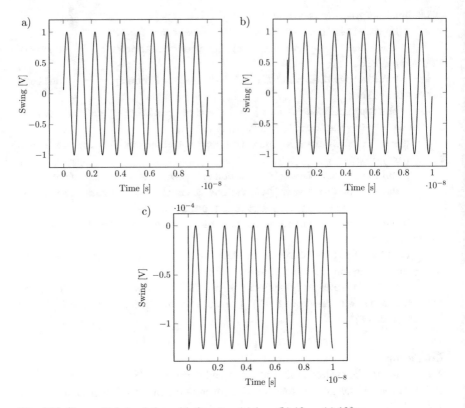

**Fig. 4.30** Trapezoidal simulation with timesteps (**a**) 1 ps, (**b**) 10 ps, (**c**) 100 ps

By following steps similar to the previous integration methods, we find matrix implementation will now look like

$$
\begin{array}{c}
\quad v_a \quad v_b \\
\begin{array}{c} v_a \\ v_b \\ i_C \end{array}
\left(
\begin{array}{ccc}
 & & 1 \\
 & & -1 \\
-\dfrac{3C}{2\Delta t} & \dfrac{3C}{2\Delta t} & 1
\end{array}
\right)
\end{array}
=
\left(
\begin{array}{c}
\\
\\
\dfrac{C}{\Delta t}\left(-2u(t_n)+\dfrac{1}{2}u(t_{n-1})\right)
\end{array}
\right)
\tag{4.64}
$$

A pseudocode can look like

```
if DeviceType = 'Capacitor':
    Matrix[DeviceBranchIndex][DeviceNode1Index] = 2*DeviceValue
    Matrix[DeviceBranchIndex][DeviceNode2Index] = -2*DeviceValue
    Matrix[DeviceBranchIndex][DeviceBranchIndex] = 1
    rhs[DeviceBranchIndex] =2*(V[DeviceNode1Index]-V[DeviceNode2
Index])+deltaT*I[DeviceBranchIndex]/DeviceValue
```

```
// This is the branch equation row, indicated by the variable
DeviceBranchRow,
    // The columns are set by the DeviceNode1Columns and
DeviceNode2Column variables
        Matrix[Nodes.index[DeviceNode1Index][DeviceBranchIndex] = 1
        Matrix[Nodes.index[DeviceNode2Index][DeviceBranchIndex] = -1
    // These are the two KCL equation entries
```

Likewise for the inductor,

$$u(t_{n+1}) = \frac{L}{\Delta t}\left(\frac{3}{2}i(t_{n+1}) - 2i(t_n) + \frac{1}{2}i(t_{n-1})\right) \tag{4.65}$$

$$i(t_{n+1}) = u(t_{n+1})\frac{2\Delta t}{3L} + \frac{4}{3}i(t_n) - \frac{1}{3}i(t_{n-1}) \tag{4.66}$$

And the matrix implementation will look like

$$
\begin{array}{cc} & v_a \quad v_b \end{array}
\begin{array}{c} v_a \\ v_b \\ i_L \end{array}
\left(
\begin{array}{ccc}
 & & 1 \\
 & & -1 \\
-\dfrac{2\Delta t}{3L} & \dfrac{2\Delta t}{3L} & 1
\end{array}
\right)
\left(
\begin{array}{c}
\\
\\
\dfrac{4}{3}i_L(t_n) - \dfrac{1}{3}i_L(t_{n-1})
\end{array}
\right)
= \tag{4.67}
$$

Notice in these formulae we need information from two timesteps back!
A pseudocode can look like

```
if DeviceType = 'Inductor':
        Matrix[DeviceBranchIndex][DeviceBranchIndex] = 1
        Matrix[DeviceBranchIndex][DeviceNode1Index]    = -deltaT/
(2*DeviceValue)
        Matrix[DeviceBranchIndex][DeviceNode2Index]    =    deltaT/
(2*DeviceValue)
        rhs[DeviceBranchIndex] = I[DeviceBransIndex]+deltaT*(V[Dev
iceNode1]-V[DeviceNode2])/(2*DeviceValue)
    // This is the branch equation row, indicated by the variable
DeviceBranchIndex,
        Matrix[Nodes.index[DeviceNode1Index][DeviceBranchIndex]=1
        Matrix[Nodes.index[DeviceNode2Index][DeviceBranchIndex]=-1
    // These are the two KCL equation entries
```

## Simulation
The difference in the code is the build routine again (Sect. 4.9.8).

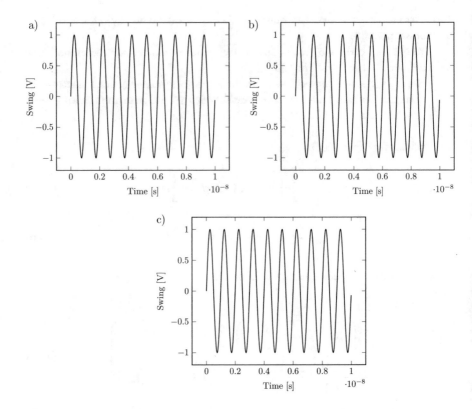

**Fig. 4.31** Gear2 simulation with timesteps (**a**) 1 ps, (**b**) 10 ps, (**c**) 100 ps

Figure 4.31 shows the result of simulating netlist 4.7 with various timesteps.

These examples were mainly made to demonstrate the last three methods are stable with respect to timestep in contrast with forward Euler that has some real problems with stability and large timesteps.

### 4.6.5   Stiff Circuits

The kinds of circuits where there are time constants, like $R\,C$, that are small compared to the timestep are called stiff circuits. In electrical systems, it is very common to have many high-frequency poles due to small capacitors, for example. Normally a timestep would be much larger than what is needed to determine the characteristics of such time constants, mostly because they are way beyond the bandwidth of the system and will play negligible role in the dynamical evolution of the system. We noticed in Sect. 4.6.1 the forward Euler method has some real problems and needs to take really small timesteps in order to converge. This is very characteristic for this method. The other three integration methods we are using are stable when used on stiff systems. For the circuit we have used so far, we notice the following time constants:

$$\tau_{c1} = \frac{L}{R_1} = 10^{-11}\,[s], \quad \tau_{c2} = R_1 C = 1.2 \cdot 10^{-10}\,[s], \quad \tau_{c3} = \sqrt{LC} = 3.5 \cdot 10^{-11}\,[s]$$

It is then clear the smallest time scale is set by $\tau_{c1}$. We thus need a timestep <10 ps to make sure forward Euler converges which is confirmed by our experiment in Sect. 4.6.1.

### 4.6.6   Local Truncation Error

The fact that we approximate the differential equation with a truncated difference scheme causes LTEs, in which higher-order derivatives are discarded [25]. This error needs to be estimated somehow, and there are many ways one can do this and exercise some control of the timestep the simulator is taking. One way is to use a fixed step size and run through the simulation and compare it to a simulation with smaller step size and see to what degree the two simulations agree. This is a cumbersome way to approach the problem and is rarely used practically. A better way would be to estimate the LTE. We can do this by extrapolating the previous few timesteps assuming the solution follows a parabola [25]. Let us call this extrapolated voltage, $v_{n,\,pred}(t)$. The difference between the solved values $v_n(t)$ and $v_{n,\,pred}(t)$ can be viewed as an estimate of the LTE. We can then exercise a timestep control by forcing this estimated LTE to be less than a certain criterion (see [25]) at every time point:

$$v_n(t) - v_{npred}(t) < \alpha \; (\text{reltol}\, v_{nmax} + \text{vabstol})$$

Here, $\alpha$ is some user-controlled factor. $v_{nmax}$ can be chosen in different ways as

$$v_{nmax} = \begin{cases} \max \forall t, \forall v_n \\ \max \forall t, v_n \\ |v_n(t)| \end{cases}$$

The first criterion is often called a global error criteria, the next a local error since it concerns the max value over past time for node $n$. The last is often referred to as the point local error criteria. If the LTE is greater than this criterion, one needs to take a smaller timestep. Let us implement these criteria in our simulator and take a look at the convergence rate. Depending on the situation, either of these $v_{nmax}$ settings is relevant. Sometimes the point local criterion is simply too tight and does not provide any benefits, and sometimes it is necessary. We will look into these choices in more detail in Chap. 5. The $v_{n,\,pred}$ can be calculated as follows:

Let us assume the solution follows a second-order polynomial in time:

$$v_{n,pred} = a + bt_n + ct_n^2$$

We know the previous time values are $t_{n-1}$, $t_{n-2}$, $t_{n-3}$. It is convenient to shift the times such that $\tau = t - t_{n-3}$. We then find

$$\tau_0 = 0, \tau_1 = t_{n-2} - t_{n-3}, \tau_2 = t_{n-1} - t_{n-3}, \tau_3 = t_n - t_{n-3}$$

and we can write

$$v_{n-1} = a + b\tau_2 + c\tau_2^2$$

$$v_{n-2} = a + b\tau_1 + c\tau_1^2$$

$$v_{n-3} = a$$

$$\begin{cases} v_{n-1} = a + b\tau_2 + c\tau_2^2 \\ v_{n-2} = a + b\tau_1 + c\tau_1^2 \\ v_{n-3} = a \end{cases}$$

We can now clearly solve for the coefficients, $a$, $b$, $c$ and calculate the predicted value as

$$v_{n,pred} = a + b\tau_3 + c\tau_3^2$$

Verify this and the python implementation again.

```
Python code
        LTEConverged=True
        while LTEConverged:
            for i in range(NumberOfNodes):
                    # Assume  parabola  v=a+b*t+c*t^2  v[t_solm1
(t=0)]=solm1[i] -> a=solm1[i], a+b*t_nm1+c*t_nm1^2=solold[i]
                    # a+b*tn+c*t_n^2=sol[i]. A simple 2x2 matrix
                    #
                    # |t_nm1     t_nm1^2| |b| =solold[i]-solm1[i]
                    # |t_n       t_n^2|  |c| =sol[i]-solm1[i]
                    #
                    # This gives
                    #
                    # |b|=1/(t_nm1*t_n^2-t_nm1^2*t_n) | t_n^2      -t_nm
1^2||solold[i]-solm1[i]|
```

```
                        #  |c|=          | -t_n           t_nm1^2||sol[i]-solm1[i]|
                        #
                        PredMatrix[0][0]=(timeVector[iter-1]-timeVector
[iter-2])
                        PredMatrix[0, 1]=(timeVector[iter-1]-timeVector
[iter-2])*(timeVector[iter-1]-timeVector[iter-2])
                        PredMatrix[0, 1]=(SimTime-timeVector[iter-2])
                        PredMatrix[1]=(SimTime-timeVector[iter-2])*(Sim
Time-timeVector[iter-2])
                        Predrhs[0]=solold[i]-solm1[i]
                        Predrhs[1]=sol[i]-solm1[i]
              Predsol=numpy.matmul(numpy.linalg.inv(PredMatrix),Predrhs)
                        vpred=solm1[i]+Predsol[0]*SimTime+Predsol[1]*SimT
ime*SimTime
                     if PointLocal:
                          for node in range(NumberOfNodes):
                              vkmax=max(abs(sol[node]),abs(sol[node]-Solut
ionCorrection[node]))
                              if abs(vpred-sol[i])>vkmax*reltol+vabstol:
                                  LTEConverged=False
                     elif GlobalTruncation:
                          for node in range(NumberOfNodes):
                              if abs(vpred-sol[i])>vkmax*reltol+vabstol:
                                  LTEConverged=False
                     else:
                          print('Error: Unknown truncation error')
                          sys.exit()
End Python
```

With this code, we can now proceed with simulations having an estimate of the local truncation error. Let us look at the simple circuit we just investigated with this code added to the forward Euler implementation we discussed in Sect. 4.6.1 (see Code 4.10 Sect. 4.9.9). The result is shown in Table 4.1.

We notice the LTE algorithm flags the result and stops the simulation if the timestep is too large. In the exercises, the reader is encouraged to investigate this type of algorithm further.

Interestingly, at times the circuit convergence can be really difficult to achieve, and the simulator can be forced to take increasingly smaller and smaller timesteps. In fact, the step can get so small the whole simulation progress crawls to a virtual halt. These situations are almost always due to some modeling issue in some of the devices used in the circuit. In some simulators, there is a trick where the minimum timestep can be defined so that the LTE criterion is temporarily ignored and simulator can bypass the tough spot. It is called LTEminstep or other similar name.

**Table 4.1** The result of running a forward Euler implementation with an LTE algorithm

| Timestep | LTE OK |
|----------|--------|
| 1e-12 | Yes |
| 1e-11 | No |
| 1e-10 | No |

### 4.6.7  Odd Cases: Things to Watch out for

**Trapezoidal Ringing**

The "trapezoidal ringing" is a fairly known weakness of the trapezoidal method [25]. It is quite characteristic, and one can easily recognize it in the simulator output. It tends to show up around steps, which is why a simulator will often shift to backward Euler around such events.

Since we are now armed with a range of integration methods, let us illustrate this ringing with a very simple test case. Let us use a single capacitor to ground with a voltage source across it, according to the following netlist:

```
*
vdd vdd 0 0
c1 vdd 0 1e-12
netlist 4p8
```

We induce a step by having the initial voltage at the cap be 1 V, by taking advantage of the initial condition feature of the code we have developed. Adding

```
.ic v(a)=1
```

to the netlist will accomplish this, and the next timestep we force it to zero with the voltage source. We use both trap and Gear2. We find the result in Fig. 4.32.

For these simulations, we used 10 ps timestep and 10 steps in time.

**Numerical Loss**

The Gear2 implementation does not suffer from this odd ringing featured with the trap method. This can be a relief. However, it's doing so because of artificial dissipation. The method adds artificial resistors to the circuit, and any ringing will quickly roll-off. We can illustrate this with the following classic example [25] shown in Fig. 4.33.

It shows an LC tank with a shunting resistor. If one perturbs it, by adding

```
.ic v(in)=1
```

to the netlist

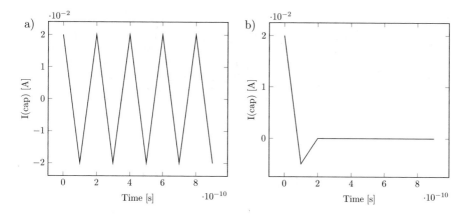

**Fig. 4.32** Trapezoidal ringing. (**a**) Trap method, (**b**) Gear2 method. The oscillatory behavior for trap is very characteristic

**Fig. 4.33** Idealized LC
tank oscillator

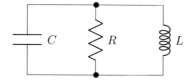

as an initial condition, like we just did in the previous example; it will start oscillating with an amplitude that will decrease with time due to the shunting resistor. The analytical solution is

$$v(t) = \exp(-t/(RC))\sin\sqrt{LC - \frac{1}{4R^2C^2}}t \qquad (4.68)$$

where $R$ represents a real loss in the tank. If $R$ is equal to infinity, the oscillation will go on forever with an amplitude set by the initial conditions. Let us simulate this with our transient simulators we have just developed using the netlist (Figs. 4.34, 4.35, and 4.36):

```
*ideal LC
l1 a vs 1e-9
c1 a vs 1e-12
vs vs 0 0
end
```
**netlist 4p9**

As we just mentioned about the Gear2 method, the amplitude decreases with time due to the "artificial" resistor that in this case acts as a shunting resistor. The backward Euler is even worse! However, the trapezoidal just tugs along with no attenuation in time.

Gear2 and Euler method results in an artificial resistor being inserted across inductors and capacitors. The trapezoidal method does not. Bear this in mind when employing a particular integration method. The trapezoidal has the ringing effect as an implementation side effect.

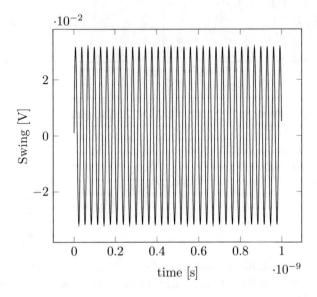

**Fig. 4.34** Ideal LC simulation with trapezoidal method, the output has been subsampled to cover the roll-off of the Gear2 method

This should not discourage the user too much from using backward Euler. Sometimes, when convergence is a real problem, the "resistor" effect with Euler can really help quiet down nodes that misbehave for whatever reason. All these methods have their uses.

## 4.7  Simulator Options Considered

- Forward Euler
- Backward Euler
- Trapezoidal
- Gear2
- reltol
- iabstol
- vntol
- Iteratio
- LTEminstep

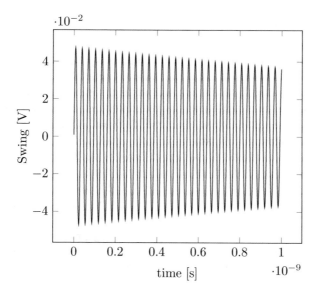

**Fig. 4.35** Ideal LC simulation with Gear2 method. The output has been subsampled to cover the roll-off of the Gear2 method

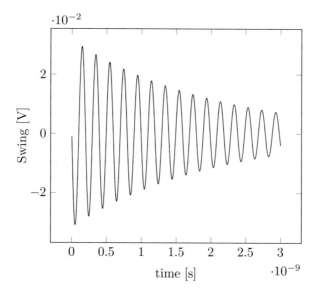

**Fig. 4.36** Ideal LC simulation with time backward Euler. The implementation rolls off quickly so there is no subsampling of the output waveform

## 4.8 Summary

We have looked at a linear simulator for the three basic cases of AC, DC, and transient simulations. The idea of circuit simulation is about the same age as the inventions of the computer. It has been a crucial development tool in the technology revolution, and a good understanding of such tools and what their weaknesses and strengths are is critical to the modern circuit designer. We have in this chapter discussed the basic ideas behind such simulators, and we have studied linear simulators from AC to transient simulations in quite some detail and shown the pros and cons of various integration methods. The belief is the more one is aware of what is going "under the hood," the better equipped one is to handle a design project. Fundamentally the simulator is just a very simple test bench where the user can explore different effects directly. A word of caution, the reader must not fall into the temptation that the suggested code example is the beginning of a full-fledged simulator. Modern simulators are a whole different ball game where things like convergence aids are often proprietary and the result of many years of intense research.

## 4.9 Codes

### 4.9.1 Code 4.2

```
#!/usr/bin/env python3
# -*- coding: utf-8 -*-
"""
Created on Thu Feb 28 22:33:04 2019

@author: mikael
"""
import sys
import numpy
import matplotlib.pyplot as plt
import analogdef as ana
#
# Initialize Variables
#
MaxNumberOfDevices=100
DevType=[0*i for i in range(MaxNumberOfDevices)]
DevLabel=[0*i for i in range(MaxNumberOfDevices)]
DevNode1=[0*i for i in range(MaxNumberOfDevices)]
DevNode2=[0*i for i in range(MaxNumberOfDevices)]
DevNode3=[0*i for i in range(MaxNumberOfDevices)]
DevModel=[0*i for i in range(MaxNumberOfDevices)]
```

```
DevValue=[0*i for i in range(MaxNumberOfDevices)]
Nodes=[]
#
# Read modelfile
#
modeldict=ana.readmodelfile('models.txt')
ICdict={}
Plotdict={}
Printdict={}
Optionsdict={}
SetupDict={}
SimDict={}
#
# Read the netlist
#
DeviceCount=ana.readnetlist('netlist_4p2.txt',modeldict,ICdict,
Plotdict,Printdict,Optionsdict,DevType,DevValue,DevLabel,DevNode1
,DevNode2,DevNode3,DevModel,Nodes,MaxNumberOfDevices)
#
# Create Matrix based on circuit size. We do not implement strict
Modified Nodal Analysis. We keep instead all currents
# but keep referring to the voltages as absolute voltages. We
believe this will make the operation clearer to the user.
#
MatrixSize=DeviceCount+len(Nodes)
#
# The number of branch equations are given by the number of
devices
# The number of KCL equations are given by the number of nodes
in the netlist.
# Hence the matrix size if set by the sum of DeviceCount and
len(Nodes)
#
STA_matrix=[[0 for i in range(MatrixSize)] for j in
range(MatrixSize)]
STA_rhs=[0 for i in range(MatrixSize)]
sol=[0 for i in range(MatrixSize)]
#
# Loop through all devices and create matrix/rhs entries accord-
ing to signature
#
NumberOfNodes=len(Nodes)

for i in range(DeviceCount):
    if DevType[i]=='capacitor' or DevType[i]=='inductor':
```

```
       DevValue[i]*=(0+1j)
    if DevType[i] == 'resistor' or DevType[i] == 'inductor':
      STA_matrix[NumberOfNodes+i][NumberOfNodes+i]=-DevValue[i]
      STA_matrix[NumberOfNodes+i][Nodes.index(DevNode1[i])]=1
      STA_matrix[Nodes.index(DevNode1[i])][NumberOfNodes+i]=1
      STA_matrix[NumberOfNodes+i][Nodes.index(DevNode2[i])]=-1
      STA_matrix[Nodes.index(DevNode2[i])][NumberOfNodes+i]=-1
    if DevType[i]=='capacitor':
      STA_matrix[NumberOfNodes+i][NumberOfNodes+i]=1
      STA_matrix[Nodes.index(DevNode1[i])][NumberOfNodes+i]=1
      STA_matrix[Nodes.index(DevNode2[i])][NumberOfNodes+i]=-1
      STA_matrix[NumberOfNodes+i][Nodes.index(DevNode1[i])]=-Dev
Value[i]
       STA_matrix[NumberOfNodes+i][Nodes.index(DevNode2[i])]=+Dev
Value[i]
     if DevType[i]=='VoltSource':
       if DevNode1[i]!= '0':
         STA_matrix[NumberOfNodes+i][Nodes.index(DevNode1[i])]=1
         STA_matrix[Nodes.index(DevNode1[i])][NumberOfNodes+i]=1
       if DevNode2[i] != '0':
         STA_matrix[NumberOfNodes+i][Nodes.index(DevNode2[i])]=-1
         STA_matrix[Nodes.index(DevNode2[i])][NumberOfNodes+i]=-1
       STA_matrix[NumberOfNodes+i][NumberOfNodes+i]=0
       STA_rhs[NumberOfNodes+i]=DevValue[i]
     if DevType[i]=='CurrentSource':
       if DevNode1[i] != '0' :
         STA_matrix[Nodes.index(DevNode1[i])][NumberOfNodes+i]=1
       if DevNode2[i] != '0' :
         STA_matrix[Nodes.index(DevNode2[i])][NumberOfNodes+i]=-1
       STA_matrix[NumberOfNodes+i][NumberOfNodes+i]=1
       STA_rhs[NumberOfNodes+i]=0 # No AC source, so put to zero
     if DevType[i]=='transistor':
       STA_matrix[NumberOfNodes+i][NumberOfNodes+i]=1
       STA_matrix[NumberOfNodes+i][Nodes.index(DevNode2[i])]=1/
DevValue[i]
       STA_matrix[NumberOfNodes+i][Nodes.index(DevNode3[i])]=-1
/DevValue[i]
       STA_matrix[Nodes.index(DevNode1[i])][NumberOfNodes+i]=1
       STA_matrix[Nodes.index(DevNode3[i])][NumberOfNodes+i]=-1
  #
  #Loop through frequency points
  #
  val=[[0 for i in range(100)] for j in range(MatrixSize)]
  freqpnts=[0 for i in range(100)]
  for iter in range(100):
```

```
        omega=iter*1e8*2*3.14159265
        for i in range(DeviceCount):
            if DevType[i]=='capacitor':
                if DevNode1[i] != '0' :
                    STA_matrix[NumberOfNodes+i][Nodes.index(DevNode1
[i])]=DevValue[i]*omega
                if DevNode2[i] != '0' :
                    STA_matrix[NumberOfNodes+i][Nodes.index(DevNode2
[i])]=-DevValue[i]*omega
            if DevType[i]=='inductor':
                STA_matrix[NumberOfNodes+i][NumberOfNodes+i]=DevVal
ue[i]*omega
        STA_inv=numpy.linalg.inv(STA_matrix)
        sol=numpy.matmul(STA_inv,STA_rhs)
        freqpnts[iter]=iter*1e8
        for j in range(MatrixSize):
            val[j][iter]=abs(sol[j])

    ana.plotdata(Plotdict,NumberOfNodes,freqpnts,val,Nodes)
    if len(Printdict)> 0:
        ana.printdata(Printdict,NumberOfNodes,freqpnts,val,Nodes)
```

## *4.9.2   Code 4.3*

```
#!/usr/bin/env python3
# -*- coding: utf-8 -*-
"""
Created on Thu Feb 28 22:33:04 2019

@author: mikael
"""
import sys
import numpy as np
import matplotlib.pyplot as plt
import analogdef as ana
import math
#
# Initialize Variables
#
MaxNumberOfDevices=100
DevType=[0*i for i in range(MaxNumberOfDevices)]
DevLabel=[0*i for i in range(MaxNumberOfDevices)]
DevNode1=[0*i for i in range(MaxNumberOfDevices)]
```

```
DevNode2=[0*i for i in range(MaxNumberOfDevices)]
DevNode3=[0*i for i in range(MaxNumberOfDevices)]
DevModel=[0*i for i in range(MaxNumberOfDevices)]
DevValue=[0*i for i in range(MaxNumberOfDevices)]
Nodes=[]
k=1.3823e-23 #Avogadro's constant
Temperature=300
#
# Read modelfile
#
modeldict=ana.readmodelfile('models.txt')
ICdict={}
Plotdict={}
Printdict={}
Optdict={}
#
# Read the netlist
#
 DeviceCount=ana.readnetlist('netlist_noise2_4p4.txt',modeldic-
t,ICdict,Plotdict,Printdict,Optdict,DevType,DevValue,DevLabel,Dev
Node1,DevNode2,DevNode3,DevModel,Nodes,MaxNumberOfDevices)
#
# Create Matrix based on circuit size. We do not implement strict
Modified Nodal Analysis. We keep instead all currents
# but keep referring to the voltages as absolute voltages. We
believe this will make the operation clearer to the user.
#
#
# We will add a new current souce
#
DeviceCount=DeviceCount+1
MatrixSize=DeviceCount+len(Nodes)
#
# The number of branch equations are given by the number of
devices
# The number of KCL equations are given by the number of nodes
in the netlist.
# Hence the matrix size if set by the sum of DeviceCount and
len(Nodes)
#
    STA_matrix=[[0  for  i  in  range(MatrixSize)]  for  j  in
range(MatrixSize)]
 STA_rhs=[0 for i in range(MatrixSize)]
#
```

```
# Loop through all devices and create matrix/rhs entries accord-
ing to signature
#
NumberOfNodes=len(Nodes)
for i in range(DeviceCount-1):# We will not do the added current
source yet
    if DevType[i]=='capacitor' or DevType[i]=='inductor':
        DevValue[i]*=(0+1j)
    if DevType[i] == 'resistor' or DevType[i] == 'inductor':
        STA_matrix[NumberOfNodes+i][NumberOfNodes+i]=-DevValue[i]
        STA_matrix[NumberOfNodes+i][Nodes.index(DevNode1[i])]=1
        STA_matrix[Nodes.index(DevNode1[i])][NumberOfNodes+i]=1
        STA_matrix[NumberOfNodes+i][Nodes.index(DevNode2[i])]=-1
        STA_matrix[Nodes.index(DevNode2[i])][NumberOfNodes+i]=-1
    if DevType[i]=='capacitor':
        STA_matrix[NumberOfNodes+i][NumberOfNodes+i]=1
        STA_matrix[Nodes.index(DevNode1[i])][NumberOfNodes+i]=1
        STA_matrix[Nodes.index(DevNode2[i])][NumberOfNodes+i]=-1
        STA_matrix[NumberOfNodes+i][Nodes.index(DevNode1[i])]=-Dev
Value[i]
        STA_matrix[NumberOfNodes+i][Nodes.index(DevNode2[i])]=+D
evValue[i]
    if DevType[i]=='VoltSource':
        if DevNode1[i]!= '0':
         STA_matrix[NumberOfNodes+i][Nodes.index(DevNode1[i])]=1
         STA_matrix[Nodes.index(DevNode1[i])][NumberOfNodes+i]=1
        if DevNode2[i] != '0':
         STA_matrix[NumberOfNodes+i][Nodes.index(DevNode2[i])]=-1
         STA_matrix[Nodes.index(DevNode2[i])][NumberOfNodes+i]=-1
        STA_matrix[NumberOfNodes+i][NumberOfNodes+i]=0
        STA_rhs[NumberOfNodes+i]=0
    if DevType[i]=='CurrentSource':
        if DevNode1[i] != '0' :
        STA_matrix[Nodes.index(DevNode1[i])][NumberOfNodes+i]=1
        if DevNode2[i] != '0' :
        STA_matrix[Nodes.index(DevNode2[i])][NumberOfNodes+i]=-1
        STA_matrix[NumberOfNodes+i][NumberOfNodes+i]=1
        STA_rhs[NumberOfNodes+i]=0

val=[[0 for j in range(100)] for i in range(DeviceCount)]
freqpnts=[i*1e8 for i in range(100)]
for NoiseSource in range(DeviceCount):
    if DevType[NoiseSource]=='resistor':# Here we add the current
noise source in parallel with the resistor
```

```
        if DevNode1[NoiseSource] != '0' :
            STA_matrix[Nodes.index(DevNode1[NoiseSource])][Numb
erOfNodes+DeviceCount-1]=1
        if DevNode2[NoiseSource] != '0' :
            STA_matrix[Nodes.index(DevNode2[NoiseSource])][Numb
erOfNodes+DeviceCount-1]=-1
            STA_matrix[NumberOfNodes+DeviceCount-1][NumberOfNodes
+DeviceCount-1]=1
    #
    #Loop through frequency points
    #
        for iter in range(100):
            omega=iter*1e8*2*3.14159265
            for i in range(DeviceCount):
                if DevType[i]=='capacitor':
                    if DevNode1[i] != '0' :
                        STA_matrix[NumberOfNodes+i][Nodes.index
(DevNode1[i])]=DevValue[i]*omega
                    if DevNode2[i] != '0' :
                        STA_matrix[NumberOfNodes+i][Nodes.index
(DevNode2[i])]=-DevValue[i]*omega
                if DevType[i]=='inductor':
                    STA_matrix[NumberOfNodes+i][NumberOfNodes+i
]=DevValue[i]*omega
                if DevType[i]=='resistor' and i==NoiseSource:
                    STA_rhs[NumberOfNodes+DeviceCount-1]=math.
sqrt(4*k*Temperature/DevValue[i])
            sol=np.matmul(np.linalg.inv(STA_matrix),STA_rhs)
            val[NoiseSource][iter]=abs(sol[Nodes.index('out')])
        if DevNode1[NoiseSource] != '0' :
            STA_matrix[Nodes.index(DevNode1[NoiseSource])][Numb
erOfNodes+DeviceCount-1]=0
        if DevNode2[NoiseSource] != '0' :
            STA_matrix[Nodes.index(DevNode2[NoiseSource])][Numb
erOfNodes+DeviceCount-1]=0
  TotalNoiseSpectrum=[0 for i in range(100)]
  for NoiseSource in range(DeviceCount):
      if DevType[NoiseSource]=='resistor':
          for i in range(100):
              TotalNoiseSpectrum[i]+=abs(val[NoiseSource][i])*abs
(val[NoiseSource][i])

  plt.plot(freqpnts,TotalNoiseSpectrum)
  plt.title('Noise Power vs frequency')
  plt.xlabel('frequency [Hz]')
```

```
plt.ylabel('Noise Power [V^2/Hz]')
#fp=open('../pictures/Noisedata_4p4.csv',"w+")
#fp.write('frequency noise')
#for i in range(100):
#    fp.write("%g " % freqpnts[i])
#    fp.write("%g \n" % TotalNoiseSpectrum[i])
#fp.close()
```

### 4.9.3   Code 4.4

```
#!/usr/bin/env python3
# -*- coding: utf-8 -*-
"""
Created on Thu Feb 28 22:33:04 2019

@author: mikael
"""
import numpy as np
import matplotlib.pyplot as plt
import analogdef as ana
#
# Initialize Variables
#
MaxNumberOfDevices=100
DevType=[0*i for i in range(MaxNumberOfDevices)]
DevLabel=[0*i for i in range(MaxNumberOfDevices)]
DevNode1=[0*i for i in range(MaxNumberOfDevices)]
DevNode2=[0*i for i in range(MaxNumberOfDevices)]
DevNode3=[0*i for i in range(MaxNumberOfDevices)]
DevModel=[0*i for i in range(MaxNumberOfDevices)]
DevValue=[0*i for i in range(MaxNumberOfDevices)]
Nodes=[]
FreqStep=3e7
#
# Read modelfile
#
modeldict=ana.readmodelfile('models.txt')
ICdict={}
Plotdict={}
Printdict={}
Optionsdict={}

#
```

```
# Read the netlist
#
DeviceCount=ana.readnetlist('netlist_ac_stab_4p5.txt',modeldict
,ICdict,Plotdict,Printdict,Optionsdict,DevType,DevValue,DevLabel,
DevNode1,DevNode2,DevNode3,DevModel,Nodes,MaxNumberOfDevices)
#
# Create Matrix based on circuit size. We do not implement strict
Modified Nodal Analysis. We keep instead all currents
# but keep referring to the voltages as absolute voltages. We
believe this will make the operation clearer to the user.
#
MatrixSize=DeviceCount+len(Nodes)
#
# The number of branch equations are given by the number of
devices
# The number of KCL equations are given by the number of nodes
in the netlist.
# Hence the matrix size if set by the sum of DeviceCount and
len(Nodes)
#
    STA_matrix=[[0    for    i    in    range(MatrixSize)]    for    j    in
range(MatrixSize)]
    STA_rhs=[0 for i in range(MatrixSize)]
#
# Loop through all devices and create matrix/rhs entries accord-
ing to signature
#
NumberOfNodes=len(Nodes)
for i in range(DeviceCount):
    if DevType[i]=='capacitor' or DevType[i]=='inductor':
        DevValue[i]*=(0+1j)
    if DevType[i] == 'resistor' or DevType[i] == 'inductor':
        STA_matrix[NumberOfNodes+i][NumberOfNodes+i]=-DevValue[i]
        STA_matrix[NumberOfNodes+i][Nodes.index(DevNode1[i])]=1
        STA_matrix[Nodes.index(DevNode1[i])][NumberOfNodes+i]=1
        STA_matrix[NumberOfNodes+i][Nodes.index(DevNode2[i])]=-1
        STA_matrix[Nodes.index(DevNode2[i])][NumberOfNodes+i]=-1
    if DevType[i]=='capacitor':
        STA_matrix[NumberOfNodes+i][NumberOfNodes+i]=1
        STA_matrix[Nodes.index(DevNode1[i])][NumberOfNodes+i]=1
        STA_matrix[Nodes.index(DevNode2[i])][NumberOfNodes+i]=-1
        STA_matrix[NumberOfNodes+i][Nodes.index(DevNode1[i])]=-Dev
Value[i]
        STA_matrix[NumberOfNodes+i][Nodes.index(DevNode2[i])]=+Dev
Value[i]
```

```
    if DevType[i]=='VoltSource':
      if DevNode1[i]!= '0':
        STA_matrix[NumberOfNodes+i][Nodes.index(DevNode1[i])]=1
        STA_matrix[Nodes.index(DevNode1[i])][NumberOfNodes+i]=1
      if DevNode2[i] != '0':
        STA_matrix[NumberOfNodes+i][Nodes.index(DevNode2[i])]=-1
        STA_matrix[Nodes.index(DevNode2[i])][NumberOfNodes+i]=-1
        STA_matrix[NumberOfNodes+i][NumberOfNodes+i]=0
        STA_rhs[NumberOfNodes+i]=DevValue[i]
    if DevType[i]=='CurrentSource':
      if DevNode1[i] != '0' :
        STA_matrix[Nodes.index(DevNode1[i])][NumberOfNodes+i]=1
      if DevNode2[i] != '0' :
        STA_matrix[Nodes.index(DevNode2[i])][NumberOfNodes+i]=-1
        STA_matrix[NumberOfNodes+i][NumberOfNodes+i]=1
        STA_rhs[NumberOfNodes+i]=0 # No AC source, so put to zero
    if DevType[i]=='transistor':
        STA_matrix[NumberOfNodes+i][NumberOfNodes+i]=1
        STA_matrix[NumberOfNodes+i][Nodes.index(DevNode2[i])]=1/
DevValue[i]
        STA_matrix[NumberOfNodes+i][Nodes.index(DevNode3[i])]=-1/
DevValue[i]
        STA_matrix[Nodes.index(DevNode1[i])][NumberOfNodes+i]=1
        STA_matrix[Nodes.index(DevNode3[i])][NumberOfNodes+i]=-1
  #
  # Neutralize all voltage sources and turn on vstab
  #
  # For this to work properly the current, istab, needs to shoot
into the positive end of vstab. Then the voltage
  # ve is at the positive terminal of vstab and the current out of
the positive terminal of vstab that counts in the
  # second stage
  #
  print('Setting up stab run 1')
  for i in range(DeviceCount):
     if DevType[i]=='VoltSource':
         if DevLabel[i]=='vstab':
             STA_rhs[NumberOfNodes+i]=1
             VElabel=DevNode1[i]
             StabProbeIndex=i
             print('Found stability probe')
         else:
             STA_rhs[NumberOfNodes+i]=0
  #
  #Loop through frequency points
```

```
#
val=[[0 for i in range(100)] for j in range(MatrixSize)]
freqpnts=[0 for i in range(100)]
D=[0+0j for i in range(100)]
B=[0+0j for i in range(100)]
for iter in range(100):
    omega=iter*FreqStep*2*3.14159265
    for i in range(DeviceCount):
        if DevType[i]=='capacitor':
            if DevNode1[i] != '0' :
                STA_matrix[NumberOfNodes+i][Nodes.index(DevNode
1[i])]=DevValue[i]*omega
            if DevNode2[i] != '0' :
                            STA_matrix[NumberOfNodes+i][Nodes.
index(DevNode2[i])]=-DevValue[i]*omega
        if DevType[i]=='inductor':
            STA_matrix[NumberOfNodes+i][NumberOfNodes+i]=DevVal
ue[i]*omega
    STA_inv=np.linalg.inv(STA_matrix)
    sol=np.matmul(STA_inv,STA_rhs)
    freqpnts[iter]=iter*FreqStep
    for j in range(MatrixSize):
        val[j][iter]=abs(sol[j])
        if j<NumberOfNodes:
            if Nodes[j]==VElabel:
                D[iter]=sol[j]
        if j==NumberOfNodes+StabProbeIndex:
            B[iter]=sol[j]

#plt.plot(freqpnts,20*numpy.log10(D))#20*numpy.log10(val[7]))


print('Setting up stab run 2')
for i in range(DeviceCount):
    if DevType[i]=='VoltSource':
        STA_rhs[NumberOfNodes+i]=0
    if DevType[i]=='CurrentSource':
        if DevLabel[i]=='istab':
            STA_rhs[NumberOfNodes+i]=1
            print('Found stability current probe')
#
#Loop through frequency points
#
val=[[0 for i in range(100)] for j in range(MatrixSize)]
freqpnts=[0 for i in range(100)]
```

```
   C=[0+0j for i in range(100)]
   A=[0+0j for i in range(100)]
   for iter in range(100):
        omega=iter*FreqStep*2*3.14159265
        for i in range(DeviceCount):
             if DevType[i]=='capacitor':
                  if DevNode1[i] != '0' :
                       STA_matrix[NumberOfNodes+i][Nodes.index(DevNode
1[i])]=DevValue[i]*omega
                  if DevNode2[i] != '0' :
                       STA_matrix[NumberOfNodes+i][Nodes.index(DevNode
2[i])]=-DevValue[i]*omega
             if DevType[i]=='inductor':
                  STA_matrix[NumberOfNodes+i][NumberOfNodes+i]=DevVal
ue[i]*omega
        STA_inv=np.linalg.inv(STA_matrix)
        sol=np.matmul(STA_inv,STA_rhs)
        freqpnts[iter]=iter*FreqStep
        for j in range(MatrixSize):
             val[j][iter]=abs(sol[j])
             if j<NumberOfNodes:
                  if Nodes[j]==VElabel:
                       C[iter]=sol[j]
             if j==NumberOfNodes+StabProbeIndex:
                  A[iter]=-sol[j] # its the current out of the positive
end that counts

   T=[0+0j for i in range(100)]
   magdB=[0 for i in range(100)]
   phasedegree=[0 for i in range(100)]
   for iter in range(100):
    T[iter]=(2*(A[iter]*D[iter]-B[iter]*C[iter])-A[iter]+D[iter])/
(2*(B[iter]*C[iter]-A[iter]*D[iter])+A[iter]-D[iter]+1)
        magdB[iter]=20*np.log10(np.abs(T[iter]))
    if (180/np.pi*np.arctan(np.imag(T[iter])/np.real(T[iter])))>=0:
        phasedegree[iter]=180-180/np.pi*np.arctan(np.imag(T[iter])/
np.real(T[iter]))
        else:
             phasedegree[iter]=180-(180+180/np.pi*np.arctan(np.imag
(T[iter])/np.real(T[iter])))

   for i in range(100-1):
        if magdB[i]*magdB[i+1]<0:
             print('Phasemargin: ',phasedegree[i])
             print('UGF ',freqpnts[i])
```

```
for i in range(100-1):
    if phasedegree[i]*phasedegree[i+1]<0:
        print('Gainmargin: ',-magdB[i])

#
plt.xscale('log')
plt.title('Gain/Phase vs Frequency')
plt.xlabel('Frequency [Hz]')
plt.ylabel('Phase [degrees], Gain [dB]')
#
plt.plot(freqpnts,magdB,label='Gain')
plt.plot(freqpnts,phasedegree,label='Phase')
plt.subplot(111).legend(loc='upper center', bbox_to_anchor=(0.8,
0.97), shadow=True)
plt.show()
#
if len(Printdict)> 0:
    ana.printdata(Printdict,NumberOfNodes,freqpnts,val,Nodes)
```

### 4.9.4    Code 4.5

```
#!/usr/bin/env python3
# -*- coding: utf-8 -*-
"""
Created on Thu Feb 28 22:33:04 2019

@author: mikael
"""
import sys
import numpy as np
import matplotlib.pyplot as plt
import analogdef as ana
import math
#
# Initialize Variables
#
MaxNumberOfDevices=100
DevType=[0*i for i in range(MaxNumberOfDevices)]
DevLabel=[0*i for i in range(MaxNumberOfDevices)]
DevNode1=[0*i for i in range(MaxNumberOfDevices)]
DevNode2=[0*i for i in range(MaxNumberOfDevices)]
DevNode3=[0*i for i in range(MaxNumberOfDevices)]
DevModel=[0*i for i in range(MaxNumberOfDevices)]
```

```
DevValue=[0*i for i in range(MaxNumberOfDevices)]
Nodes=[]
Ports=['vp1','vp2']
PortNodes=['in','out']
NPorts=2
#
# Read modelfile
#
modeldict=ana.readmodelfile('models.txt')
ICdict={}
Plotdict={}
Printdict={}
Optionsdict={}
#sys.exit()

#
# Read the netlist
#
DeviceCount=ana.readnetlist('netlist_sp2_4p6.txt',modeldict,ICd
ict,Plotdict,Printdict,Optionsdict,DevType,DevValue,DevLabel,DevN
ode1,DevNode2,DevNode3,DevModel,Nodes,MaxNumberOfDevices)
#
# Create Matrix based on circuit size. We do not implement strict
Modified Nodal Analysis. We keep instead all currents
# but keep referring to the voltages as absolute voltages. We
believe this will make the operation clearer to the user.
#
MatrixSize=DeviceCount+len(Nodes)
#
# The number of branch equations are given by the number of
devices
# The number of KCL equations are given by the number of nodes
in the netlist.
# Hence the matrix size if set by the sum of DeviceCount and
len(Nodes)
#
STA_matrix=[[0 for i in range(MatrixSize)] for j in
range(MatrixSize)]
STA_rhs=[0 for i in range(MatrixSize)]
#
# Loop through all devices and create matrix/rhs entries accord-
ing to signature
#
NumberOfNodes=len(Nodes)
for i in range(DeviceCount):
```

```
    if DevType[i]=='capacitor' or DevType[i]=='inductor':
      DevValue[i]*=(0+1j)
    if DevType[i] == 'resistor' or DevType[i] == 'inductor':
      STA_matrix[NumberOfNodes+i][NumberOfNodes+i]=-DevValue[i]
      STA_matrix[NumberOfNodes+i][Nodes.index(DevNode1[i])]=1
      STA_matrix[Nodes.index(DevNode1[i])][NumberOfNodes+i]=1
      STA_matrix[NumberOfNodes+i][Nodes.index(DevNode2[i])]=-1
      STA_matrix[Nodes.index(DevNode2[i])][NumberOfNodes+i]=-1
    if DevType[i]=='capacitor':
      STA_matrix[NumberOfNodes+i][NumberOfNodes+i]=1
      STA_matrix[Nodes.index(DevNode1[i])][NumberOfNodes+i]=1
      STA_matrix[Nodes.index(DevNode2[i])][NumberOfNodes+i]=-1
      STA_matrix[NumberOfNodes+i][Nodes.index(DevNode1[i])]=-Dev
Value[i]
        STA_matrix[NumberOfNodes+i][Nodes.index(DevNode2[i])]=+
DevValue[i]
   if DevType[i]=='VoltSource':
     if DevNode1[i]!= '0':
       STA_matrix[NumberOfNodes+i][Nodes.index(DevNode1[i])]=1
        STA_matrix[Nodes.index(DevNode1[i])][NumberOfNodes+i]=1
     if DevNode2[i] != '0':
       STA_matrix[NumberOfNodes+i][Nodes.index(DevNode2[i])]=-1
       STA_matrix[Nodes.index(DevNode2[i])][NumberOfNodes+i]=-1
       STA_matrix[NumberOfNodes+i][NumberOfNodes+i]=0
       STA_rhs[NumberOfNodes+i]=DevValue[i]
   if DevType[i]=='CurrentSource':
     if DevNode1[i] != '0' :
       STA_matrix[Nodes.index(DevNode1[i])][NumberOfNodes+i]=1
     if DevNode2[i] != '0' :
       STA_matrix[Nodes.index(DevNode2[i])][NumberOfNodes+i]=-1
       STA_matrix[NumberOfNodes+i][NumberOfNodes+i]=1
       STA_rhs[NumberOfNodes+i]=0 # No AC source, so put to zero
  #
  #Loop through frequency points
  #
  val=[[[0 for i in range(1000)] for j in range(NPorts)] for k in
range(NPorts)]
  for port in range(NPorts):
     for iter in range(1000):
        omega=iter*1e8*2*3.14159265
        for i in range(DeviceCount):
           if DevType[i]=='capacitor':
              if DevNode1[i] != '0' :
                 STA_matrix[NumberOfNodes+i][Nodes.index(Dev
Node1[i])]=DevValue[i]*omega
```

```
                    if DevNode2[i] != '0' :
                        STA_matrix[NumberOfNodes+i][Nodes.index(Dev
Node2[i])]=-DevValue[i]*omega
                if DevType[i]=='inductor':
                    STA_matrix[NumberOfNodes+i][NumberOfNodes+i]=De
vValue[i]*omega
                if DevLabel[i]==Ports[port]:
                    print('Exciting port:',DevLabel[i])
                    STA_rhs[NumberOfNodes+i]=2
                else:
                    STA_rhs[NumberOfNodes+i]=0
            STA_inv=np.linalg.inv(STA_matrix)
            sol=np.matmul(STA_inv,STA_rhs)
            for j in range(NPorts):
                print('Sniffing port: ',PortNodes[j])
                if port != j :
                    val[port][j][iter]=20*math.log10(abs(sol[Nodes.
index(PortNodes[j])]))
                else :
                    val[port][j][iter]=20*math.log10(abs(sol[Nodes.
index(PortNodes[j])]-1))
    plt.plot(val[0][0])
    plt.plot(val[0, 1])
    plt.plot(val[0, 1])
    plt.plot(val[1])
```

### 4.9.5   Code 4.6

```python
#!/usr/bin/env python3
# -*- coding: utf-8 -*-
"""
Created on Thu Feb 28 22:33:04 2019

@author: mikael
"""
import numpy as np
import matplotlib.pyplot as plt
import math
import analogdef as ana


#
# Read netlist
#
```

```
MaxNumberOfDevices=100
DevType=[0*i for i in range(MaxNumberOfDevices)]
DevLabel=[0*i for i in range(MaxNumberOfDevices)]
DevNode1=[0*i for i in range(MaxNumberOfDevices)]
DevNode2=[0*i for i in range(MaxNumberOfDevices)]
DevNode3=[0*i for i in range(MaxNumberOfDevices)]
DevModel=[0*i for i in range(MaxNumberOfDevices)]
DevValue=[0*i for i in range(MaxNumberOfDevices)]
Nodes=[]
#
# Read modelfile
#
modeldict=ana.readmodelfile('models.txt')
ICdict={}
Plotdict={}
Printdict={}
Optionsdict={}
Optionsdict['deltaT']=1e-12
Optionsdict['NIterations']=200
#
# Read the netlist
#
DeviceCount=ana.readnetlist('netlist_4p7.txt',modeldict,ICdict,
Plotdict,Printdict,Optionsdict,DevType,DevValue,DevLabel,DevNode1
,DevNode2,DevNode3,DevModel,Nodes,MaxNumberOfDevices)
#
# Create Matrix based on circuit size. We do not implement strict
Modified Nodal Analysis. We keep instead all currents
# but keep referring to the voltages as absolute voltages. We
believe this will make the operation clearer to the user.
#
MatrixSize=DeviceCount+len(Nodes)
    STA_matrix=[[0  for  i  in  range(MatrixSize)]  for  j  in
range(MatrixSize)]
 STA_rhs=[0 for i in range(MatrixSize)]
 sol=[0 for i in range(MatrixSize)]
#
# Create sim parameters
#
deltaT=Optionsdict['deltaT']
NIterations=int(Optionsdict['NIterations'])
#
# Loop through all devices and create matrix/rhs entries accord-
ing to signature
#
```

```
NumberOfNodes=len(Nodes)
for i in range(DeviceCount):
  if DevType[i] != 'VoltSource' and DevType[i] != 'CurrentSource':
      STA_matrix[NumberOfNodes+i][NumberOfNodes+i]=-DevValue[i]
    if DevNode1[i] != '0' :
      STA_matrix[NumberOfNodes+i][Nodes.index(DevNode1[i])]=1
      STA_matrix[Nodes.index(DevNode1[i])][NumberOfNodes+i]=1
    if DevNode2[i] != '0' :
      STA_matrix[NumberOfNodes+i][Nodes.index(DevNode2[i])]=-1
      STA_matrix[Nodes.index(DevNode2[i])][NumberOfNodes+i]=-1
    if DevType[i]=='capacitor':
      STA_matrix[NumberOfNodes+i][NumberOfNodes+i]=0
      if DevNode1[i] != '0' : STA_matrix[NumberOfNodes+i][Nodes.
index(DevNode1[i])]=+1
      if DevNode2[i] != '0' : STA_matrix[NumberOfNodes+i][Nodes.
index(DevNode2[i])]=-1
      if DevNode1[i] != '0' : STA_matrix[Nodes.index(DevNode1[i])]
[NumberOfNodes+i]=+1
      if DevNode2[i] != '0' : STA_matrix[Nodes.index(DevNode2[i])]
[NumberOfNodes+i]=-1
        STA_rhs[NumberOfNodes+i]=sol[Nodes.index(DevNode1[i])]-
sol[Nodes.index(DevNode2[i])]+deltaT*sol[NumberOfNodes+i]/
DevValue[i]
    if DevType[i]=='inductor':
        STA_matrix[NumberOfNodes+i][NumberOfNodes+i]=1
      if DevNode1[i] != '0' : STA_matrix[NumberOfNodes+i][Nodes.
index(DevNode1[i])]=0
      if DevNode2[i] != '0' : STA_matrix[NumberOfNodes+i][Nodes.
index(DevNode2[i])]=0
      if DevNode1[i] != '0' : STA_matrix[Nodes.index(DevNode1[i])]
[NumberOfNodes+i]=1
      if DevNode2[i] != '0' : STA_matrix[Nodes.index(DevNode2[i])]
[NumberOfNodes+i]=-1
        STA_rhs[NumberOfNodes+i]=sol[NumberOfNodes+i]+(sol[Nodes.
index(DevNode1[i])]-sol[Nodes.index(DevNode2[i])])*deltaT/
DevValue[i]
    if DevType[i]=='VoltSource':
        STA_matrix[NumberOfNodes+i][NumberOfNodes+i]=0
      STA_rhs[NumberOfNodes+i]=ana.getSourceValue(DevValue[i],0)
  #
  #Loop through frequency points
  #
  val=[[0 for i in range(NIterations)] for j in range(MatrixSize)]
  timeVector=[0 for i in range(NIterations)]
  for iter in range(NIterations):
```

```
        SimTime=iter*deltaT
        STA_inv=np.linalg.inv(STA_matrix)
        sol=np.matmul(STA_inv,STA_rhs)
        timeVector[iter]=SimTime
        for j in range(MatrixSize):
            val[j][iter]=sol[j]
        for i in range(DeviceCount):
            if DevType[i]=='capacitor':
            STA_rhs[NumberOfNodes+i]=sol[Nodes.index(DevNode1[i])]-
sol[Nodes.index(DevNode2[i])]+deltaT*sol[NumberOfNodes+i]/
DevValue[i]
            if DevType[i]=='inductor':
                STA_rhs[NumberOfNodes+i]=sol[NumberOfNodes+i]+(sol[
Nodes.index(DevNode1[i])]-sol[Nodes.index(DevNode2[i])])*deltaT/
DevValue[i]
            if DevType[i]=='VoltSource':
                STA_rhs[NumberOfNodes+i]=ana.getSourceValue(DevValu
e[i],SimTime)

    ana.plotdata(Plotdict,NumberOfNodes,timeVector,val,Nodes)
```

### 4.9.6  Code 4.7

```python
#!/usr/bin/env python3
# -*- coding: utf-8 -*-
"""
Created on Thu Feb 28 22:33:04 2019

@author: mikael
"""
import numpy as np
import matplotlib.pyplot as plt
import math
import analogdef as ana

MaxNumberOfDevices=100
DevType=[0*i for i in range(MaxNumberOfDevices)]
DevLabel=[0*i for i in range(MaxNumberOfDevices)]
DevNode1=[0*i for i in range(MaxNumberOfDevices)]
DevNode2=[0*i for i in range(MaxNumberOfDevices)]
DevNode3=[0*i for i in range(MaxNumberOfDevices)]
DevModel=[0*i for i in range(MaxNumberOfDevices)]
DevValue=[0*i for i in range(MaxNumberOfDevices)]
```

```
Nodes=[]
#
# Read modelfile
#
modeldict=ana.readmodelfile('models.txt')
ICdict={}
Plotdict={}
Printdict={}
Optionsdict={}
Optionsdict['deltaT']=1e-12
Optionsdict['NIterations']=200
#
# Read the netlist
#
DeviceCount=ana.readnetlist('netlist_4p7.txt',modeldict,ICdict,
Plotdict,Printdict,Optionsdict,DevType,DevValue,DevLabel,DevNode1
,DevNode2,DevNode3,DevModel,Nodes,MaxNumberOfDevices)
#
# Create Matrix based on circuit size. We do not implement strict
Modified Nodal Analysis. We keep instead all currents
# but keep referring to the voltages as absolute voltages. We
believe this will make the operation clearer to the user.
#
MatrixSize=DeviceCount+len(Nodes)
    STA_matrix=[[0    for    i    in    range(MatrixSize)]    for    j    in
range(MatrixSize)]
  STA_rhs=[0 for i in range(MatrixSize)]
  sol=[0 for i in range(MatrixSize)]
#
# update initial conditions if present
#
NumberOfNodes=len(Nodes)
if len(ICdict)>0:
    for i in range(len(ICdict)):
        for j in range(NumberOfNodes):
            if Nodes[j]==ICdict[i]['NodeName']:
                sol[j]=ICdict[i]['Value']
                print('Setting ',Nodes[j],' to ',sol[j])
#
# Create sim parameters
#
deltaT=Optionsdict['deltaT']
NIterations=int(Optionsdict['NIterations'])
```

```
        #
        # Loop through all devices and create matrix/rhs entries accord-
ing to signature
        #
        for i in range(DeviceCount):
          if DevType[i] != 'VoltSource' and DevType[i] != 'CurrentSource':
              STA_matrix[NumberOfNodes+i][NumberOfNodes+i]=-DevValue[i]
          if DevNode1[i] != '0' :
              STA_matrix[NumberOfNodes+i][Nodes.index(DevNode1[i])]=1
              STA_matrix[Nodes.index(DevNode1[i])][NumberOfNodes+i]=1
          if DevNode2[i] != '0' :
              STA_matrix[NumberOfNodes+i][Nodes.index(DevNode2[i])]=-1
              STA_matrix[Nodes.index(DevNode2[i])][NumberOfNodes+i]=-1
          if DevType[i]=='capacitor':
              STA_matrix[NumberOfNodes+i][NumberOfNodes+i]=1
              if DevNode1[i] != '0' : STA_matrix[NumberOfNodes+i][Nodes.
index(DevNode1[i])]=-DevValue[i]/deltaT
              if DevNode2[i] != '0' : STA_matrix[NumberOfNodes+i][Nodes.
index(DevNode2[i])]=DevValue[i]/deltaT
              if DevNode1[i] != '0' : STA_matrix[Nodes.index(DevNode1[i])]
[NumberOfNodes+i]=1
              if DevNode2[i] != '0' : STA_matrix[Nodes.index(DevNode2[i])]
[NumberOfNodes+i]=-1
                  if DevNode1[i] != '0' and DevNode2[i] != '0':
                  STA_rhs[NumberOfNodes+i]=-DevValue[i]/deltaT*(sol[Nodes.
index(DevNode1[i])]-sol[Nodes.index(DevNode2[i])])
                  if DevNode1[i] == '0':
                          STA_rhs[NumberOfNodes+i]=-DevValue[i]/deltaT*(-
sol[Nodes.index(DevNode2[i])])
                  if DevNode2[i] == '0':
                  STA_rhs[NumberOfNodes+i]=-DevValue[i]/deltaT*(sol[Nodes.
index(DevNode1[i])])
              if DevType[i]=='inductor':
                  STA_matrix[NumberOfNodes+i][NumberOfNodes+i]=1
                  if DevNode1[i] != '0' : STA_matrix[NumberOfNodes+i][Nodes.
index(DevNode1[i])]=-deltaT/DevValue[i]
                  if DevNode2[i] != '0' : STA_matrix[NumberOfNodes+i][Nodes.
index(DevNode2[i])]=deltaT/DevValue[i]
              if DevNode1[i] != '0' : STA_matrix[Nodes.index(DevNode1[i])]
[NumberOfNodes+i]=1
              if DevNode2[i] != '0' : STA_matrix[Nodes.index(DevNode2[i])]
[NumberOfNodes+i]=-1
                  STA_rhs[NumberOfNodes+i]=sol[NumberOfNodes+i]
              if DevType[i]=='VoltSource':
```

```
            STA_matrix[NumberOfNodes+i][NumberOfNodes+i]=0
        STA_rhs[NumberOfNodes+i]=ana.getSourceValue(DevValue[i],0)
    #
    #Loop through frequency points
    #
    val=[[0 for i in range(NIterations)] for j in range(MatrixSize)]
    timeVector=[0 for i in range(NIterations)]
    for iter in range(NIterations):
        SimTime=iter*deltaT
        STA_inv=np.linalg.inv(STA_matrix)
        sol=np.matmul(STA_inv,STA_rhs)
        timeVector[iter]=SimTime
        for j in range(MatrixSize):
            val[j][iter]=sol[j]
        for i in range(DeviceCount):
            if DevType[i]=='capacitor':
                if DevNode1[i] != '0' and DevNode2[i] != '0':
                    STA_rhs[NumberOfNodes+i]=-DevValue[i]/deltaT*(sol
[Nodes.index(DevNode1[i])]-sol[Nodes.index(DevNode2[i])])
                if DevNode1[i] == '0':
                    STA_rhs[NumberOfNodes+i]=-DevValue[i]/deltaT*(-sol
[Nodes.index(DevNode2[i])])
                if DevNode2[i] == '0':
                    STA_rhs[NumberOfNodes+i]=-DevValue[i]/deltaT*(sol
[Nodes.index(DevNode1[i])])
            if DevType[i]=='inductor':
                STA_rhs[NumberOfNodes+i]=sol[NumberOfNodes+i]
            if DevType[i]=='VoltSource':
                STA_rhs[NumberOfNodes+i]=ana.getSourceValue(DevValu
e[i],SimTime)
    #
    ana.plotdata(Plotdict,NumberOfNodes,timeVector,val,Nodes)
```

### 4.9.7   Code 4.8

```
#!/usr/bin/env python3
# -*- coding: utf-8 -*-
"""
Created on Thu Feb 28 22:33:04 2019

@author: mikael
"""
import numpy as np
```

```
import matplotlib.pyplot as plt
import math
import analogdef as ana
#
# Initialize Variables
#
MaxNumberOfDevices=100
DevType=[0*i for i in range(MaxNumberOfDevices)]
DevLabel=[0*i for i in range(MaxNumberOfDevices)]
DevNode1=[0*i for i in range(MaxNumberOfDevices)]
DevNode2=[0*i for i in range(MaxNumberOfDevices)]
DevNode3=[0*i for i in range(MaxNumberOfDevices)]
DevModel=[0*i for i in range(MaxNumberOfDevices)]
DevValue=[0*i for i in range(MaxNumberOfDevices)]
Nodes=[]
#
# Read modelfile
#
modeldict=ana.readmodelfile('models.txt')
ICdict={}
Plotdict={}
Printdict={}
Optionsdict={}
Optionsdict['deltaT']=1e-12
Optionsdict['NIterations']=200
#
# Read the netlist
#
DeviceCount=ana.readnetlist('netlist_4p9.txt',modeldict,ICdict,
Plotdict,Printdict,Optionsdict,DevType,DevValue,DevLabel,DevNode1
,DevNode2,DevNode3,DevModel,Nodes,MaxNumberOfDevices)
#
# Create Matrix based on circuit size. We do not implement strict
Modified Nodal Analysis. We keep instead all currents
# but keep referring to the voltages as absolute voltages. We
believe this will make the operation clearer to the user.
#
MatrixSize=DeviceCount+len(Nodes)
    STA_matrix=[[0 for i in range(MatrixSize)] for j in
range(MatrixSize)]
STA_rhs=[0 for i in range(MatrixSize)]
sol=[0 for i in range(MatrixSize)]
#
# update initial conditions if present
#
```

```
NumberOfNodes=len(Nodes)
if len(ICdict)>0:
    for i in range(len(ICdict)):
        for j in range(NumberOfNodes):
            if Nodes[j]==ICdict[i]['NodeName']:
                sol[j]=ICdict[i]['Value']
                print('Setting ',Nodes[j],' to ',sol[j])
#
# Create sim parameters
#
deltaT=Optionsdict['deltaT']
NIterations=int(Optionsdict['NIterations'])
#
# Loop through all devices and create matrix/rhs entries accord-
ing to signature
#
for i in range(DeviceCount):
  if DevType[i] != 'VoltSource' and DevType[i] != 'CurrentSource':
      STA_matrix[NumberOfNodes+i][NumberOfNodes+i]=-DevValue[i]
  if DevNode1[i] != '0' :
      STA_matrix[NumberOfNodes+i][Nodes.index(DevNode1[i])]=1
      STA_matrix[Nodes.index(DevNode1[i])][NumberOfNodes+i]=1
  if DevNode2[i] != '0' :
      STA_matrix[NumberOfNodes+i][Nodes.index(DevNode2[i])]=-1
      STA_matrix[Nodes.index(DevNode2[i])][NumberOfNodes+i]=-1
    if DevType[i]=='capacitor':
        STA_matrix[NumberOfNodes+i][NumberOfNodes+i]=1
      if DevNode1[i] != '0' : STA_matrix[NumberOfNodes+i][Nodes.
index(DevNode1[i])]=-2*DevValue[i]/deltaT
        if DevNode2[i] != '0' : STA_matrix[NumberOfNodes+i][Nodes.
index(DevNode2[i])]=2*DevValue[i]/deltaT
        if DevNode1[i] != '0' : STA_matrix[Nodes.index(DevNode1[i])]
[NumberOfNodes+i]=1
        if DevNode2[i] != '0' : STA_matrix[Nodes.index(DevNode2[i])]
[NumberOfNodes+i]=-1
          if DevNode1[i] != '0' and DevNode2[i] != '0':
          STA_rhs[NumberOfNodes+i]=-2*DevValue[i]/deltaT*(sol[Nodes.
index(DevNode1[i])]-sol[Nodes.index(DevNode2[i])])-sol
[NumberOfNodes+i]
            if DevNode1[i] == '0':
                STA_rhs[NumberOfNodes+i]=-2*DevValue[i]/deltaT*(-sol
[Nodes.index(DevNode2[i])])-sol[NumberOfNodes+i]
            if DevNode2[i] == '0':
            STA_rhs[NumberOfNodes+i]=-2*DevValue[i]/deltaT*(sol[Nodes.
index(DevNode1[i])])-sol[NumberOfNodes+i]
```

```
    if DevType[i]=='inductor':
        STA_matrix[NumberOfNodes+i][NumberOfNodes+i]=1
      if DevNode1[i] != '0' : STA_matrix[NumberOfNodes+i][Nodes.
index(DevNode1[i])]=-deltaT/(2*DevValue[i])
        if DevNode2[i] != '0' : STA_matrix[NumberOfNodes+i][Nodes.
index(DevNode2[i])]=deltaT/(2*DevValue[i])
        if DevNode1[i] != '0' : STA_matrix[Nodes.index(DevNode1[i])]
[NumberOfNodes+i]=1
        if DevNode2[i] != '0' : STA_matrix[Nodes.index(DevNode2[i])]
[NumberOfNodes+i]=-1
        if DevNode1[i] != '0' and DevNode2[i] != '0':
            STA_rhs[NumberOfNodes+i]=sol[NumberOfNodes+i]+delta
T*(sol[Nodes.index(DevNode1[i])]-sol[Nodes.index(DevNode2[i])])/
(2*DevValue[i])
        if DevNode1[i] == '0':
            STA_rhs[NumberOfNodes+i]=sol[NumberOfNodes+i]+del
taT*(-sol[Nodes.index(DevNode2[i])])/(2*DevValue[i])
        if DevNode2[i] == '0':
            STA_rhs[NumberOfNodes+i]=sol[NumberOfNodes+i]+delta
T*(sol[Nodes.index(DevNode1[i])])/(2*DevValue[i])
    if DevType[i]=='VoltSource':
        STA_matrix[NumberOfNodes+i][NumberOfNodes+i]=0
      STA_rhs[NumberOfNodes+i]=ana.getSourceValue(DevValue[i],0)
 #
 #Loop through frequency points
 #
 val=[[0 for i in range(NIterations)] for j in range(MatrixSize)]
 timeVector=[0 for i in range(NIterations)]
 for iter in range(NIterations):
     SimTime=iter*deltaT
     STA_inv=np.linalg.inv(STA_matrix)
     sol=np.matmul(STA_inv,STA_rhs)
     timeVector[iter]=SimTime
     for j in range(MatrixSize):
         val[j][iter]=sol[j]
     for i in range(DeviceCount):
         if DevType[i]=='capacitor':
             if DevNode1[i] != '0' and DevNode2[i] != '0':
                 STA_rhs[NumberOfNodes+i]=-2*DevValue[i]/deltaT*
(sol[Nodes.index(DevNode1[i])]-sol[Nodes.index(DevNode2[i])])-
sol[NumberOfNodes+i]
             if DevNode1[i] == '0':
                 STA_rhs[NumberOfNodes+i]=-2*DevValue[i]/deltaT*(-
sol[Nodes.index(DevNode2[i])])-sol[NumberOfNodes+i]
             if DevNode2[i] == '0':
                 STA_rhs[NumberOfNodes+i]=-2*DevValue[i]/deltaT*
```

```
(sol[Nodes.index(DevNode1[i])])-sol[NumberOfNodes+i]
            if DevType[i]=='inductor':
                if DevNode1[i] != '0' and DevNode2[i] != '0':
                    STA_rhs[NumberOfNodes+i]=sol[NumberOfNodes+i]+d
eltaT*(sol[Nodes.index(DevNode1[i])]-sol[Nodes.index
(DevNode2[i])])/(2*DevValue[i])
                if DevNode1[i] == '0':
                    STA_rhs[NumberOfNodes+i]=sol[NumberOfNodes+i]+d
eltaT*(-sol[Nodes.index(DevNode2[i])])/(2*DevValue[i])
                if DevNode2[i] == '0':
                    STA_rhs[NumberOfNodes+i]=sol[NumberOfNodes+i]+d
eltaT*(sol[Nodes.index(DevNode1[i])])/(2*DevValue[i])
            if DevType[i]=='VoltSource':
                STA_rhs[NumberOfNodes+i]=ana.getSourceValue(DevValu
e[i],SimTime)
    #
    ana.plotdata(Plotdict,NumberOfNodes,timeVector,val,Nodes)
```

## 4.9.8   Code 4.9

```
#!/usr/bin/env python3
# -*- coding: utf-8 -*-
"""
Created on Thu Feb 28 22:33:04 2019

@author: mikael
"""
import numpy as np
import matplotlib.pyplot as plt
import math
import analogdef as ana
#
# Initialize Variables
#
MaxNumberOfDevices=100
DevType=[0*i for i in range(MaxNumberOfDevices)]
DevLabel=[0*i for i in range(MaxNumberOfDevices)]
DevNode1=[0*i for i in range(MaxNumberOfDevices)]
DevNode2=[0*i for i in range(MaxNumberOfDevices)]
DevNode3=[0*i for i in range(MaxNumberOfDevices)]
DevModel=[0*i for i in range(MaxNumberOfDevices)]
DevValue=[0*i for i in range(MaxNumberOfDevices)]
Nodes=[]
```

```
#
# Read modelfile
#
modeldict=ana.readmodelfile('models.txt')
ICdict={}
Plotdict={}
Printdict={}
Optionsdict={}
Optionsdict['deltaT']=1e-12
Optionsdict['NIterations']=200
#
# Read the netlist
#
DeviceCount=ana.readnetlist('netlist_4p9.txt',modeldict,ICdict,
Plotdict,Printdict,Optionsdict,DevType,DevValue,DevLabel,DevNode1
,DevNode2,DevNode3,DevModel,Nodes,MaxNumberOfDevices)
#
# Create Matrix based on circuit size. We do not implement strict
Modified Nodal Analysis. We keep instead all currents
# but keep referring to the voltages as absolute voltages. We
believe this will make the operation clearer to the user.
MatrixSize=DeviceCount+len(Nodes)
    STA_matrix=[[0 for i in range(MatrixSize)]  for  j  in
range(MatrixSize)]
STA_rhs=[0 for i in range(MatrixSize)]
sol=[0 for i in range(MatrixSize)]
solm1=[0 for i in range(MatrixSize)]
#
# Create sim parameters
#
deltaT=Optionsdict['deltaT']
NIterations=int(Optionsdict['NIterations'])
#
# update initial conditions if present
#
NumberOfNodes=len(Nodes)
if len(ICdict)>0:
    for i in range(len(ICdict)):
        for j in range(NumberOfNodes):
            if Nodes[j]==ICdict[i]['NodeName']:
                sol[j]=ICdict[i]['Value']
                print('Setting ',Nodes[j],' to ',sol[j])
#
# Loop through all devices and create matrix/rhs entries accord-
ing to signature
```

```
     #
     for i in range(DeviceCount):
       if DevType[i] != 'VoltSource' and DevType[i] != 'CurrentSource':
           STA_matrix[NumberOfNodes+i][NumberOfNodes+i]=-DevValue[i]
       if DevNode1[i] != '0' :
           STA_matrix[NumberOfNodes+i][Nodes.index(DevNode1[i])]=1
           STA_matrix[Nodes.index(DevNode1[i])][NumberOfNodes+i]=1
       if DevNode2[i] != '0' :
           STA_matrix[NumberOfNodes+i][Nodes.index(DevNode2[i])]=-1
           STA_matrix[Nodes.index(DevNode2[i])][NumberOfNodes+i]=-1
       if DevType[i]=='capacitor':
           STA_matrix[NumberOfNodes+i][NumberOfNodes+i]=1
           if DevNode1[i] != '0' : STA_matrix[NumberOfNodes+i][Nodes.
index(DevNode1[i])]=-3/2.0*DevValue[i]/deltaT
           if DevNode2[i] != '0' : STA_matrix[NumberOfNodes+i][Nodes.
index(DevNode2[i])]=3/2.0*DevValue[i]/deltaT
           if DevNode1[i] != '0' : STA_matrix[Nodes.index(DevNode1[i])]
[NumberOfNodes+i]=1
           if DevNode2[i] != '0' : STA_matrix[Nodes.index(DevNode2[i])]
[NumberOfNodes+i]=-1
             if DevNode1[i] != '0' and DevNode2[i] != '0':
                 STA_rhs[NumberOfNodes+i]=DevValue[i]/deltaT*(-2*(sol
[Nodes.index(DevNode1[i])]-sol[Nodes.index(DevNode2[i])])+1/2*(so
lm1[Nodes.index(DevNode1[i])]-solm1[Nodes.index(DevNode2[i])]) )
             if DevNode1[i] == '0':
                 STA_rhs[NumberOfNodes+i]=DevValue[i]/deltaT*(-2*(-sol
[Nodes.index(DevNode2[i])])+1/2*(-solm1[Nodes.index(DevNode2[i])] ))
             if DevNode2[i] == '0':
                 STA_rhs[NumberOfNodes+i]=DevValue[i]/deltaT*(-2*(sol
[Nodes.index(DevNode1[i])])+1/2*(solm1[Nodes.index(DevNode1[i])] ))
       if DevType[i]=='inductor':
           STA_matrix[NumberOfNodes+i][NumberOfNodes+i]=1
           if DevNode1[i] != '0' : STA_matrix[NumberOfNodes+i][Nodes.
index(DevNode1[i])]=-2/3*deltaT/DevValue[i]
           if DevNode2[i] != '0' : STA_matrix[NumberOfNodes+i][Nodes.
index(DevNode2[i])]=2/3*deltaT/DevValue[i]
           if DevNode1[i] != '0' : STA_matrix[Nodes.index(DevNode1[i])]
[NumberOfNodes+i]=1
           if DevNode2[i] != '0' : STA_matrix[Nodes.index(DevNode2[i])]
[NumberOfNodes+i]=-1
             STA_rhs[NumberOfNodes+i]=4/3*sol[NumberOfNodes+i]-1/3*
solm1[NumberOfNodes+i]
       if DevType[i]=='VoltSource':
           STA_matrix[NumberOfNodes+i][NumberOfNodes+i]=0
           STA_rhs[NumberOfNodes+i]=ana.getSourceValue(DevValue[i],0)
```

```
#
#Loop through frequency points
#
val=[[0 for i in range(NIterations)] for j in range(MatrixSize)]
timeVector=[0 for i in range(NIterations)]
for iter in range(NIterations):
    SimTime=iter*deltaT
    STA_inv=np.linalg.inv(STA_matrix)
    solm1=sol[:]
    sol=np.matmul(STA_inv,STA_rhs)
    timeVector[iter]=SimTime
    for j in range(MatrixSize):
        val[j][iter]=sol[j]
    for i in range(DeviceCount):
        if DevType[i]=='capacitor':
            if DevNode1[i] != '0' and DevNode2[i] != '0':
                STA_rhs[NumberOfNodes+i]=DevValue[i]/deltaT*(-
2*(sol[Nodes.index(DevNode1[i])]-sol[Nodes.index(DevNode2[i])])+1
/2*(solm1[Nodes.index(DevNode1[i])]-solm1[Nodes.index
(DevNode2[i])]) )
            if DevNode1[i] == '0':
                STA_rhs[NumberOfNodes+i]=DevValue[i]/deltaT*(-
2*(-sol[Nodes.index(DevNode2[i])])+1/2*(-solm1[Nodes.index
(DevNode2[i])] ))
            if DevNode2[i] == '0':
                STA_rhs[NumberOfNodes+i]=DevValue[i]/deltaT*(-
2*(sol[Nodes.index(DevNode1[i])])+1/2*(solm1[Nodes.
index(DevNode1[i])] ))
        if DevType[i]=='inductor':
                STA_rhs[NumberOfNodes+i]=4/3*sol[NumberOfNode
s+i]-1/3*solm1[NumberOfNodes+i]
        if DevType[i]=='VoltSource':
            STA_rhs[NumberOfNodes+i]=ana.getSourceValue(DevValu
e[i],SimTime)
    #
  ana.plotdata(Plotdict,NumberOfNodes,timeVector,val,Nodes)
```

## 4.9.9  Code 4.10

```
#!/usr/bin/env python3
# -*- coding: utf-8 -*-
"""
Created on Thu Feb 28 22:33:04 2019
```

```
@author: mikael
"""
import numpy
import matplotlib.pyplot as plt
import math
import sys
import analogdef as ana

#
# Read netlist
#
MaxNumberOfDevices=100
DevType=[0*i for i in range(MaxNumberOfDevices)]
DevLabel=[0*i for i in range(MaxNumberOfDevices)]
DevNode1=[0*i for i in range(MaxNumberOfDevices)]
DevNode2=[0*i for i in range(MaxNumberOfDevices)]
DevNode3=[0*i for i in range(MaxNumberOfDevices)]
DevModel=[0*i for i in range(MaxNumberOfDevices)]
DevValue=[0*i for i in range(MaxNumberOfDevices)]
Nodes=[]
vkmax=0
#
# Read modelfile
#
modeldict=ana.readmodelfile('models.txt')
ICdict={}
Plotdict={}
Printdict={}
Optionsdict={}
Optionsdict['reltol']=1e-2
Optionsdict['iabstol']=1e-7
Optionsdict['vabstol']=1e-2
Optionsdict['lteratio']=2
Optionsdict['deltaT']=1e-12
Optionsdict['NIterations']=200
Optionsdict['GlobalTruncation']=True
#
# Read the netlist
#
DeviceCount=ana.readnetlist('netlist_4p7.txt',modeldict,ICdict,
Plotdict,Printdict,Optionsdict,DevType,DevValue,DevLabel,DevNode1
,DevNode2,DevNode3,DevModel,Nodes,MaxNumberOfDevices)
#
# Create Matrix based on circuit size. We do not implement strict
Modified Nodal Analysis. We keep instead all currents
```

```
    # but keep referring to the voltages as absolute voltages. We
believe this will make the operation clearer to the user.
    #
    MatrixSize=DeviceCount+len(Nodes)
    STA_matrix=[[0   for   i   in   range(MatrixSize)]   for   j   in
range(MatrixSize)]
    STA_rhs=[0 for i in range(MatrixSize)]
    sol=[0 for i in range(MatrixSize)]
    solold=[0 for i in range(MatrixSize)]
    solm1=[0 for i in range(MatrixSize)]
    solm2=[0 for i in range(MatrixSize)]
    #
    # Create sim parameters
    #
    deltaT=Optionsdict['deltaT']
    NIterations=int(Optionsdict['NIterations'])
    GlobalTruncation=Optionsdict['GlobalTruncation']
    PointLocal=not GlobalTruncation
    reltol=Optionsdict['reltol']
    iabstol=Optionsdict['iabstol']
    vabstol=Optionsdict['vabstol']
    lteratio=Optionsdict['lteratio']
    #
    # Loop through all devices and create matrix/rhs entries accord-
ing to signature
    #
    NumberOfNodes=len(Nodes)
    for i in range(DeviceCount):
      if DevType[i] != 'VoltSource' and DevType[i] != 'CurrentSource':
        STA_matrix[NumberOfNodes+i][NumberOfNodes+i]=-DevValue[i]
      if DevNode1[i] != '0' :
        STA_matrix[NumberOfNodes+i][Nodes.index(DevNode1[i])]=1
        STA_matrix[Nodes.index(DevNode1[i])][NumberOfNodes+i]=1
      if DevNode2[i] != '0' :
        STA_matrix[NumberOfNodes+i][Nodes.index(DevNode2[i])]=-1
        STA_matrix[Nodes.index(DevNode2[i])][NumberOfNodes+i]=-1
      if DevType[i]=='capacitor':
        STA_matrix[NumberOfNodes+i][NumberOfNodes+i]=0
          if DevNode1[i] != '0' : STA_matrix[NumberOfNodes+i][Nodes.
index(DevNode1[i])]=+1
          if DevNode2[i] != '0' : STA_matrix[NumberOfNodes+i][Nodes.
index(DevNode2[i])]=-1
          if DevNode1[i] != '0' : STA_matrix[Nodes.index(DevNode1[i])]
[NumberOfNodes+i]=+1
          if DevNode2[i] != '0' : STA_matrix[Nodes.index(DevNode2[i])]
```

```
[NumberOfNodes+i]=-1
            STA_rhs[NumberOfNodes+i]=sol[Nodes.index(DevNode1[i])]-
sol[Nodes.index(DevNode2[i])]+deltaT*sol[NumberOfNodes+i]/
DevValue[i]
        if DevType[i]=='inductor':
          STA_matrix[NumberOfNodes+i][NumberOfNodes+i]=1
            if DevNode1[i] != '0' : STA_matrix[NumberOfNodes+i][Nodes.
index(DevNode1[i])]=0
            if DevNode2[i] != '0' : STA_matrix[NumberOfNodes+i][Nodes.
index(DevNode2[i])]=0
            if DevNode1[i] != '0' : STA_matrix[Nodes.index(DevNode1[i])]
[NumberOfNodes+i]=1
            if DevNode2[i] != '0' : STA_matrix[Nodes.index(DevNode2[i])]
[NumberOfNodes+i]=-1
            STA_rhs[NumberOfNodes+i]=sol[NumberOfNodes+i]+(sol[Nodes.
index(DevNode1[i])]-sol[Nodes.index(DevNode2[i])])*deltaT/
DevValue[i]
        if DevType[i]=='VoltSource':
            STA_matrix[NumberOfNodes+i][NumberOfNodes+i]=0
          STA_rhs[NumberOfNodes+i]=ana.getSourceValue(DevValue[i],0)
  #
  #Loop through frequency points
  #
  val=[[0 for i in range(NIterations)] for j in range(MatrixSize)]
  timeVector=[0 for i in range(NIterations)]
  PredMatrix=[[0 for i in range(2)] for j in range(2)]
  Predrhs=[0 for i in range(2)]
  for iter in range(NIterations):
      SimTime=iter*deltaT
      STA_inv=numpy.linalg.inv(STA_matrix)
      solm2=[solm1[i] for i in range(MatrixSize)]
      solm1=[solold[i] for i in range(MatrixSize)]
      solold=[sol[i] for i in range(MatrixSize)]
      sol=numpy.matmul(STA_inv,STA_rhs)
      for node in range(NumberOfNodes):
          vkmax=max(vkmax,abs(sol[node]))
      timeVector[iter]=SimTime
      for j in range(MatrixSize):
          val[j][iter]=sol[j]
      for i in range(DeviceCount):
          if DevType[i]=='capacitor':
          STA_rhs[NumberOfNodes+i]=sol[Nodes.index(DevNode1[i])]-
sol[Nodes.index(DevNode2[i])]+deltaT*sol[NumberOfNodes+i]/
DevValue[i]
```

```
        if DevType[i]=='inductor':
            STA_rhs[NumberOfNodes+i]=sol[NumberOfNodes+i]+(sol[
Nodes.index(DevNode1[i])]-sol[Nodes.index(DevNode2[i])])*deltaT/
DevValue[i]
        if DevType[i]=='VoltSource':
            STA_rhs[NumberOfNodes+i]=ana.getSourceValue(DevValu
e[i],SimTime)
    if iter>2:
        LTEConverged=True
        for i in range(NumberOfNodes):
            tau1=(timeVector[iter-2]-timeVector[iter-3])
            tau2=(timeVector[iter-1]-timeVector[iter-3])
            PredMatrix[0][0]=tau2
            PredMatrix[0, 1]=tau2*tau2
            PredMatrix[0, 1]=tau1
            PredMatrix[1]=tau1*tau1
            Predrhs[0]=solold[i]-solm2[i]
            Predrhs[1]=solm1[i]-solm2[i]
            Predsol=numpy.matmul(numpy.linalg.inv(PredMatrix),
Predrhs)
            vpred=solm2[i]+Predsol[0]*(SimTime-timeVector[iter-
3])+Predsol[1]*(SimTime-timeVector[iter-3])*(SimTime-
timeVector[iter-3])
            if PointLocal:
                for node in range(NumberOfNodes):
                    vkmax=max(abs(sol[node]),abs(solm1[node]))
                    print('Is vkmax correct here?')
                    if abs(vpred-sol[i])> lteratio*(vkmax*relto
l+vabstol):
                        LTEConverged=False
            elif GlobalTruncation:
                for node in range(NumberOfNodes):
                    if abs(vpred-sol[i])> lteratio*(vkmax*relto
l+vabstol):
                        LTEConverged=False
            else:
                print('Error: Unknown truncation error')
                sys.exit()
        if not LTEConverged:
            print('LTE NOT converging, change time step')
            sys.exit(0)

  ana.plotdata(Plotdict,NumberOfNodes,timeVector,val,Nodes)
```

## 4.10   Exercises

1. Play with the codes and change the netlists provided. Can you run the code on large networks? What is the limitation? Any errors please go to www.fastictechniques.com and report!
2. Build a DC linear simulator where the netlist consists of

   (a) Resistors and capacitors only.
   (b) Inductors that should be shorted.

3. Implement a mutual inductor element in the simulator.
4. Build a transfer function analysis routine that simulates the AC response from all independent sources to a user-defined output.
5. Build a sensitivity analysis routine that simulates a user-defined output response to properties like resistor sizes.
6. Add the LTE code to the rest of the difference approximations. Investigate LTE convergence as a function of reltol, iabstol, vntol, etc. What is the significance of Iteratio?
7. In your favorite simulator, setup and run the ideal LC tank we showcased in Sect. 4.6.7. Investigate how the various difference schemes affect the response. Is it the same we found here or is it better or worse?

## References

1. Berry, R. D. (1971). An optimal ordering of electronic circuit equations for a sparse matrix solution. *IEEE Transactions on Circuit Theory, 18*, 40–50.
2. Calahan, D. A. (1972). *Computer-aided network design* (Rev ed.). McGraw-Hill, New York.
3. Chua, L. O., & Lin, P.-M. (1975). *Computer-aided analysis of electronic circuits.* New York: McGraw-Hill.
4. Ho, C.-W., Zein, A., Ruehli, A. E., & Brennan, P. A. (1975). The modified nodal approach to network analysis. *IEEE Transactions on Circuits and Systems, 22*, 504–509.
5. Hachtel, G. D., Brayton, R. K., & Gustavson, F. G. (1971). The sparse tableau approach to network analysis and design. *IEEE Transactions on Circuit Theory, 18*, 101–113.
6. Nagel, L. W. (1975). *SPICE2: A computer program to simulate semiconductor circuits.* PhD thesis, University of California Berkeley. Memorandum No ERL-M520.
7. Nagel, L. W., & Pederson, D. O. (1973). *Simulation program with integrated circuit emphasis.* Waterloo: In Proceedings of the Sixteenth Midwest Symposium on Circuit Theory.
8. Ho, C. W., Zein, D. A., Ruehli, A. E., & Brennan, P. A. (1977). An algorithm for DC solutions in an experimental general purpose interactive circuit design program. *IEEE Transactions on Circuits and Systems, 24*, 416–422.
9. Saad, Y. (2003). *Iterative method for sparse linear systems* (2nd ed.). Philadelphia: Society for Industrial and Applied Mathematics.
10. Najm, F. N. (2010). *Circuit simulation.* Hobroken: Wiley.
11. Kundert, K., White, J., & Sangiovanni-Vicentelli, A. (1990). *Steady-state methods for simulating analog and microwave circuits.* Norwell: Kluwer Academic Publications.
12. Ogrodzki, J. (1994). *Circuit simulation methods and algorithms.* Boca Raton: CRC Press.

13. Vlach, J., & Singhai, K. (1994). *Computer methods for circuit analysis and design* (2nd ed.). New York: Van Nostrand Reinhold.
14. McCalla, W. J. (1988). *Fundamentals of computer-aided circuit simulation*. Norwell: Kluwer Academic Publishers.
15. Ruehli, A. E., editor (1986). *Circuit analysis, simulation and design – Part 1*, North-Holland, Amsterdam published as Volume 3 of Advances in CAD for VLSI.
16. Ruehli, A. E., editor (1987). *Circuit analysis, simulation and design – Part 2*, North-Holland, Amsterdam published as Volume 3 of Advances in CAD for VLSI.
17. Vladimirescu, A. (1994). *The spice book*. New York: Wiley.
18. Nyquist, H. (1932). Regeneration theory. *Bell System Technical Journal, 11*(1), 126–147.
19. Franklin, F. F., Emami-Naeini, A., & Powell, J. D. (2005). *Feedback control of dynamic systems*. Englewood Cliffs: Prentice Hall.
20. Lee, T. (2003). *The design of CMOS radio-frequency integrated circuits* (2nd ed.). Cambridge: Cambridge University Press.
21. Middlebrook, R. D. (1975). Measurement of loop gain in feedback systems. *International Journal of Electronics, 38*(4), 485–512.
22. Visvanathan, V., Hantgan, J., & Kundert, K. (2001). Striving for small signal stability. *IEEE Circuits and Devices, 17*, 31–40.
23. Sahrling, M. (2019). *Fast techniques for integrated circuit design*. Cambridge: Cambridge University Press.
24. Posar, D. (2012). *Microwave engineering* (4th ed.). Wiley and Sons: Hoboken.
25. Kundert, K. (1995). *The designers guide to spice and spectre*. Norwell: Kluwer Academic Press.

# Chapter 5
# Circuit Simulation: Nonlinear Case

**Abstract** After the chapter on linear simulators, we will venture into the nonlinear aspects of circuit simulators by first looking at nonlinear DC operating point simulations. This was historically perhaps the most difficult problem to solve since one usually does not have an actual starting point unlike a transient solution where one often starts from the DC point and look at small changes with the new timestep. After the DC discussion, we will do full nonlinear transient simulations with active devices. The chapter ends with exploring other developments in circuit simulation such as periodic steady-state simulators. They have been extraordinarily helpful with circuit development, and we will examine several implementations using working code examples.

## 5.1 Introduction

As we already mentioned in Chap. 2 in the discussion on nonlinear solvers, they invariably involve an iterative step most commonly implemented as a version of the Newton-Raphson method. In this chapter, we will discuss this in more detail with respect to circuit simulators [1–21]. Perhaps most notably, the first difficulty one runs into is a DC convergence; it is one of the most challenging problems in circuit simulations with nonlinear elements, and we will discuss this in the first section. The next section will cover how nonlinear issues are tackled in a transient simulation. We will follow this by a detailed section on periodic steady-state-solvers (PSS) where we discuss the most popular implementations, harmonic balance, and shooting methods. Once a PSS solution has been found, we can use perturbation techniques to find solutions to periodic AC (PAC) and periodic noise (PNOISE) problems, and these are discussed in the subsequent sections. We end this chapter discussing a handful of common circuit example simulations where the steady-state approach is helpful.

It should be noted the material in this chapter is very dense. Many complex issues are merely hinted at through brief and very simple examples. Things like convergence issues are extraordinarily difficult to address thoroughly in such small space as this chapter. The reader is therefore highly encouraged to explore outside the examples provided. It is not difficult to construct circuitry that has serious convergence difficulties, and the reader should consider what one can do about them and

© Springer Nature Switzerland AG 2021

M. Sahrling, *Analog Circuit Simulators for Integrated Circuit Designers*,

https://doi.org/10.1007/978-3-030-64206-8_5

experiment on his or her own using the Python codes as a base. Modern simulators are a result of a multitude of researchers working over many decades to address these types of problems, and it is the hope the reader will learn to appreciate the difficulties these developers have overcome in modern simulators. Updates to the codes and a place for healthy discussions will be provided at www.fastictechniques.com.


## 5.2   DC Nonlinear Simulation

In Chap. 4, we showed a few implementations of linear simulators, and we discussed the fact that characterizing circuit elements as having a certain matrix signature makes it really easy to build up the basic matrix equation including the known right-hand side (rhs) column matrix. Here we will discuss a DC simulator and the key difficulties in implementing such a simulator when the branch equations are nonlinear and common remedies to these problems.

As a circuit designer, it can often seem like the DC simulation is trivial in that it is only solving for the operating points of the circuit; no capacitors or inductors or other frequency dependent objects are used and can be ignored. We can use the scheme developed earlier and simply short all inductors and remove all capacitors, and we would be done. The problems come when we have nonlinear elements. Many solutions can then exist, and if a starting point for the DC iterations is not good, it can take a long time to find one of them, if at all feasible in a reasonable time [4, 18]. We will here outline some of these difficulties, and in doing so, we will encounter common terminology and explain their technical meaning.


### 5.2.1   Solution Techniques

The study of nonlinear equations has a long and rich history. We mentioned in Chap. 2 the dominant solution method as being the Newton-Raphson algorithm. In Chap. 2, we discussed a one-dimensional derivation of the method. This section will introduce a multidimensional version, and we will show how one can implement it numerically. It is one of the work horses in nonlinear numerical codes, and a good understanding of it and its convergence properties can be a really helpful tool in one's toolbox. There are also some interesting pathologies stemming from not being "close enough" to the correct solution initially, and we will mention these briefly also.


#### 5.2.1.1   Nonlinear Model Implementations

We will discuss the DC problem by looking at a specific example (see Fig. 5.1). The new element is a nonlinear transistor we will model in three different ways following our analytical outline in Sect. 3.4. A transistor can be seen as a VCCS where an input voltage between gate and source (or base-emitter) controls the current through the drain (or collector). There is a lot more to modeling a transistor as we discussed already in Chap. 3, but for our purposes, we will keep the models really simple so various aspects of analog simulators can be showcased. We will start with the simplest CMOS model from Sect. 3.4. This model shows the drain current varies like $I_d \sim V_{gs}^2$. This is perhaps the simplest nonlinearity one can think of. In order to proceed, we need to take a look at the definitions of the direction of currents as they apply to NMOS and PMOS transistors, respectively. Let us look one more time at Fig. 3.8 where the directions of the source and drain currents are indicated. The current direction convention is that a current going into a device (or subcircuit) is positive; outgoing currents are negative. For the NMOS transistor, we have then that the drain current is

$$I_d = K\left(V_g - V_s\right)^2 = KV_{gs}^{\,2} \tag{5.1}$$

For the PMOS, $I_d$ becomes negative:

$$I_d = -K\left(V_s - V_g\right)^2 = -KV_{sg}^{\,2} \tag{5.2}$$

Note the subtle change of subscript for the PMOS. For this model, it has no significance, but it will play a role later on with more sophisticated implementations.

With these sign conventions in mind, let us now discuss solution techniques for such nonlinear devices.

**Fig. 5.1** Simple network with two nonlinear elements (diff pair)

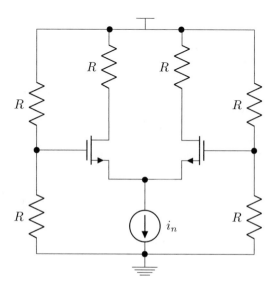

Let us look at this model with our circuit in Fig. 5.1. The reader will realize the solution is fairly straightforward analytically, but let us go through Newton-Raphson iterations to see how it works. We find after some trivial algebra the equations we are trying to solve are

$$\begin{cases} I_p + I_n = I_{sink} \\ I_p = K(V_a - V_s)^2 \\ I_n = K(V_b - V_s)^2 \end{cases} \rightarrow \begin{cases} f_0 : I_p + I_n - I_{sink} = 0 \\ f_1 : I_p - K(V_a - V_s)^2 = 0 \\ f_2 : I_n - K(V_b - V_s)^2 = 0 \end{cases} \rightarrow f(v) = 0, v = \begin{pmatrix} V_s \\ I_p \\ I_n \end{pmatrix} \quad (5.3)$$

where we have explicitly solved for the voltages $V_A = V_{dd}R_1/(R_1 + R_2)$ and $V_B = V_{dd}R_3/(R_3 + R_4)$ to simplify the math and keep the network equations manageable for analytical calculations. The drain voltages follow trivially from the currents, $I_p$, $I_n$, and we will not add them directly to the network equations. $v$ is here the vector of unknown quantities. The function $f$ is a common symbol in the simulation literature [3, 4, 10, 11] when discussing network equations. There the full MNA description is often used, and no explicit current branches are included; it is therefore common to see $f$ simply as the equation for KCL. In this book, we explicitly include all branch equations, so $f$ will here also include those. The reader should keep this in mind when reading the common reference literature [1–21]. When we explicitly refer to the KCL equations only, we denote $f$ as $f_{KCL}$.

These set of equations contain linear terms and nonlinear terms. We can separate these out such that we have a set of equations that look like

$$f(v) = \begin{pmatrix} 0 & 1 & 1 \\ 0 & 1 & 0 \\ 0 & 0 & 1 \end{pmatrix} \begin{pmatrix} V_s \\ I_p \\ I_n \end{pmatrix} + \begin{pmatrix} 0 \\ -K(V_a - V_s)^2 \\ -K(V_b - V_s)^2 \end{pmatrix} - \begin{pmatrix} I_{sink} \\ 0 \\ 0 \end{pmatrix} = 0 \quad (5.4)$$

The second term in the equation is known as the nonlinear term. Is it simply added as a separate column matrix to the matrix equation. The last term is the right-hand side of the matrix equation we have just moved over as part of the definition of $f$. This way of separating out the linear and nonlinear parts of the network equations is standard practice.

Following the Newton-Raphson recipe we discussed in Chap. 2, we can now write this equation as a Taylor expansion in multidimensions around some point $v_0$; for a detailed derivation, see Chap. 7:

$$f(v) = f(v_0) + \frac{\partial f}{\partial v}\bigg|_{v_0} (v - v_0) \quad (5.5)$$

Let us now calculate the matrix in the second term, known as the *Jacobian*:

$$J_{ij} = \frac{\partial f_i}{\partial v_j} \tag{5.6}$$

It is fairly straightforward to calculate for this case, and we find the following expression:

$$J = \left( \frac{\partial f}{\partial V_s} \; \frac{\partial f}{\partial I_p} \; \frac{\partial f}{\partial I_n} \right) = \begin{pmatrix} 0 & 1 & 1 \\ 2K(V_a - V_s) & 1 & 0 \\ 2K(V_a - V_s) & 0 & 1 \end{pmatrix} \tag{5.7}$$

This matrix is evaluated at the previous iteration so all the entities are known.

Newton's method now tells us we can iterate the solution in the following way (see Chap. 7 and [1, 3, 15–20]):

$$v_{k+1} = v_k - (J)^{-1} f(v_k) \tag{5.8}$$

This is very similar in structure to the one-dimensional case we presented in Chap. 2, but now it is all in matrix annotation. As should be clear to the reader, the difficulty here is mainly finding the Jacobian; luckily most of the time, it is already known through the device models. We find the following Python code example:

```
Python code
.if DevType[i]=='transistor':
    STA_matrix[NumberOfNodes+i][NumberOfNodes+i]=DevValue[i]
    STA_matrix[NumberOfNodes+i][Nodes.index(DevNode1[i])]=0
    STA_matrix[Nodes.index(DevNode1[i])][NumberOfNodes+i]=1
    STA_matrix[NumberOfNodes+i][Nodes.index(DevNode3[i])]=0
    STA_matrix[Nodes.index(DevNode3[i])][NumberOfNodes+i]=-1
    STA_nonlinear[NumberOfNodes+i]=(sol[Nodes.index(DevNode2[i]
)] - sol[ Nodes.index( DevNode3[i])])**2

#
# Now we need the Jacobian, the transistors look like VCCS with
a specific gain = 2 K (Vg-Vs) in our case
#
    for i in range(MatrixSize):
        for j in range(MatrixSize):
            Jacobian[i][j]=STA_matrix[i][j]
        for i in range(DeviceCount):
            if DevType[i]=='transistor':
                Jacobian[NumberOfNodes+i][NumberOfNodes+i]=DevValue
[i] # due to derfivative leading to double gain
```

```
        Jacobian[ NumberOfNodes+I ][ Nodes.index(DevNode2[i])
] = 2*( sol[Nodes.index( DevNode2[i] ) ]- sol[Nodes.index
(DevNode3[i])])+0.01
        Jacobian[NumberOfNodes+i][Nodes.index(DevNode3[i])]= - 2
*( sol[ Nodes.index( DevNode2 [i] )] - sol[Nodes.index
(DevNode3[i])])-0.01
        Jacobian[Nodes.index(DevNode1[i])][NumberOfNodes+i]=1
        Jacobian[Nodes.index(DevNode3[i])][NumberOfNodes+i]=-1
        sol=sol-numpy.matmul(numpy.linalg.inv(Jacobian),f)
    for i in range(MatrixSize):
        solm1[i]=solold[i]
    f=numpy.matmul(STA_matrix,sol)-STA_rhs+STA_nonlinear
    val[iter]=sol[5]#2*(sol[Nodes.index(DevNode2[12])]-sol[Nodes.
index(DevNode3[12])])#sol[3]#sol[1]-sol[3]#0.9-K*(sol[3]-
sol[7])**2*100
  #plt.plot(vin)
  plt.plot(val)
  Here python implementation of Newton's method will be presented
  .
  End Python
  Code 5.1
```

Armed with this implementation, we can simulate the circuit in Fig. 5.1 with the netlist:

```
*
vdd vdd 0 0.9
vss vss 0 0
r1 vdd inp 30
r2 inp vss 150
r3 vdd inn 30
r4 inn vss 150
i1 vs vss 1e-3
m1 vdd inp vs nch1
m2 vdd inn vs nch1
.options MaxNewtonIterations=10
.ic v(vdd)=1 v(vss)=0 v(inp)=0 v(vs)=-2
.plot v(vs)
netlist 5.1
```

After executing the code, we can now observe the node "vs" in terms of iteration in Table 5.1 for our transistor model.

**Table 5.1** Node voltage value vs iteration number for model1

| Iteration # | Value of node "vs" | Error (%) |
|---|---|---|
| 0 | −2.0 | 4762.7 |
| 1 | −0.375 | 974.3 |
| 2 | −0.034722 | 180.9 |
| 3 | 0.039054 | 8.949 |
| 4 | 0.042883 | 0.024 |
| 5 | 0.042893 | 0 |

We can now follow the same steps to derive matrix elements for the more realistic model in Sect. 3.4. Here among other things, we need to take the finite output conductance into account and add equations for the drains of the transistors. We find for the case where $V_{gs} - V_T < 0$:

$$
f(v_k) = \begin{pmatrix} 0 & 0 & 0 & 1 & 0 & -1 & 0 \\ 0 & 0 & 0 & 0 & 1 & 0 & -1 \\ 0 & 0 & 0 & 1 & 1 & 0 & 0 \\ 0 & 0 & 0 & 1 & 0 & 0 & 0 \\ 0 & 0 & 0 & 0 & 1 & 0 & 0 \\ 0 & 0 & 0 & 0 & 0 & 1 & 0 \\ 0 & 0 & 0 & 0 & 0 & 0 & 1 \end{pmatrix} \begin{pmatrix} V_{op} \\ V_{on} \\ V_{s} \\ I_{p} \\ I_{n} \\ I_{R,p} \\ I_{R,n} \end{pmatrix} + \begin{pmatrix} 0 \\ 0 \\ 0 \\ 0 \\ 0 \\ 0 \\ 0 \end{pmatrix} - \begin{pmatrix} 0 \\ 0 \\ I_{sink} \\ 0 \\ 0 \\ 0 \\ 0 \end{pmatrix} = 0 \qquad (5.9)
$$

The Jacobian is simply the derivative of this equation with respect to the unknowns, and we find

$$
J = \begin{pmatrix} 0 & 0 & 0 & 1 & 0 & -1 & 0 \\ 0 & 0 & 0 & 0 & 1 & 0 & -1 \\ 0 & 0 & 0 & 1 & 1 & 0 & 0 \\ 0 & 0 & 0 & 1 & 0 & 0 & 0 \\ 0 & 0 & 0 & 0 & 1 & 0 & 0 \\ 0 & 0 & 0 & 0 & 0 & 1 & 0 \\ 0 & 0 & 0 & 0 & 0 & 0 & 1 \end{pmatrix}, \qquad (5.10)
$$

$V_{ds} < V_{gs} - V_T:$

$$
f(v_k) = \begin{pmatrix} 0 & 0 & 0 & 1 & 0 & -1 & 0 \\ 0 & 0 & 0 & 0 & 1 & 0 & -1 \\ 0 & 0 & 0 & 1 & 1 & 0 & 0 \\ 0 & 0 & 0 & 1 & 0 & 0 & 0 \\ 0 & 0 & 0 & 0 & 1 & 0 & 0 \\ 0 & 0 & 0 & 0 & 0 & 1 & 0 \\ 0 & 0 & 0 & 0 & 0 & 0 & 1 \end{pmatrix} \begin{pmatrix} V_{op} \\ V_{on} \\ V_s \\ I_p \\ I_n \\ I_{R,p} \\ I_{R,n} \end{pmatrix} +
$$

$$
+ \begin{pmatrix} 0 \\ 0 \\ 0 \\ -2K(V_a - V_s - V_T)^2 (1 + \lambda(V_{op} - V_s)) \\ -2K(V_a - V_s - V_T)^2 (1 + \lambda(V_{on} - V_s)) \\ 0 \\ 0 \end{pmatrix} - \begin{pmatrix} 0 \\ 0 \\ I_{sink} \\ 0 \\ 0 \\ 0 \\ 0 \end{pmatrix} = 0 \qquad (5.11)
$$

Likewise for the Jacobian (the explicit matrix entries are not shown for clarity)

$$
J = \left( \frac{\partial I_d}{\partial V_g}, \frac{\partial I_d}{\partial V_s}, \frac{\partial I_d}{\partial V_d} \right) = K(V_d - V_s), -2K(V_g - V_s - V_T),
$$

$$
K\big((V_g - V_s - V_T) - (V_d - V_s)\big) = K(V_g - V_d - V_T) \qquad (5.12)
$$

$V_{ds} > V_{gs} - V_T:$

$$
f(v_k) = \begin{pmatrix} 0 & 0 & 0 & 1 & 0 & -1 & 0 \\ 0 & 0 & 0 & 0 & 1 & 0 & -1 \\ 0 & 0 & 0 & 1 & 1 & 0 & 0 \\ 0 & 0 & 0 & 1 & 0 & 0 & 0 \\ 0 & 0 & 0 & 0 & 1 & 0 & 0 \\ 0 & 0 & 0 & 0 & 0 & 1 & 0 \\ 0 & 0 & 0 & 0 & 0 & 0 & 1 \end{pmatrix} \begin{pmatrix} V_{op} \\ V_{on} \\ V_s \\ I_p \\ I_n \\ I_{R,p} \\ I_{R,n} \end{pmatrix} +
$$

$$
+\begin{pmatrix}
0 \\
0 \\
0 \\
-2K\left(\left(V_a-V_s-V_T\right)\left(V_{op}-V_s\right)-\dfrac{1}{2}\left(V_{op}-V_s\right)^2\right) \\
-2K\left(\left(V_a-V_s-V_T\right)\left(V_{on}-V_s\right)-\dfrac{1}{2}\left(V_{on}-V_s\right)^2\right) \\
0 \\
0
\end{pmatrix}
-\begin{pmatrix}
0 \\
0 \\
I_{sink} \\
0 \\
0 \\
0 \\
0
\end{pmatrix}=0 \qquad (5.13)
$$

With the Jacobian given by

$$
J=\left(\frac{\partial I_d}{\partial V_g},\frac{\partial I_d}{\partial V_s},\frac{\partial I_d}{\partial V_d}\right)=2K\left(V_g-V_s-V_T\right)\left(1+\lambda V_{ds}\right),
$$

$$
-2K\left(V_g-V_s-V_T\right)\left(1+\lambda V_{ds}\right)-\lambda K\left(V_g-V_s-V_T\right)^2,
$$

$$
K\lambda\left(V_g-V_s-V_T\right)^2 \qquad (5.14)
$$

To reiterate our earlier statement, this matrix is evaluated at the previous iteration, so all elements are known.

Implementing this model is now straightforward, and we find the code snippets:

```
Python code
   .    if DevType[i]=='transistor':
        STA_matrix[NumberOfNodes+i][NumberOfNodes+i]=DevValue[i]
        STA_matrix[NumberOfNodes+i][Nodes.index(DevNode1[i])]=0
        STA_matrix[Nodes.index(DevNode1[i])][NumberOfNodes+i]=1
        STA_matrix[NumberOfNodes+i][Nodes.index(DevNode3[i])]=0
        STA_matrix[Nodes.index(DevNode3[i])][NumberOfNodes+i]=-1
        VD=sol[Nodes.index(DevNode1[i])]
        VG=sol[Nodes.index(DevNode2[i])]
        VS=sol[Nodes.index(DevNode3[i])]
        Vgs=VG-VS
        Vds=VD-VS
        if Vds < Vgs-VT :
             STA_nonlinear[NumberOfNodes+i]=2*((Vgs-VT)*Vds-0.5*
Vds**2)
           else :
             STA_nonlinear[NumberOfNodes+i]=(Vgs-VT)**2*(1+lambda
T*Vds)
```

```
    #
    # Now we need the Jacobian, the transistors look like VCCS with
a specific gain = 2 K (Vg-Vs) in our case
    #
        for i in range(MatrixSize):
            for j in range(MatrixSize):    .
                Jacobian[i][j]=STA_matrix[i][j]
        for i in range(DeviceCount):
            if DevType[i]=='transistor':
                Jacobian[NumberOfNodes+i][NumberOfNodes+i]=DevValue[i]
# due to derfivative leading to double gain
                VD=sol[Nodes.index(DevNode1[i])]
                VG=sol[Nodes.index(DevNode2[i])]
                VS=sol[Nodes.index(DevNode3[i])]
                Vgs=VG-VS
                Vds=VD-VS
                Vgd=VG-VD
                if Vgs<VT :
                    Jacobian[NumberOfNodes+i][Nodes.index(DevNode1[i])]=
1e-5
                    Jacobian[NumberOfNodes+i][Nodes.index(DevNode2[i])]=
1e-5
                    Jacobian[NumberOfNodes+i][Nodes.index(DevNode3[i])]=
-2e-5
                    Jacobian[Nodes.index(DevNode1[i])][NumberOfNodes+i]=1
                    Jacobian[Nodes.index(DevNode3[i])][NumberOfNodes+i]=-1
                elif Vds <= Vgs-VT:
                    Jacobian[NumberOfNodes+i][Nodes.index(DevNode1[i])]=
2*(Vgd-VT)
                    Jacobian[NumberOfNodes+i][Nodes.index(DevNode2[i])]=
2*Vds
                    Jacobian[NumberOfNodes+i][Nodes.index(DevNode3[i])]=
-2*(Vgs-VT)
                    Jacobian[Nodes.index(DevNode1[i])][NumberOfNodes+i]=1
                    Jacobian[Nodes.index(DevNode3[i])][NumberOfNodes+i]=-1
                else :
                    Jacobian[NumberOfNodes+i][Nodes.index(DevNode1[i])]=
lambdaT*(Vgs-VT)**2
                    Jacobian[NumberOfNodes+i][Nodes.index(DevNode2[i])]=
2*(Vgs-VT)*(1+lambdaT*Vds)
                    Jacobian[NumberOfNodes+i][Nodes.index(DevNode3[i])]=
-2*(Vgs-VT)*(1+lambdaT*Vds)-lambdaT*(Vgs-VT)**2
```

```
        Jacobian[Nodes.index(DevNode1[i])][NumberOfNodes+i]=1
        Jacobian[Nodes.index(DevNode3[i])][NumberOfNodes+i]=-1
     sol=sol-numpy.matmul(numpy.linalg.inv(Jacobian),f)
  for i in range(MatrixSize):
     sol1[i]=solold[i]
  f=numpy.matmul(STA_matrix,sol)-STA_rhs+STA_nonlinear
Here python implementation of Newton's method will be presented
.

End Python
```
**Code 5.2**

```
vss vss 0 0
vdd vdd 0 0.9
r1 vdd inp 30
r2 inp vss 150
r3 vdd inn 30
r4 inn vss 150
i1 vs vss 1e-3
m1 vdd inp vs nch
m2 vdd inn vs nch
.options MaxNewtonIterations=10
.ic v(vdd)=1 v(vss)=0 v(inp)=0 v(vs)=-2
.plot v(vs)
```
**Netlist 5.2** The only difference compared to netlist 5.1 is the transistor model used.

After executing the code with netlist 5.2, we can now observe the node "vs" in terms of iteration in Table 5.2.

**Table 5.2** Node voltage value vs iteration number model2. Exact value is −0.1736067 V

| Iteration # | Value of node "vs" | Error (%) |
|---|---|---|
| 0 | -2 | 1682 |
| 1 | −0.465637606 | 468 |
| 2 | −0.08847079 | 170 |
| 3 | 0.073748 | 41.6 |
| 4 | 0.12137 | 3.97 |
| 5 | 0.12634 | 0.0007 |

The final bipolar model in Sect. 3.4. will now look like the following set of matrix equations:

$V_{be} < 0$:

$$f(v_k) = \begin{pmatrix} 0 & 0 & 0 & 1 & 0 & -1 & 0 \\ 0 & 0 & 0 & 0 & 1 & 0 & -1 \\ 0 & 0 & 0 & 1 & 1 & 0 & 0 \\ 0 & 0 & 0 & 1 & 0 & 0 & 0 \\ 0 & 0 & 0 & 0 & 1 & 0 & 0 \\ 0 & 0 & 0 & 0 & 0 & 1 & 0 \\ 0 & 0 & 0 & 0 & 0 & 0 & 1 \end{pmatrix} \begin{pmatrix} V_{op} \\ V_{on} \\ V_s \\ I_p \\ I_n \\ I_{R,p} \\ I_{R,n} \end{pmatrix} + \begin{pmatrix} 0 \\ 0 \\ 0 \\ 0 \\ 0 \\ 0 \\ 0 \end{pmatrix} - \begin{pmatrix} 0 \\ 0 \\ I_{sink} \\ 0 \\ 0 \\ 0 \\ 0 \end{pmatrix} = 0 \qquad (5.15)$$

With the Jacobian

$$J = \begin{pmatrix} 0 & 0 & 0 & 1 & 0 & -1 & 0 \\ 0 & 0 & 0 & 0 & 1 & 0 & -1 \\ 0 & 0 & 0 & 1 & 1 & 0 & 0 \\ 0 & 0 & 0 & 1 & 0 & 0 & 0 \\ 0 & 0 & 0 & 0 & 1 & 0 & 0 \\ 0 & 0 & 0 & 0 & 0 & 1 & 0 \\ 0 & 0 & 0 & 0 & 0 & 0 & 1 \end{pmatrix}$$

$V_{be} > 0$:

$$f(v_k) = \begin{pmatrix} 0 & 0 & 0 & 1 & 0 & -1 & 0 \\ 0 & 0 & 0 & 0 & 1 & 0 & -1 \\ 0 & 0 & 0 & 1 & 1 & 0 & 0 \\ 0 & 0 & 0 & 1 & 0 & 0 & 0 \\ 0 & 0 & 0 & 0 & 1 & 0 & 0 \\ 0 & 0 & 0 & 0 & 0 & 1 & 0 \\ 0 & 0 & 0 & 0 & 0 & 0 & 1 \end{pmatrix} \begin{pmatrix} V_{op} \\ V_{on} \\ V_s \\ I_p \\ I_n \\ I_{R,p} \\ I_{R,n} \end{pmatrix} +$$

$$+ \begin{pmatrix} 0 \\ 0 \\ 0 \\ -I_0 \exp\!\left(\dfrac{V_{BE}\,q}{kT}\right)\!\left(1+\dfrac{V_{CE}}{V_A}\right) \\ -I_0 \exp\!\left(\dfrac{V_{BE}\,q}{kT}\right)\!\left(1+\dfrac{V_{CE}}{V_A}\right) \\ 0 \\ 0 \end{pmatrix} - \begin{pmatrix} 0 \\ 0 \\ I_{sink} \\ 0 \\ 0 \\ 0 \\ 0 \end{pmatrix} = 0 \qquad (5.16)$$

With the Jacobian

$$J = \left(\frac{\partial I_c}{\partial V_b}, \frac{\partial I_c}{\partial V_e}, \frac{\partial I_c}{\partial V_c}\right) = -I_0 \frac{q}{kT}\exp\!\left(\frac{V_{be}\,q}{kT}\right)\!\left(1+\frac{V_{ce}}{V_A}\right),$$

$$I_0 \frac{q}{kT}\exp\!\left(\frac{V_{be}\,q}{kT}\right)\!\left(\frac{q}{kT}\!\left(1+\frac{V_{ce}}{V_A}\right)+\frac{1}{V_A}\right),$$

$$-I_0 \exp\!\left(\frac{V_{be}\,q}{kT}\right)\frac{1}{V_A} \qquad (5.17)$$

Let us implement this model in a code, and we find

```
Python code
.    if DevType[i]=='bipolar':
        STA_matrix[NumberOfNodes+i][NumberOfNodes+i]=DevValue[i]
        STA_matrix[NumberOfNodes+i][Nodes.index(DevNode1[i])]=0
        STA_matrix[Nodes.index(DevNode1[i])][NumberOfNodes+i]=1
        STA_matrix[NumberOfNodes+i][Nodes.index(DevNode3[i])]=0
        STA_matrix[Nodes.index(DevNode3[i])][NumberOfNodes+i]=-1
        VC=sol[Nodes.index(DevNode1[i])]
        VB=sol[Nodes.index(DevNode2[i])]
        VE=sol[Nodes.index(DevNode3[i])]
        Vbe=VB-VE
        Vce=VC-VE
        if Vbe < 0 :
            STA_nonlinear[NumberOfNodes+i]=0
        else :
            STA_nonlinear[NumberOfNodes+i]=math.exp(Vbe/Vthermal)
* (1+Vce/VEarly)
```

```
    #
    # Now we need the Jacobian, the transistors look like VCCS with
a specific gain = 2 K (Vg-Vs) in our case
    #
    for i in range(MatrixSize):
        for j in range(MatrixSize):
            Jacobian[i][j]=STA_matrix[i][j]
    for i in range(DeviceCount):
        if DevType[i]=='bipolar':
            Jacobian[NumberOfNodes+i][NumberOfNodes+i]=DevValue[i]
# due to derfivative leading to double gain
            VC=sol[Nodes.index(DevNode1[i])]
            VB=sol[Nodes.index(DevNode2[i])]
            VE=sol[Nodes.index(DevNode3[i])]
            Vbe=VB-VE
            Vce=VC-VE
            Vbc=VB-VC
            if Vbe<=0 :
                Jacobian[NumberOfNodes+i][Nodes.index(DevNode1[i])]=
1e-5
                Jacobian[NumberOfNodes+i][Nodes.index(DevNode2[i])]=
1e-5
                Jacobian[NumberOfNodes+i][Nodes.index(DevNode3[i])]=
-1e-5
                Jacobian[Nodes.index(DevNode1[i])][NumberOfNodes+i]=1
                Jacobian[Nodes.index(DevNode3[i])][NumberOfNodes+i]=-1


            else :
                Jacobian[ NumberOfNodes+ I ][ Nodes.index( DevNode1
[i]) ] = math.exp( Vbe/ Vthermal ) /VEarly
                Jacobian[NumberOfNodes+i][Nodes.index(DevNode2[i])]=
math.exp(Vbe/Vthermal)*(1+Vce/VEarly)/Vthermal
                Jacobian[NumberOfNodes+i][Nodes.index(DevNode3[i])]=
(-math.exp(Vbe/Vthermal)/VEarly-math.exp(Vbe/Vthermal)*(1+Vce/
VEarly)/Vthermal)
                Jacobian[Nodes.index(DevNode1[i])][NumberOfNodes+i]=1
                Jacobian[Nodes.index(DevNode3[i])][NumberOfNodes+i]=-1
        sol=sol-numpy.matmul(numpy.linalg.inv(Jacobian),f)

        Jac_inv=numpy.linalg.inv(Jacobian)
    for i in range(MatrixSize):
        solm1[i]=solold[i]
    f=numpy.matmul(STA_matrix,sol)-STA_rhs+STA_nonlinear
Here python implementation of Newton's method will be presented
    .

End Python
```
**Code 5p3**

**Table 5.3** Node voltage value vs iteration number for the bipolar transistor model

| Iteration # | Value of node "vs" | Error (%) |
|---|---|---|
| 0 | −0.3 | 6542 |
| 1 | 0.022582378 | 58.2 |
| 2 | 0.040781412 | 24.58 |
| 3 | 0.051170494 | 5.363 |
| 4 | 0.053913564 | 0.2900 |
| 5 | 0.054069913 | 0.00088 |

After executing the code, we can now observe the node "vs" in terms of iteration in Table 5.3 for our simple bipolar transistor model.

Note the use of initial conditions here is critical. With an exponential dependency, the routine can and will quickly completely miss the real solution and go haywire unless the initial conditions are close. We will see this again when we look at a simple bandgap. Exponential dependency in the model is very difficult to handle without a lot of help from initial conditions. Often this kind of model has some kind of exponential limit built into it to avoid "exploding" solutions.

*PMOS Transistor*
A PMOS transistor implementation will look very similar to the NMOS but with some sign differences. We find if we, compared to NMOS, do

$$K \rightarrow -K, \ \left(V_g - V_s\right) \rightarrow -\left(V_g - V_s\right) \tag{5.18}$$

we can use the same coding implementation as we did for NMOS. Let us illustrate with the example in Fig. 5.2

When using the CMOS approximation outlined in Sect. 3.4, model1, the network matrix, will look like

$$\begin{cases} V_g = V_{in} \\ I_d = -K\left(V_s - V_g\right)^2 \\ I_{source} = -I_d \end{cases} \rightarrow f\left(v_k\right) = \begin{cases} V_g - V_{in} = 0 \\ I_d + K\left(V_s - V_g\right)^2 = 0 \\ I_{source} + I_d = 0 \end{cases} \tag{5.19}$$

With unknowns, $v_k = (V_g, I_d, V_s)$. And the accompanying Jacobian matrix becomes

$$J = \begin{pmatrix} 1 & 0 & 0 \\ -2K\left(V_s - V_g\right) & 1 & 2K\left(V_s - V_g\right) \\ 0 & 1 & 0 \end{pmatrix} \tag{5.20}$$

**Fig. 5.2** Simple PMOS
follower

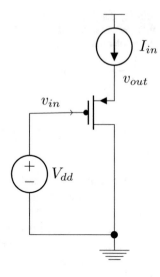

A python implementation can look like

```
Python code
        .
              if DevModel[i][0]=='p':
                  PFET=-1
                  Vgs=-Vgs
                  Vds=-Vds
                  Vgd=-Vgd
              else:
                  PFET=1
For the rest of the details see Code 5.2
        .
End Python
```

We now run the code with the netlist

```
*
vdd vdd 0 1
vss vss 0 0
m1 vss in out pch
vin in vss 0.2
i1 vdd out 1e-3
netlist 5.3
```

and find the result in Fig. 5.3.

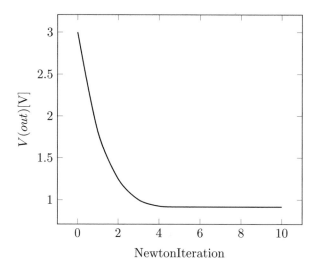

**Fig. 5.3** Source voltage of PMOS follower as a function of iteration

We see the iterations converge really quickly like we now would expect with Newton-Raphson.

**Model Implementation Verification**
Let us verify our model implementations with some DC sweeps.

*Gate/Base Sweep*
As a final check on our transistor implementations, let us simulate the drain and collector currents as a function of gate and drain (base and collector) voltage and compare with the exact analytical formula we devised in Sect. 3.4. Figures 5.4, 5.5, and 5.6 show the results.

The resulting error can be adjusted with the number of Newton-Raphson iterations.

*Drain/Collector Sweep*
The drain and collector voltages can also be swept, and we can compare again with the analytical result.

We find again the accuracy improves with the number of iterations just like one would expect.

### 5.2.1.2 Bandgap Circuit

A bandgap is a classic analog circuit. We will use the circuit in Fig. 5.7 as an example and see how convergence is affected. It also shows a case where both bipolar and CMOS transistors coexist and can have some interesting consequences. We will in this example simulate the startup of such a circuit where the power supplies ramp up emulating turning the circuit on.

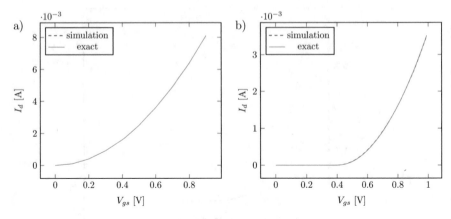

**Fig. 5.4** Drain current vs gate voltage for (**a**) CMOS model1, (**b**) CMOS model2 compared to exact calculation

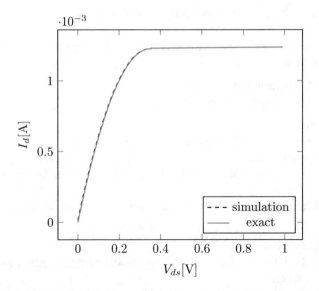

**Fig. 5.5** Drain current vs drain voltage for CMOS model2 compared to exact calculation

This circuit although simple looking has some real convergence issues. First of all, there are at least two stable modes: one where there is no current going any-where and $v(out) \approx 0$ and then there is the other desired one. Without help, the simu-lator easily finds one or the other, mostly the off state. Our netlist is

```
*
vdd vdd 0 2
vss vss 0 0
m1 out a vdd pch
```

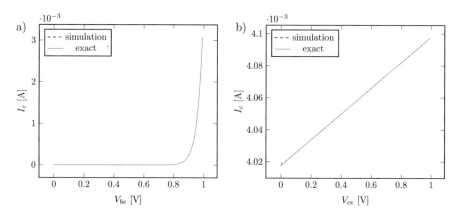

**Fig. 5.6** Collector current vs (**a**) base voltage (**b**) collector voltage for model3 compared to exact calculation

```
m2 a a vdd pch
q1 out out vss npn
q2 a out d npn
q3 a out d npn
q4 a out d npn
q5 a out d npn
q6 a out d npn
q7 a out d npn
q8 a out d npn
q9 a out d npn
q10 a out d npn
q11 a out d npn
r2 d vss 59.51
.options MaxNewtonIterations=40
.plot v(out) v(a) v(d) v(vdd)
```
**netlist 5.4**

Let us first run this circuit without any specific initial conditions (see Fig. 5.8).

We see circuit simulator finds the off mode. This is not the desired solution, but we can find that one by using initial condition settings for the circuits. In reality a startup circuit would be used to ensure the desired solution is reached. First let us calculate the desired bias voltages. We find when the current in the two legs are equal

$$I_s e^{V_{out}/V_{th}} = 10 I_s e^{(V_{out}-V_d)/V_{th}} \rightarrow V_d = V_{th} \ln 10 = 59.51 mV$$

For a 1 mA current, we see $R_1 = \dfrac{V_d}{1m} = 59.51 ohm$ and $V_{out} = V_{th} \ln (10^{-3}/I_s) = 0.71406 \, V$. We know $Vdd - V_a = \sqrt{10^{-3}/K} + V_T = \sqrt{0.1} + 0.4 \rightarrow V_a = 1.28377 V$

**Fig. 5.7**  A simple bandgap
circuit

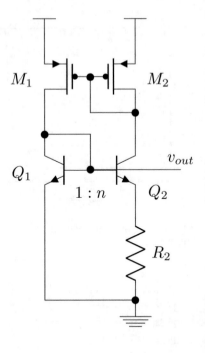

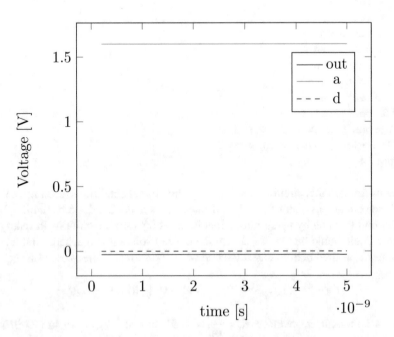

**Fig. 5.8**  Bandgap circuit without any specific initial conditions. The result is a state where no current is running through the devices. It is a valid solution but not the desired one

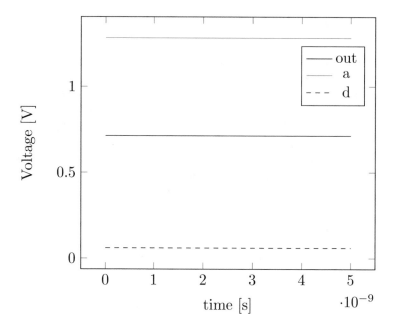

**Fig. 5.9** Bandgap output voltage as a function of time where the initial condition statement has been used to set the desired output voltage

We plug in these numbers in the simulator by adding the line

```
.ic v(a)=1.5 v(out)=0.9 v(d)=0.06 v(vdd)=2 v(vss)=0
```

to the netlist, and we can now simulate and find the plot in Fig. 5.9.

This is now the desired result and the simulator confirms our calculations.

This circuit can be very difficult to get to converge properly. Try various combinations of initial conditions to see where the difficulty might be. Some settings will cause the solution to blow up due to the strong exponential in the bipolar transistor; other settings will find the before-mentioned zero solution. This serves as an example the Newton-Raphson will work if the starting point is close enough to the final solution. A more sophisticated model will prevent the blowup solution from appearing. What can you think of that will limit this problem?

The bandgap has been biased at some arbitrary point since the purpose was just to illustrate the simulator performance. Normally, such a circuit will need to have its temperature dependence flat at the operating temperature. As an exercise, the reader is encouraged to size the circuit properly.

For each of these examples of transistor models, try experimenting with different initial conditions. Can you think of a way to break the code so it will not converge? The bipolar transistor model can easily blow up due to the exponential; try various combinations and explore! What could be done differently? Can you think of a way to improve convergence?

## 5.2.2   Convergence Criteria

As we can see from the iterations in Table 5.1 to Table 5.3, the total error goes down
fairly rapidly for each iteration, and we simply need to define what being done
means. In other words, decide what accuracy we desire. One might think a constraint on the KCL, $f_{KCL}(v_k) < \varepsilon$, would be an obvious choice, and in some simulators, it is. One refers to this constraint as the residue constraint. However, it is not
always implemented, and the user should be aware of what convergence criterions
are implemented in the specific simulator used. The most heavily used criteria rely
on the difference of the consecutive iterations $v_k$ and $v_{k+1}$, called the update criterion, in addition to the residue convergence criterion see for example [4].

For the residue, $f_{KCL}(v_k)$, we say the Newton-Raphson iteration residue has converged if

$$\left| f_{KCL}\left(v_k\right)\right| < \text{reltol}\, i_{n,\max} + \text{iabstol} \tag{5.21}$$

where $i_{n,\,max}$ is the largest absolute current entering node $n$.

The update convergence criterion for a Newton-Raphson iteration is

$$\left| v_k - v_{k-1}\right| < \text{reltol}\, v_{k,\max} + \text{vabstol} \tag{5.22}$$

where $v_{k,\,max}$ is usually $\max(|v_k|, |v_{k-1}|)$.

In most simulators, both of these needs to be fulfilled for the Newton-Raphson
iteration to be considered converged. We can implement this in our Python environment in the following way:

```
Python code
    .

    .

    .

        ResidueConverged=True
        node=0
        while ResidueConverged and node<NumberOfNodes:
# Let us find the maximum current going into node, Nodes[node]
            MaxCurrent=0
            for current in range(NumberOfCurrents):
                MaxCurrent= max(MaxCurrent,abs(STA_matrix[node]
[NumberOfNodes+current]*(
                                 sol[NumberOfNodes+current])))
            if f[node] > reltol*MaxCurrent+iabstol:
                ResidueConverged=False
            node=node+1
```

.
.
.

```
        SolutionCorrection=numpy.matmul(numpy.linalg.inv(Jacob
ian),f)
        UpdateConverged=True
        if PointLocal:
            for node in range(NumberOfNodes):
                        vkmax=max(abs(sol[node]),abs(sol[n
ode]-SolutionCorrection[node]))
            if abs(SolutionCorrection[node])>vkmax*reltol+vabstol:
                    UpdateConverged=False
        elif GlobalTruncation:
            for node in range(NumberOfNodes):
                if abs(SolutionCorrection[node])>vkmax*reltol+v
abstol:
                    UpdateConverged=False
    End Python code
```

We can now run the previous examples with this convergence check and compare the number of iterations needed for a given tolerance (Codes 5.3.1) and find the results in Table 5.4.

As one would expect, the number of iterations increases with the setting of reltol. The more complicated bandgap circuit takes a few more iterations for the given tolerance. The number of iterations are also highly dependent on the initial conditions which should come as no surprise.

### 5.2.2.1   Convergence Issues

For real circuits, there can be millions of nonlinear devices, and finding the correct operating point (there can be many) can be a difficult task. In response to this, several methods were developed. For one thing, it should be clear that if all voltage and current sources are zero, a solution presents itself, namely, zero! One can then simply ramp up the various voltage and current sources to their nominal value, and if one does this slowly enough, the circuitry should be able to tag along, and we will

**Table 5.4**  Number of Newton-Raphson iterations versus reltol for the examples in Figs. 5.2, 5.3, and 5.4

| reltol | 5p2 | 5p3 | 5p4 |
|--------|-----|-----|-----|
| 1e-1   | 5   | 5   | 3   |
| 1e-3   | 6   | 6   | 18  |
| 1e-7   | 7   | 7   | 19  |

have a realistic DC solution. Oftentimes the solution at the end of this ramp up can be used as a starting point for another kind of simulation, like a transient simulation. There are other methods we will discuss in the section, and choosing one is often referred to as choosing a homotopy (the term comes from the mathematical discipline topology where a homotopy is a relationship between two functions where one function is continuously changed into the other through some parameter, loosely speaking). For the ramp-up case we just discussed, we often see ramp up, or source stepping or some such option for the simulator.

**Homotopy Methods**
In this section, we will discuss the most common homotopy methods, source stepping, $g_{min}$ stepping, and (briefly) pseudo-transient methods. We will be quite brief and highlight what often works best in practice.

*Source Stepping*
Most circuits have the convenient property that if the voltage sources are at 0 V and the current sources are at 0 A, the circuit has the solution = 0; all other voltages and currents are zero. We mentioned this case briefly in the introduction to this section. In the case where there is a problem with DC convergence, one method would then be to start everything off at zero and slowly ramp all the sources up to their final value. This is clearly a continuous change of the circuit from one state to the other, so it is classified as a homotopy. It is called source stepping in most commercial simulators. It is often a very convenient way to achieve DC convergence. The trick is often in the rate of the ramp where one needs to adjust the next value of the sources if a new step fails. It is not always guaranteed to converge, and there is a lot of theoretical work that has been done to investigate these problems. Often the Achilles heel is in the severe nonlinearity in switching-type circuits where the gain around the switch point can be large.

Let us take a look at a cross-coupled inverter circuit where we can showcase a way to implement the stepping algorithm Code5.4. The circuit can be found in Fig. 5.10, and the netlist is

**Fig. 5.10** Cross-coupled inverters with pull-up/down resistors

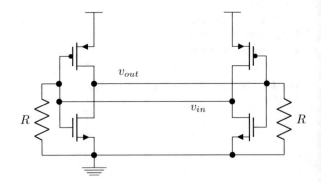

```
*
M1 out in vss nch
M2 out in vdd pch
M3 in out vss nch
M4 in out vss pch
R1 in vss 10000
R2 out vss 10000
Vdd vdd 0 1
Vss vss 0 0
```
**netlist 5.5**

For the stepping algorithm, we simply take the supply voltage, Vdd, and change its value from 0 to full scale with a certain number of steps. Remember, this is a DC simulation, so all capacitors and inductors are removed or shorted as needed.

The Python snippet will be

```
NumberOfSourceSteps=100
For step in range(NumberOfSteps):
    ... Do Newton-Raphson iteration
    ...
        If DevLabel[i]='Vdd':
            STA_rhs[NumberOfNodes+i]=DevValue[i]*steps/
NumberOfSourceSteps
    ...
```

where the rhs line change the source value as the steps progress. The result of such a simulation can be found in Fig. 5.11. It is evident the algorithm picks the midpoint solution. Is there a way to make the algorithm pick another solution? If one change the pull-down resistors to pull-up and also a mix of pull-up and pull-down, one can, depending on the resistor value, see several different outcomes of this implementation of the algorithm.

A natural difficulty is when a circuit comes to a bifurcation, where there are several paths available to several solutions. The algorithm can then get stuck and fail to converge to either of the solutions. Often one can affect this by changing the circuit conditions. We can change the resistors to pull-up/pull-down and step the supply voltage as before. Figure 5.12 shows the result for the four combinations of resistor pull-up/ pull-down.

We see here that given different combinations of pull-up/pull-down resistors, the response changes as a function of ramp voltage. We can also see the convergence is having a hard time around vdd = 0.4 corresponding to VT for the devices. This just serves as a simple illustration to the convergence problems the method can sometimes run into.

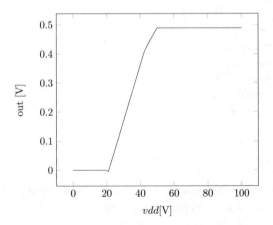

**Fig. 5.11** Source stepping response with both resistors at 100kohm and pull down to gnd

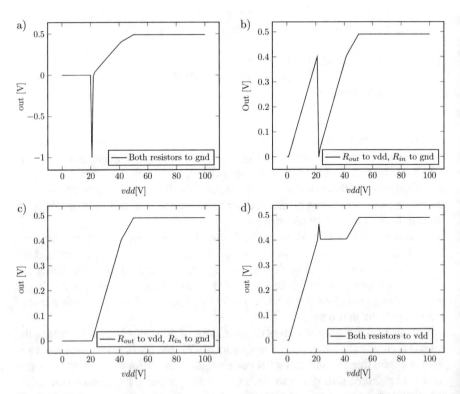

**Fig. 5.12** Node out as a function of ramp voltage for 10Mohm resistors. (**a**) both resistors are pulled down to vss; (**b, c**) one resistor ties to vss, the other to vdd; (**d**) both resistors tie to vdd

*gmin Stepping*

Another commonly used method is adding a large resistor, $1/g_{min}$, between every node and ground and slowly make the resistor larger and larger, $g_{min} \rightarrow 0$, and solve the matrix equation for each step. When the value of $g_{min}$ has reached some small number, where the voltages and currents are not changing beyond some tolerance, the circuit can be considered as having converged. In practical situations with real silicon, there is always some resistance between every node and ground due to leakage and other factors, so having a $g_{min} \neq 0$ is not necessarily a bad approximation. This method is clearly also a homotopy since it is continuously changing the circuit from one to another. We can illustrate this with a really simple example using one of our CMOS transistor models we just developed, model1. Let us use it to investigate the inverter circuit in Fig. 5.13.

The netlist we will use is as follows:

```
*

M1 Vo Vi vss nch
M2 Vo Vi vdd pch
Vdd vdd vss 0.9
Vss vss 0 0
Vin Vi 0 0.2
    netlist 5.6
```

When one attempts to run this netlist, the result is a singularity error. To investigate further, let us write the matrix equation (skipping the trivial supply and ground current equation for clarity):

**Fig. 5.13** A CMOS inverter with a DC input voltage

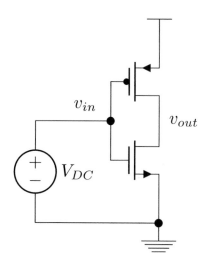

$$\begin{pmatrix} 0 & 1 & 1 \\ 0 & 1 & 0 \\ 0 & 0 & 1 \end{pmatrix}\begin{pmatrix} V_o \\ i_p \\ i_n \end{pmatrix} = \begin{pmatrix} 0 \\ -K\left(V_{in}-V_{dd}\right)^2 \\ KV_{in}^{\,2} \end{pmatrix} \qquad (5.23)$$

The matrix is clearly singular in that its determinant is equal to zero. The root cause of this is the output $V_o$ does not appear in any equation and is unrestricted. Clearly removing it altogether from the system of equation means KCL will not be met at that node, so that is not an option. In addition, there is also the problem with the output currents from the PMOS, and CMOS transistors do not match in general, only when $V_{in} = V_{dd}/2$. These problems are both due to the fact the output conductance is zero with the very simplified transistor model we are using. Shortly we will discuss examples with transistor model2 from Chap. 3, but let us stay with this one a little bit longer and illustrate the gmin stepping method. Let us add a resistor with value $R = 1/g_{min}$ between the output and *vss* and see what happens (Fig. 5.14). Such a resistor will add another equation to our matrix which now becomes

$$\begin{pmatrix} 0 & 1 & 1 & 1 \\ 0 & 1 & 0 & 0 \\ 0 & 0 & 1 & 0 \\ -1/g_m & 0 & 0 & 1 \end{pmatrix}\begin{pmatrix} V_o \\ i_p \\ i_n \\ i_o \end{pmatrix} = \begin{pmatrix} 0 \\ -K\left(V_{in}-V_{dd}\right)^2 \\ KV_{in}^{\,2} \\ 0 \end{pmatrix} \qquad (5.24)$$

If nothing else now, the system of equation is solvable. Let us look at the output node as a function of $g_m$. The result is in Table 5.5.

We note that as $g_m$ is changing, the output voltage is changing with it, so there is no convergence to a specific output voltage. The reason is the difference in current between the two MOS transistor output flows into this resistor, and the output voltage will scale directly with it. In a professional simulator, such a situation will result

**Fig. 5.14** A resistor added to output of inverter model

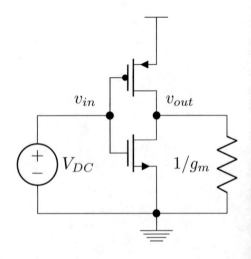

**Table 5.5** Output voltage as
a function of $g_m$ for $V_{in} = 0.2V$

| $g_m$ | V(out) |
|-------|--------|
| 1e-5 | 45 |
| 1e-7 | 4500 |
| 1e-9 | 450,000 |

**Table 5.6** Output voltage as
a function of $g_m$ for
$V_{in} = \dfrac{V_{dd}}{2} = 0.45V$

| $g_m$ | V(out) |
|-------|--------|
| 1e-5 | 0 |
| 1e-7 | 0 |
| 1e-9 | 0 |

in an error, and a convergence flag will be set. The situation is different if we set the input voltage $V_{in} = V_{dd}/2$. Now the current outputs match, and we should see a better convergence (Table 5.6). We have used **code 5.5**.

This was a trivial example of how the gmin method operates. It attempts to find a stable solution by simply adding resistors to various nodes.

*Pseudo-transient*

The pseudo-transient method, ptran for short, is perhaps the method that works best in practice. Here all the dynamic elements, inductors, and capacitors are not disabled but actually used. The basic idea of the method is to use *time* as the continuation variable, and if one can think of the circuit as being simulated "forever" clearly, a DC value can be reached in principle, keeping all independent sources at their DC value. The method works by adding inductors in series with all voltage sources and some dependent sources and capacitors across all independent current sources and some dependent current sources. Naturally the STA system is now increasing with the additional nodes and currents this will provide. Furthermore as an initial condition on the circuit, all new nodes created by the insertion of a series inductor are set to the value of the voltage source. All other nodes are set to 0 V initially, so many capacitors have 0 V across them. Lastly all currents are set to 0 A initially. With this initial condition, a regular transient simulation is run. In practice this method works really well. Clearly, the transition from the initial state to the final state is of no interest in this analysis. Therefore, one can ignore the truncation error as long as the circuit converges to the correct final value. This means the step size in pseudo-time can be much larger than for regular transient, and one needs only to make sure the Newton-Raphson algorithm converges.

One of the issues with such a method is all these dynamic elements in terms of inductors and capacitors can cause resonances and oscillations can occur.

We will look at an example of this kind of algorithm in the transient section of this chapter.

*dptran*

The damped pseudo-transient method is a variety of the ptran method that improves the convergence by reducing the likelihood of an oscillation.

### 5.2.3   Odd Cases: Things to Watch Out for

*Floating Device Example*
Let us look at how a floating resistor will look like following the methodology we have outlined here. Following our outline, the matrix equation will now look like

$$
\begin{pmatrix} 0 & 0 & -1 \\ 0 & 0 & 1 \\ -1 & 1 & -R \end{pmatrix} \begin{pmatrix} v_a \\ v_b \\ i \end{pmatrix} = \begin{pmatrix} 0 \\ 0 \\ 0 \end{pmatrix}
$$

We quickly see the matrix has determinant zero (which it needs to in order to have a nontrivial solution). The upper two rows result in the same equation for the current $i = 0$, and we are left with the last row:

$$
\begin{pmatrix} -1 & 1 \end{pmatrix} \begin{pmatrix} v_a \\ v_b \end{pmatrix} = 0
$$

This is an underdetermined system of equations, one equation and two unknowns, resulting in an infinite number of solutions. A simulator will issue some complaint relating to singular matrix or some such warning. Try the netlist

```
Netlist
*
R1 a b 10
```

in the Python setup. The script will exit with a singular matrix warning.
The solution to this problem in practically all simulators is to put a resistor to ground from all nodes in the circuit, similar to the gmin dampening method we discussed in the previous section. This resistor has a resistance value of $1/g_{min}$, where $g_{min}$ is a user controlled conductance variable. In the case just discussed, we then have

$$
\begin{pmatrix} 0 & 0 & -1 & -1 & 0 \\ 0 & 0 & 1 & 0 & 1 \\ -1 & 1 & -R & 0 & 0 \\ -1 & 0 & 0 & -1/g_{min} & 0 \\ 0 & 1 & 0 & 0 & -1/g_{min} \end{pmatrix} \begin{pmatrix} v_a \\ v_b \\ i \\ i_a \\ i_b \end{pmatrix} = \begin{pmatrix} 0 \\ 0 \\ 0 \\ 0 \\ 0 \end{pmatrix}
$$

We see now the currents $i_a, i_b = -i$ and

$$
-v_a + v_b = iR
$$

$$
v_a = i / g_{min}
$$

$$v_b = -i / g_{min}$$

which has the only possible solution $i = 0$ since $g_{min}, R > 0$.

We can also simulate this with our Python setup with the netlist

```
Netlist
*
R1 a b 10
Rgmin1 a 0 10000
Rgmin2 b 0 10000
```

where we find it now converges and to the same solution we just derived.

## 5.3 Linearization Techniques

One aspect of nonlinear circuits is we can linearize the response around a particular operating point and apply our linear analysis tools (AC analysis) in a small region around this point where everything is close to linear in the response. This section will briefly discuss how simulators can accomplish this. We will use our transistor model as examples. Let us start with the simplest one, model1.

**Model1**

This model has the simple transfer function given by Eq. 3.47. To linearize we do a simple Taylor expansion around the operating point:

$$g_m = \left. \frac{\partial I_d}{\partial V_{gs}} \right|_o = 2KV_{gs} \Big|_o = 2K \left( V_{g,o} - V_{s,o} \right)$$

where the subscript $o$ indicates the DC value at the operating point.

The input impedance is infinite and the output impedance likewise. We then find the linearized version of the transistor is a VCCS with a gain of $g_m$. This can easily be accommodated in our matrix setup in Sect. 4.3.3.

**Model2**

The transfer function is here a bit more complex given the various operating regions, so we have

$$g_m = \begin{cases} 0, V_{gs} - V_T < 0 \\ 2K V_{ds}, \ V_{ds} < V_{gs} - V_T \\ 2K \left( V_{gs} - V_T \right)\left( 1 + \lambda V_{ds} \right) \end{cases}$$

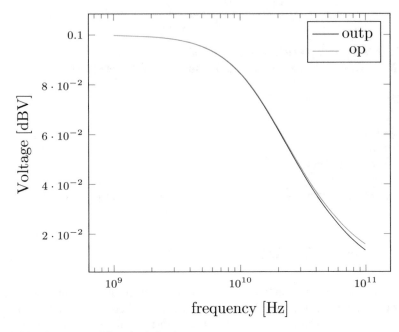

**Fig. 5.15** AC response of linearized circuit

$$g_o = \begin{cases} 0, V_{gs} - V_T < 0 \\ 2K\left((V_{gs} - V_T) - V_{ds}\right), \ V_{ds} < V_{gs} - V_T \\ K(V_{gs} - V_T)^2 \lambda \end{cases}$$

This is again a VCCS with a gain depending on the operating region, and there is also an output conductance *connecting* the drain and source terminals that is DC bias-dependent. This conductance can be modeled by an output resistor: $r_o = 1/g_o$.

The flow would then be to first run a nonlinear DC simulation to establish the operating point. Next, an AC simulation is run that replaces all nonlinear devices with their linearized version, and a new matrix system is set up.

The reader is encouraged to modify the AC simulation codes and implement this linearization scheme on the circuit we just built in Fig. 4.10. One should end up with an AC response as in Fig. 5.15

## 5.4    Transient Nonlinear Simulation

In this section, we will look at how a transient nonlinear simulator can be set up. This will involve the very important choices of timestep algorithm and integration method. We have discussed the basics of the integration methods in Chaps. 1 and 2, and here we will discuss them further. They have a significant impact on accuracy, and we will dig into the details with the help of our simulator. Furthermore, we are combining the nonlinear solver we just discussed in the DC section with a timestep algorithm where the local truncation error will be evaluated with the method outlined in Sect. 4.6.6. First, the nonlinear solver will need to find the solution with a given timestep, and then we evaluate the truncation error to see if the tolerances are met. If not, the timestep needs to be adjusted. In this chapter, we will first use a uniform timestep and take note of the number of iterations needed for each of our three transistor models for a given set of simulation parameters. Next, we discuss the idea of a variable timestep and how such can be implemented. We will then reexamine our circuits with a varying timestep algorithm and note differences in terms of number of needed iterations and compare accuracy. We will also describe the concept of global truncation error and how it depends on a given circuit and other factors. The following section will describe how nonlinear capacitors are implemented in a CMOS transistor model followed by a description of the concept of *break points*. This section ends with a detailed description of steady-state simulators including perturbation techniques.

Most of the examples will use the trapezoidal integration method. The Gear2 and backward Euler are also available in the code.

**Updated Master Transient Code**
At this point, our transient simulator has all the right pieces. We have defined a local truncation error and a code for it in Chap. 4 where we also discussed various difference equation implementations. This chapter has discussed a Newton-Raphson implementation to solve nonlinear equations. We can now incorporate all these codes into one larger code segment that includes all these different pieces. As the reader no doubt has noticed, the simulator code is getting rather large and cumbersome to read. We will therefore incorporate the various sections in subroutines defined in the file analogdef.py in appendix A. As a result, we have **Code 5.6** in Sect. 5.6.6. This code will now be used for all our subsequent simulations using transient techniques.

## 5.4.1 Fixed Timestep

Let us again look at the situation in Fig. 5.1, where there are no dynamic elements. To implement a nonlinear transient simulation, we will simply follow the recipe we outlined in the previous section and calculate the Jacobian at every timestep and do Newton Raphson iterations, and when the tolerance requirements are met, we will accept the timestep. This is called a uniform timestep simulation. The more complicated case of adjustable timesteps we will discuss in Sect. 5.4.2.

It is clear that one way to handle nonlinear situations is to take small timesteps where the devices are approximately linear and then use Newton-Raphson to nail down the error.

The flow will now be as follows:

- Guess a timestep, $\Delta t_{step}$.
- Solve for the new variables at times $N \cdot \Delta t_{step}$ using Newton-Raphson iteration. We use the convergence criteria developed in Sect. 5.2.1.
- If the Newton-Raphson method or the LTE criterion developed in Sect. 4.6.6. has not converged at a given timestep, start over from the beginning with a smaller timestep that we do by hand for this section.

This section will showcase a few examples of nonlinear transient simulations where we will examine circuits with and without dynamic components. The first two circuits will be simple amplifiers and buffers with resistive loads, and the rest will be circuitry with various dynamic elements where the memory of the circuit plays a role.

### 5.4.1.1   Example Circuits with Uniform Timesteps: Purely Resistive Case

When there are no dynamic elements like capacitors and inductors present, there are no differential equations in time that need to be solved, and so there is no memory in the circuit. The state of the circuit at time $t$ is the result of the driving source at the same time. One way to view such a transient simulation is simply as a series of DC simulations, where we can use the previous timestep as a starting point for the next step in time. Here we will show two examples of such DC-like simulations and then we introduce dynamic elements.

*Simple Buffer with CMOS Model2*
Let us simulate the buffer we had in Fig. 5.1 with transistor model2 from Sect. 3.4. The netlist we will use is

```
*
vdd vdd 0 0.9
vss vss 0 0
vinp inp 0 sin(0 1.95 1e9 0 0)
vinn inn 0 sin(0 1.95 1e9 0.5e-9 0)
r1 vdd inp 100
r2 inp vss 100
r3 vdd inn 100
r4 inn vss 100
r5 vdd outp 100
r6 vdd outn 100
i1 vs vss 1e-3
m1 outn inp vs nch1
m2 outp inn vs nch1
      .options   MaxSimTime=2e-9   reltol=1e-7   FixedTimeStep=Tue
deltaT=1e-12 iabstol=1e-12 vabstol=1e-8 lteratio=200000
   .plot v(outp) v(outn)
```
**netlist 5.7**

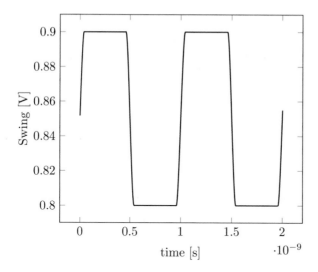

**Fig. 5.16** Buffer output with transistor model2

| **Table 5.7** Basic simulation parameters for reltol = 1e-7, iabstol = 1e-12 vabstol = 1e-8 and FixedTimeStep = True | Parameter | Value | Units |
|---|---|---|---|
| | deltaT | 1e-12 | s |
| | Number of iterations | 5719 | |

We find the output in Fig. 5.16 and Table 5.7. The model is, of course, still way unrealistic but the behavior is fairly reasonable.

The error calculation with Newton-Raphson is the same as for the DC case. For the timestep itself, there is no error estimate since there are no dynamic elements present. The main point here is to illustrate the flow at each time point. The reader will realize pretty quickly that if the solution does not converge, and one has to change the timestep and start over, it might take a while to get done. Therefore some way to have an adjustable timestep is of the essence, and we will discuss ways of doing that in Sect. 5.4.2. Many simulators use a similar algorithm on the get-go where the initial timestep is set as a fraction of the total simulation time. If error conditions are met, it just steps along. This can at times have unintended consequences as the solution the simulator traces out might be feasible in principle, think a VCO with no swing, but in practice things like noise will get the circuit going, and if the timesteps are too large, numerical noise might not be enough. In other words, the initial timestep might be too large for the circuit time scales to be noticed.

Make sure the default initial time step is small enough so that the circuit start-up properly. If not, the circuit might not respond correctly, causing confusion.

*CMOS Inverter*

The advantage with the more realistic model2 is we can also implement a PMOS transistor that we can connect with the NMOS in a digital gate fashion. We can, for example, simulate a CMOS inverter.

```
Netlist
*
vdd vdd 0 0.9
vin in 0 sin(0.45 0.45 1e9 0 0)
vss vss 0 0
mn1 out in vss nch2
mp2 out in vdd pch2
    .options   MaxSimTime=2e-9   reltol=1e-2   FixedTimeStep=True
deltaT=1e-12      iabstol=1e-3      vabstol=1e-3      lteratio=1000000
MaxNewtonIter=15
  .ic v(in)=0.45 v(out)=0.45
  .plot v(out) v(in)
Netlist 5.8
```

Let us now run a transient simulation. The input is a sine wave with amplitude 450 mV and DC offset of 450 mV, so the signal swings around the mid-point; we find the result in Fig. 5.17 and Table 5.8.

Notice, the capacitance is not modeled, and we can drive this entity with any speed and it will still work.

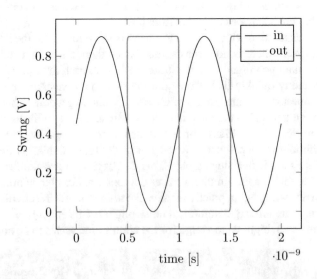

**Fig. 5.17** CMOS inverter input and output

**Table 5.8** Number of iterations for a given fixed timestep

| Parameter | Value | Units |
|---|---|---|
| deltaT | 1e-12 | s |
| Number of iterations | 2052 | |

### 5.4.1.2 Example Circuits with Uniform Timesteps: Dynamic Case

The last two examples used active devices with resistors to illustrate the nonlinear techniques. The truncation error has been nonexistent, and the LTE error-check algorithm has not been necessary. The next few circuit examples will feature both capacitors and inductors and so will be more interesting from a general case perspective. We will start with cross-coupled inverters with capacitors and follow with oscillator design and an RF mixer circuit.

*Cross-Coupled Inverters*
With the improved model of a transistor, model2, let us now look at a cross-coupled pair of inverters, similar to what we had before in the DC section. We noticed there we had three possible solutions, and in particular the mid-point solution was stable. Let us now add in some capacitors to the system with the following addition to the **netlist 5.5**:

```
C1 in vss 1e-13
C2 out vss 1e-13
.ic v(in)=0.0 v(out)=0.0
```

We can now run this as a function of time and we find the result in Table 5.9.

As we have been accustomed, the mid-point solution drives to the rails with a time scale depending directly on the capacitive load. This is in contrast to the DC solution that converged to mid supply point.

The take-home lesson from this example is that DC circuits can have some surprising properties contributing to the difficulty of finding solutions.

*Oscillator*
As another example, let us study an oscillator. It is one of the key circuit implementations in many modern chips, and a lot of efforts have been made in sorting out their behavior [22]. In a circuit development environment, a free running oscillator by itself is not that useful except perhaps for process monitoring schemes. Normally one would need a way to control the oscillation frequency, and often this is done with an input voltage controlling the frequency through a voltage-dependent capacitor. Such a device is called a voltage-controlled oscillator (VCO). Here we will look at an oscillator without such capacitors. Such an oscillator is an excellent test bench for simulator implementations since particularly the amplitude is highly dependent

**Table 5.9** Number of iterations to settle for a cross-coupled inverter and using TRAP integration method

| deltaT | Number of iterations |
|--------|----------------------|
| 1e-12  | ~10,000              |
| 1e-13  | ~125,000             |
| 1e-14  | ~125,000             |

**Fig. 5.18** Basic oscillator topology

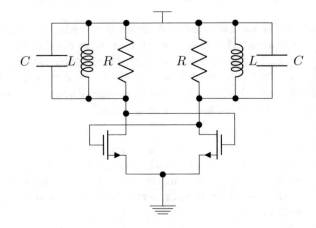

on the specific algorithm used, and we saw an example of this in Sect. 5.4. We will look at it again when discussing variable timestep implementation in Sect. 5.4.2 and in Sect. 5.5.7 when we will look at it in terms of steady-state simulators and phase-noise simulations. For now, we will just do a transient simulation, and we aim at just getting an oscillation going with the simplified MOS model2 we developed.

An oscillator is a system that exhibits a controlled instability. There is a positive feedback mechanism that causes a perturbation to grow up until nonlinearities in the transistors limit the growth. In effect the swing becomes so large the transistor pair runs out of gain. This point can be analyzed from an estimation analysis perspective and has been done elsewhere [22, 23]. There are lots of ways to achieve this feedback effect, and here we will focus on an LC oscillator consisting of an inductor and a capacitor acting as the load to a cross-coupled NMOS pair of transistors. Such a tank, in which the shunted cap and inductor is often referred to, always has some resistive loss modeled as a shunt resistor. Such a resistor can also represent the series loss leading into the tank. The point of the transistor pair is to provide negative resistance (or gain) to the system and so maintain the oscillation. Oscillator work is a rich study with many complicated issues. If the reader wants to dig through the details, [22] is a highly recommended read. Let us now look at our model of an oscillator (see Fig. 5.18). The netlist is straightforward.

```
Netlist
*
vdd vdd 0 0.9
vss vss 0 0
l1 vdd outp 1e-9
l2 vdd outn 1e-9
c1 vdd outp 1e-12
c2 vdd outn 1e-12
r1 vdd outp 1e3
r2 vdd outn 1e3
m1 outp outn vss nch1
m2 outn outp vss nch1
il outp vss pwl(0 0 1e-8 0 1.1e-8 1e-3 1.2e-8 0)
    .options   MaxSimTime=6e-8   reltol=1e-3   FixedTimeStep=True
deltaT=1e-12    iabstol=1e-6    vabstol=1e-6    MaxNewtonIter=15
ThreeLevelStep=True
   .plot v(outp) v(outn)
```
**netlist 5.9**

Note in order to get the simulation to start in a reasonable time, it is common to "jank" it with an ideal piecewise linear current source that turns on for a short time. Without this, it is possible the simulator will only find the other stable point, namely, DC particularly if the transconductor gain is weak. Firing up this netlist in our simulator with our three integration methods result in the plot in Fig. 5.19.

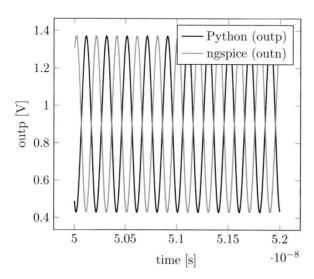

**Fig. 5.19** Oscillator output using TRAP method compared to the same method using ngspice. Note the output node is outn in ngspice and outp in Python, and not only is the amplitude and frequency the same, the phase is also the same

**Table 5.10** Number of
iterations for a 60 ns
simulation with fixed
timesteps

| Parameter | Value | Unit |
|---|---|---|
| deltaT | 5e-12 | S |
| Number of iterations | 19,402 | N/A |

An initial startup phase is followed by the steady oscillation shown in Fig. 5.25. With ngspice one can construct a transistor model that behaves just like ours; we can then use it to compare our simulator implementation to the ngspice version. The two are virtually identical as can be seen in Fig.5.25. We can use the formula derived in [23] to estimate the amplitude. We find

$$A = \sqrt{\left(g_m - \frac{1}{R}\right)\frac{4}{g_m'' 3}}\qquad(5.25)$$

And comparing to the simulation, we find a reasonable agreement, some 5% off. We can summarize the settings in Table 5.10.

*RF Mixer Circuit*
A mixer circuit is a workhorse in narrow-band radio-type circuits (Fig. 5.20). It is used to translate a frequency-modulated signal to baseband (or low frequency). It can also be used to translate a baseband frequency up to RF for use in transmitters.

The netlist is

```
*
vdd vdd 0 0.9
vss vss 0 0
vinp in1 0 sin(0 0.3 1e9 0 0)
vinn in2 0 sin(0 0.3 1e9 5e-10 0)
r1 vdd inp 100
r2 inp vss 100
r3 vdd inn 100
r4 inn vss 100
r5 vdd outp 100
r6 vdd outn 100
c1 in1 inp 1e-12
c2 in2 inn 1e-12
i1 vs vss sin(1e-3 100e-6 1.001e9 0 0)
m1 outn inp vs nch
m2 outp inn vs nch
i2 vs2 vss sin(1e-3 100e-6 1.001e9 4.995e-10 0)
m3 outp inp vs2 nch
m4 outn inn vs2 nch
    .options   MaxSimTime=2e-6   reltol=1e-3   FixedTimeStep=Tue
deltaT=2e-11   iabstol=1e-6   vabstol=1e-6   MaxNewtonIter=15
ThreeLevelStep=True
```

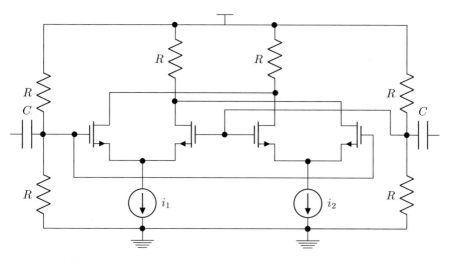

**Fig. 5.20** A Gilbert-cell mixer circuit

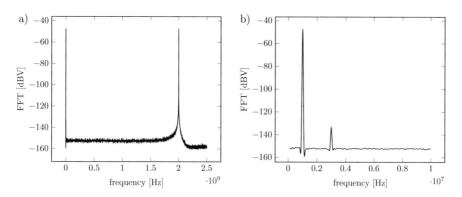

**Fig. 5.21** The figure shows the spectrum of the output differential voltage (**a**) The full spectrum up to 2.5GHz and (**b**) the zoom-in around the 1 MHz down-mixed signal

```
.plot v(outp) v(outn)
netlist 5.10
```

We can simulate this and look at the resulting baseband signal. The input RF signal we have chosen is 1GHz + 1 MHz. The LO is at 1GHz. We can then expect a baseband signal showing up at 1 MHz. This means the simulation must be >1$\mu s$ long. This is quite a stretch. Later in this chapter, Sect. 5.5.4, we will explore a similar situation with a much better type of simulation algorithm (PAC) where we will demonstrate the speed improvement with which we can achieve results. See Fig. 5.21 for the simulation result.

## 5.4.2   Adjustable Timestep

We saw in Sect. 5.4.1 it will be quite cumbersome to have a fixed timestep since the
accuracy might not be met at some point in time and one might have to redo the
simulation. Here we will demonstrate the power of timestep adjustments. There is a
bit of loss of accuracy we will look at toward the end of the section, but the speed
up with respect to simulation time is significant.

### 5.4.2.1   Timestep Adjustments

A variable timestep change is relatively easy to implement, but the question is more
what algorithm to choose. It is clear that if the Newton iteration is not converging in
some reasonable number of iterations, a common number is five, we need to take a
smaller timestep. How small this should be and what happens to the timestep after
that is something we will address in this section. One might think to try dividing the
timestep by 2 if the convergence criteria are not met and doubling the timestep if they
are met is a good idea. Let us experiment with an oscillator. Such a circuit is a good
test bench since if the timestep algorithm is wrong, it will simply stop oscillating (an
oscillator can have a DC solution in an idealized world with no noise). Let us investigate the oscillator in Sect. 5.4.1.2. We can try to implement the idea of increasing
and dividing the timestep by a factor of two and see what happens (see Fig. 5.22).

```
Python code
    if NewtonConverged:
  .


  .


  .

        if not FixedTimeStep:
            deltaT=2*deltaT
  .


  .


  .

    else:
  .


  .


  .

        SimTime=SimTime-deltaT
        deltaT=deltaT/2
End Python
```

We see the result in Fig. 5.22.

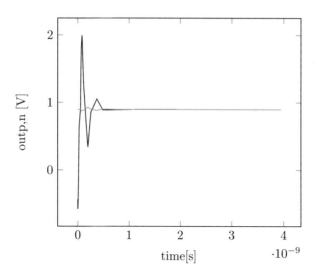

**Fig. 5.22** Simulator response with an increase and decrease in timestep a factor of 2

What is happening here? The problem is the timestep quickly becomes too large and the simulator "misses" the oscillation and finds the DC solution instead. We can try to remedy the situation with a MaxTimeStep construct.

```
Python code
     if NewtonConverged:
.
.
.
         if not FixedTimeStep:
             deltaT=min(MaxTimeStep,2*deltaT)
.
.
.
     else:
.
.
.
         SimTime=SimTime-deltaT
         deltaT=deltaT/2
End Python
```

The result is in Fig. 5.23.

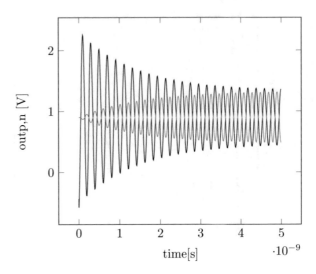

**Fig. 5.23** Simulator response with a max timestep implementation

Now the solver operates much better. Alternatively we can make the timestep increase and decrease unequal amounts as in the following code snippet:

```
Python code
    if NewtonConverged:
   .
   .
   .

        if not FixedTimeStep:
            deltaT=1.01*deltaT
   .
   .
   .

    else:
   .
   .
   .

        SimTime=SimTime-deltaT
        deltaT=deltaT/1.1
End Python
```

The result is in Fig. 5.24

This also seems to work adequately. Let us also look at the increase/decrease algorithm by changing reltol. We find the result in Table 5.11.

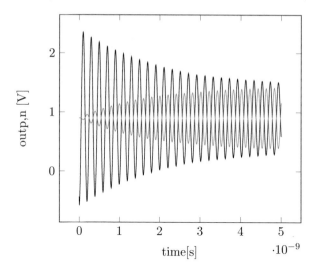

**Fig. 5.24** Simulator response with asymmetric increase and decrease in time step

**Table 5.11** Two-level timestep adjustment algorithm as a function of reltol

| reltol | Amplitude |
| --- | --- |
| 1e-1 | 0.47 |
| 1e-2 | 0.47 |
| 1e-3 | 0.47 |
| 1e-4 | 0.47 |

The simulator will always change the timestep which results in an increase in the numerical noise causing the oscillation to start even at high reltols

The fundamental difficulty with this particular implementation is that it changes the timestep every time it is updated. This will increase the accumulated error. Let us instead adjust it by adding a third, mid-level, to the adjustments. What we do is simply to keep track of the maximum ratio of the voltage difference to the error limit as in

$$v_{max} = \frac{|v_k - v_{pred}|}{reltol\, v_{k,max} + vabstol}$$

for all the voltage nodes. If this max is >0.9, we step down a percentage; if it is <0.1, we step up another percentage; and if it is between these limits, we leave the timestep unchanged. This will keep the timesteps the same for large stretch of the simulation and should improve the accuracy. Let us implement this as in the following code snippet:

**Table 5.12** Three-level timestep adjustment algorithm as a function of reltol

| reltol | Amplitude |
|--------|-----------|
| 1e-1   | 0         |
| 1e-2   | 0         |
| 1e-3   | 0.47      |
| 1e-4   | 0.47      |

Note the intermediate timestep is so long for the high reltols the simulator misses the oscillation! There is less numerical noise being generated for these cases, and the simulator finds the DC solution

**Table 5.13** Accuracy of two versions of timestep adjustments

| Property  | Fixed timestep | MaxTime, % difference | Small increase, % error |
|-----------|----------------|-----------------------|-------------------------|
| Frequency | 5.032e9        | 0.1                   | 0.1                     |
| Amplitude | 0.47           | 2                     | 1                       |

Let us run the oscillator for different reltols like we just did for the two-level algorithm. We find the result in Table 5.12.

We can use these methods and compare with the fixed timestep solution and measure frequency and amplitude using the fixed timestep as the absolute reference. We then find the result in Table 5.13.

### 5.4.2.2   Summary

Timestep control is a closely guarded secret for commercial simulators since it directly affects the simulator performance. In this section, we have demonstrated a couple of strategies one can employ and showed the possible strengths and weaknesses of each compared to a more accurate fixed timestep simulation.

The reader is strongly encouraged to explore this more on his or her own. The literature on timestep adjustments is rich, and the reader can, for example, look at [4, 15–17, 24] to get an idea of what people have tried.

The fact that we are using a variable timestep can cause problems when we are using a Gear2 implementation. If we recall from the previous sections and in Chap. 1, the Gear method relies on information from two timesteps back. If these two timesteps are different, we are not quite getting the right value, and one then needs to inter(extra)-polate to get the appropriate number. This can often result in a convergence problem unless one remedies it. Can you find a good way to circumvent this potential problem?

### 5.4.2.3 Example Circuits with Timestep Control: Purely Resistive Case

After we have examined common ways to adjust the timestep size, let us now reexamine our test circuits from Sect. 5.4.1.

*Simple Buffer with Realistic CMOS Model*
We can now resimulate the buffer we had with an adjustable timestep. The resulting number of iterations and comparison to the fixed timestep case can be found in Table 5.14.

*CMOS Inverter*
The CMOS inverter can be simulated with an adjustable timestep and be compared to the fixed timestep case.

```
Netlist
*
vdd vdd 0 0.9
vin in 0 sin(0.45 0.45 1e9 0 0)
vss vss 0 0
mn1 out in vss nch2
mp2 out in vdd pch2
    .options  MaxSimTime=2e-9  reltol=1e-2  FixedTimeStep=False
deltaT=1e-12    iabstol=1e-3    vabstol=1e-3    lteratio=1000000
MaxNewtonIter=15
  .ic v(in)=0.45 v(out)=0.45
  .plot v(out) v(in)
```

Let us now run a transient simulation. The input is a sine wave with amplitude 450 mV and and DC offset of 450 mV so the signal swings around the mid-point; we find the resulting edge rate by taking the derivative with respect to time in Fig. 5.25.

We see here the timestep adjustment algorithm we use causes errors as we step along the voltage edge. This is fundamentally why at times it is very useful to simulate a circuit with a uniform timestep.

**Table 5.14** Number of iterations for a 10 ns simulation vs reltol and fixed timestep vs three-level asymmetric timestep adjustments

| reltol | Fixed timestep | 1e-3 | 1e-4 | 1e-5 | 1e-6 | 1e-7 |
|---|---|---|---|---|---|---|
| # iterations | 5719 | 827 | 827 | 912 | 969 | 1078 |

The trend is of interest however where the tighter you set reltol, the more iterations need to be taken, and compared to the fixed timestep, there is a significant improvement in speed (number of iterations)

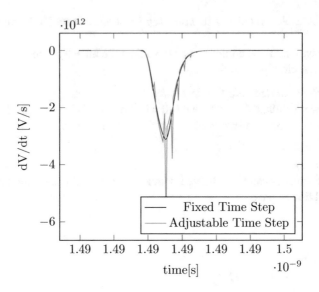

**Fig. 5.25** The derivative of the CMOS inverter output where we compare the edge rate with an adjustable timestep compared to a fixed timestep

### 5.4.2.4   Example Circuits with Timestep Control: Dynamic Case

*Cross-Coupled Inverters*
Let us run the cross-coupled inverters again with the adjustable timestep control. We can rerun the convergence simulation with various initial conditions and compare to the fixed timestep case.

Another interesting point can be made by looking at the response around the voltage edge. We find the following results in Fig. 5.26 where the regeneration time scale is simulated. See also Table 5.15, where we also compare the number of iterations to the fixed timestep case.

We can also use the result of exercise 5. (stable cross-coupled pair) to calculate the exact time constant

$$\tau = \frac{C}{g_m}$$

where for $g_m$ we use the sum of both transistor transconductance and full netlist capacitor value. We find

$$\tau = \frac{10^{-13}}{2 \cdot 2 \cdot 10^{-3}} = 2.5 \cdot 10^{-11}$$

The comparison is shown in Table 5.16. As we have expected, there are significantly fewer iterations needed to get to the final result. Note also that the difference

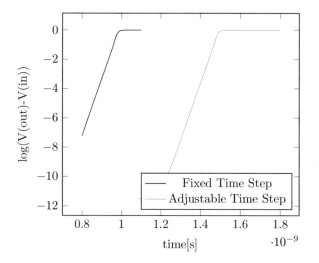

**Fig. 5.26** The regeneration time scale is simulated with and compared between fixed timestep and adjustable timestep. As one can see, the slopes are very similar for the two methods

**Table 5.15** Number of iterations to settle for a cross-coupled inverter using both fixed and adjustable timesteps

| Timestep method | Number of iterations |
|---|---|
| Three-level | No trigger |
| Two-level | ~2000 |
| Fixed (1e-12) | ~10,000 |

The adjustable timestep algorithms can easily start to take too long steps causing it to miss generating enough numerical noise to trigger the decision circuit

**Table 5.16** Comparison table for fixed and adjustable two-level timestep vs exact calculation

| Method | Fixed timestep | Adjustable timestep | Exact | Units |
|---|---|---|---|---|
| Time constant | 2.60 | 2.68 | 2.5 | $10^{-11} \ln [V]/s$ |
| Sigma | 0.28 | 0.40 | N/A | $10^{-11} \ln [V]/s$ |
| Number of iterations | 50,062 | 6221 | N/A | |

between the two methods is covered by the estimate of the slope being uncertain (sigma>slope difference). Both simulations are well within the error margin for the exact value.

*Oscillator*
The oscillator was studied in the discussion of adjustable timestep algorithms in Sect. 5.4.2.1. The number of iterations versus timestep method can be found in Table 5.17.

**Table 5.17.** The number of iterations for the oscillator in netlist 5.9 vs various timestep implementations

| Timestep method | Fixed | Two-level | Three-level |
|---|---|---|---|
| Number of iterations | 19,402 | 14,776 | 14,035 |

**Table 5.18** Comparison of number of timesteps needed for a certain accuracy per timestep between fixed and variable timestep control

| Method | Fixed timestep | Trap adjustable timestep |
|---|---|---|
| Gain error[%] | 0 | 1 |
| Number of iterations | 100,000 | 68,200 |

*RF Mixer Circuit*

As a final example of a circuit topology where we compare a fixed timestep with an adjustable one, let us look at the mixer in Fig. 5.20. We use the same tolerances and simulation stop time as before. To compare accuracy, we use the gain of the RF-baseband conversion. We find the result in Table 5.18.

The improvement in number of iterations is around 32% for a modest increase in the error.

We need to stress that due to the small circuitry, the errors are really small for all these methods, and for larger more realistic circuits, the error degradation can be higher.

### 5.4.2.5   Benefits of Timestep Control

In the previous few sections, we have looked at timesteps and how one can adjust the timestep to be smaller when the circuit activity warrants it and one can take longer steps when the activity is of lower intensity. The penalty one suffers is one of accuracy since the difference equations are less accurate when the timestep varies, in particular when it varies rapidly. The enormous benefit is one of overall simulation speed. The simple examples we looked at here had up to a factor of four increase in simulation speed. This is significant. We also noticed that the penalty in terms of error was modest. Indeed for modern simulators, the timestep algorithm is often a closely guarded secret. In practice if one is interested in high accuracy, it is very common to artificially limit the timestep by setting the MaxTimeStep variable to some small number. This usually forces the simulator to take approximately equation timesteps all through the simulation. As the reader no doubt realizes, the penalty in terms of physical time spent simulating can be drastically increased, and it is a decision a designer often makes with some reservation.

## 5.4.3 Convergence Issues

Convergence issues are a plague to the circuit designer. The good news is that modern simulators are much better in solving these problems on their own with less assistance from the designer. The days were much different way back when. A significant portion of a designer's time was spent debugging convergence problems, and one was forced to learn more about simulators than was perhaps expected. In this section, we will look at a couple of examples using our simulator where it will have some difficulties, and we will examine various attempts to work around it.

After the chosen integration method had discretized time, it is time to solve a large system of nonlinear equations timestep by timestep. We noticed in the case of an inverter with no capacitors in Sect. 5.2 that the nonlinear issues are similar to a DC analysis. In fact one can view the transient analysis as a continuation method with time as the continuation parameter. In general one will expect that if all the models are continuous and have continuous first derivatives, as long as the timestep is small enough, one can achieve convergence. This is what makes transient analysis less prone to convergence issues than DC analysis. In DC there is no previous good time point!

*Example: Jumps in Waveform*
One candidate for convergence difficulties where just shrinking the timestep is not working is when there are jumps in the waveforms. This can happen in places where there are no capacitors to ground at a particular node. This makes that node have infinite bandwidth with the resulting convergence difficulties. This is mostly an issue with poor modeling since in practice, there are no such nodes and often simulators add a *cmin* capacitor to all nodes to ensure finite bandwidth. In earlier versions of CMOS modeling, there were cases where the gate capacitance was modeled as a channel capacitance only. This could cause a sudden disappearance of capacitance when the channel disappears, but adding sidewall and overlap capacitors ensures there is always some capacitance.

### 5.4.3.1 LTE Criterion Guidelines

The LTE convergence criteria depends on a user-controlled setting called either global, local, or point local (see Sect. 4.6.6). We will discuss the usefulness of the global and local varieties in this section and give some simple arguments where each may apply.

*Global Usefulness*
The global criterion is useful in situations where there are both large and small voltages at the same time in the circuit. Let us look at the example in Fig. 5.27.

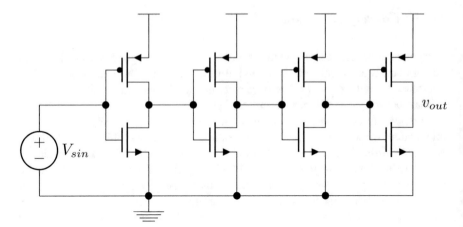

**Fig. 5.27**  Digital two-level circuit

The netlist looks like

```
*
vdd vdd 0 1
vin in 0 sin(0.45 0.45 1e9 0 0)
vss vss 0 0
mn1 o1 in vss nch
mp1 o1 in vdd pch
mn2 o2 o1 vss nch
mp2 o2 o1 vdd pch
mn3 o3 o2 vss nch
mp3 o3 o2 vdd pch
mn4 out o3 vss nch
mp4 out o3 vdd pch
c1 o1 vss 1e-15
c2 o2 vss 1e-15
c3 o3 vss 1e-15
c4 out vss 1e-15
.options reltol=1e-3 vabstol=1e-6 iabstol=1e-12 MaxSimTime=5e-9
GlobalTruncation=Tre
.ic v(in)=0 v(o1)=1 v(o2)=0 v(o3)=1 v(out)=0
.plot v(in) v(o1) v(out)
netlist 5.11
```

Let us run this with both global and point local settings. The result can be found
in Fig. 5.28.

Note that with the local setting, the ground node vss needs to settle to something
much tighter than the rest of the nodes. This is the main reason for the slow rate of
progress.

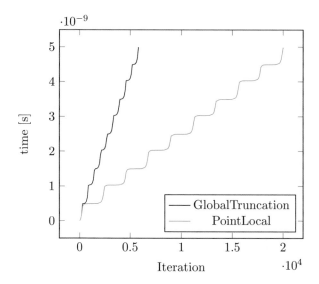

**Fig. 5.28** Global and local setting for digital circuit

*Point Local Usefulness*
The local criterion is useful in situations where there are small sensitive voltages at the same time as larger ones in the circuit. Let us look at the example in Fig. 5.29, where we have two ring oscillators off of two power rails.

The netlist looks like

```
Netlist
*
vdd vdd 0 0.9
vdd10 vdd10 0 10
vss vss 0 0
m1 int1 inp vdd10 pchp1
m2 int1 inp vss nchp1
m3 int2 int1 vdd10 pchp1
m4 int2 int1 vss nchp1
m5 inp int2 vdd10 pchp1
m6 inp int2 vss nchp1
c1 inp vss 1e-11
c2 int1 vss 1e-11
c3 int2 vss 1e-11
m1 int21 inp2 vdd pch
m2 int21 inp2 vss nch
m3 int22 int21 vdd pch
m4 int22 int21 vss nch
m5 inp2 int22 vdd pch
```

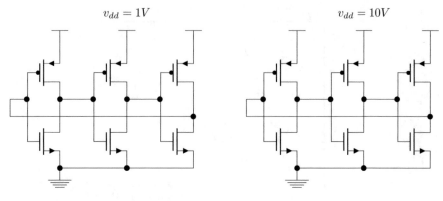

$v_{dd} = 1V$                                     $v_{dd} = 10V$

**Fig. 5.29** Multi-voltage-level circuit

```
m6 inp2 int22 vss nch
c1 inp2 vss 1e-14
c2 int21 vss 1e-14
c3 int22 vss 1e-14
.plot v(inp) v(inp2)
.ic v(inp2)=0.45 v(int21)=0.45 v(int22)=0.45
 .options  reltol=1e-3 vabstol=1e-3 iabstol=1e-3 deltaT=1e-10
MaxSimTime=1e-7 GlobalTruncation=True MaxTimeStep=1e-1End
```
**netlist 5.12**

Let us run this with both global and point local settings. The result can be found in Fig. 5.30.

One can see from the figure that the point local setting gets the oscillation going much faster for the low-voltage oscillator than for the GlobalTruncation case where it is still at its DC operating point.

This example is for illustration purposes only, and one can easily get the oscillation going some other way for the low-voltage circuit. The real difference between these two settings becomes obvious for much larger circuits beyond the capacity of our simulator.

These local and global criteria also exist in some simulators for the residue criterion in Newton-Raphson iterations, which is less tolerant to jumps. For these iterations, one should be careful using the global criteria.

### 5.4.3.2   Global Truncation Error

We have in the previous section spent a bit of time on a cross-region MOS model, and we looked at a few common circuit topologies and drew some conclusions about the need for accurate models. Errors related to the approximation of the differential equations through truncated difference schemes can be divided into two

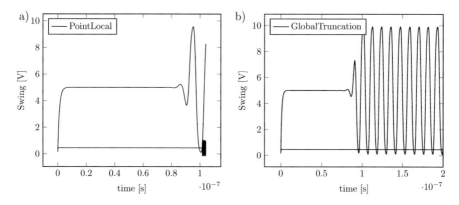

**Fig. 5.30** Global and local setting for analog multilevel circuit. (**a**) shows the PointLocal setting "pays more attention" to the small voltages, and so the low-voltage oscillator kicks in quicker. For the GlobalTruncation( **b**) case, the low voltage action is just glossed over

main groups: a local truncation error, or LTE, that we discussed in Sect. 4.6.6 and a global truncation error (GTE). Here we will now take a step back and discuss the GTE effect in more detail.

Consider, for example, a digital system where there is an error at the output of an inverter. It will quickly return to ground or supply as a function of time. These types of circuits are very insensitive to errors. Consider on the other hand an LC oscillator where we inject an error at the zero crossing. We see the resulting difference compared to the unperturbed simulation in Fig. 5.31.

The error occurs at the zero crossing and causes the phase to shift in time. This error will persist forever. In a real circuit, a noise injection at the zero crossing will contribute to phase noise (see [23]).

Consider now the error in a driven circuit like an amplifier where we perturb the output at the zero crossing with the following Python code. The error will now dissipate with time due to the driving signal as depicted in Fig. 5.32.

We have looked at two examples with varying degrees of error dissipation. Depending on the situation and the time constants involved, the GTE may or may not dissipate with time. It is then clear the global error is difficult to predict in general. In Chap. 6, we will mention a practical way to get an idea of the size of the global truncation error by changing accuracy requirements or timesteps.

### 5.4.3.3 Pseudo-Transient Method for Convergence Aid

With our final transient code, let us investigate the pseudo-transient (ptran) method for achieving convergence. Let us look at the circuit in Fig. 5.10. We noticed there that depending on the pull-down resistors and where they tie the output, the source stepping response would change. We found there are three different possible outcomes for DC source stepping. In this section, we will look at the same circuit, but

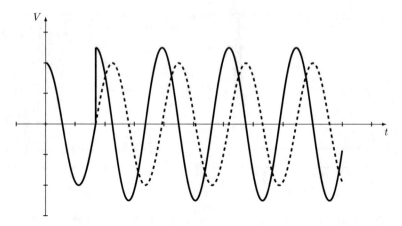

**Fig. 5.31** Oscillator-phase error, the error will persist indefinitely © [2019] Cambridge University Press. Reprinted, with permission, from Cambridge University Press

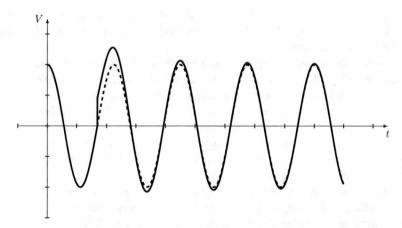

**Fig. 5.32** Amplifier step error, an error at a certain time point which will dissipate with time due to the driving signal © [2019] Cambridge University Press. Reprinted, with permission, from Cambridge University Press

this time we will imitate a pseudo-transient solution by having an inductor in series with the supply voltage source and have two capacitors shunting the outputs of the inverters to ground (Fig. 5.33). We will simulate this for a long "pseudo"-time and see what the solution becomes. Importantly we will keep the pull-up/pull-down resistors. As the reader no doubt realizes, the L and C's will form a resonant block and a certain oscillation is likely to occur. The loss provided by the resistors will shut the oscillation down as time goes by. There are obviously many ways to do this in practice, and this type of oscillation is difficult with the algorithm and there dampened versions that have been developed, often referred to as dptran methods. We will not go into that further, but merely note that in our simple example, the resistors we have added will provide the needed dampening.

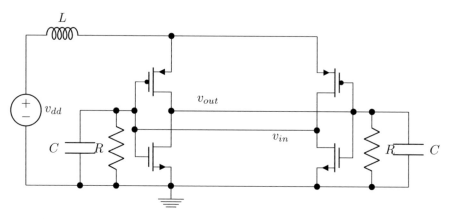

**Fig. 5.33** Circuit to exemplify ptran with the added inductors and capacitors; note the initial condition on the inductor needs to be such the voltage across it is zero; in other words the voltage on the PMOS source side is equal to vdd initially

```
Netlist
.  *
vdd vdd1 0 1
vss vss 0 0
mn1 out in vss nch
mp1 out in vdd pch
mn2 in out vss nch
mp2 in out vdd pch
r1 in vdd 1000000
r2 out vdd 1000000
l1 vdd1 vdd 1e-12
c1 in vss 1e-13
c2 out vss 1e-13
End
```
**netlist 5.13**

The result is presented in Fig. 5.34. What is noteworthy here is the relative ease of convergence. For the source stepping case we investigated in Sect. 5.2.2.1, we found severe convergence issues for some of the initial conditions, using resistors to loosely tie nodes in that case Using ptran we see the same stable operating points but with much less convergence problems. In practice this is often the case also for larger circuitry; the ptran method and its cousin dampened ptran, or dptran, have much better convergence properties. One should not draw too far-reaching conclusions from the relatively simple examples here, but the results we have discussed do resemble what can happen in day-to-day engineering development situations.

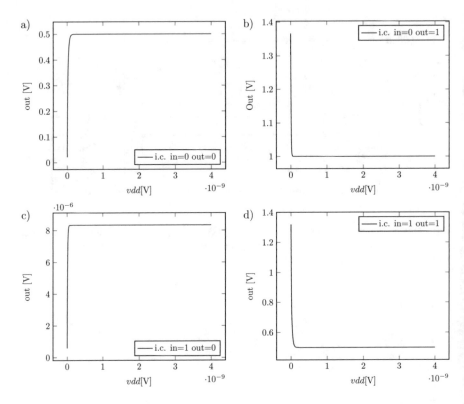

**Fig. 5.34** Result of simulating the circuit in "pseudo"-time, (**a**) shows the result using both in and out node at gnd initially; (**b**) shows the result when "in" is at gnd and "out" is at vdd; (**c**) shows the response with the initial conditions "in" is at vdd and "out" is at gnd; (**d**) finally displays the simulation response to "in" and "out" is at vdd initially

## 5.4.4   Nonlinear Capacitors

We described how one can handle nonlinear active devices in the previous section. Here we will go into nonlinear capacitors. Such devices occur naturally as we saw in the Chap. 2 where the MOSFET gate capacitance varied in a strongly nonlinear fashion with gate voltage, and it is therefore beneficial to show how one can deal with such devices. The behavior of capacitors as we saw in Chap. 2 and in Chap. 4 is governed by a differential equation, and for numerical purposes such equations need to be approximated by difference equations with the resulting accuracy difficulties. If we now add nonlinear response to the capacitance itself as a function of the voltage across it, naturally the problems become worse. We will give a brief initial motivation why this can be a problem and show various ways simulators often deal with it.

*Nonlinear Capacitors and Charge Conservation*
There is a simple argument why this causes problems, and it can be found described in more details in [4]. It goes like this:

Connect a linear 1 pF capacitor to a voltage source. Assume the voltage source change from 0 to 1 mV in one timestep and then back to zero on the next. Let both timesteps be 1 ps. The current computed on the first step is $i = CdV/dt = 10^{-12}10^{-3}/10^{-12} = 1mA$, and the charge is $Q = C U = 1fC$. On the next timestep, the charge is the opposite, and so for a linear capacitor, there is no charge conservation problem.

Consider then a nonlinear capacitor that has a capacitance of 1 pF at 0 V and 1.1 pF at 1 mV. Then as before, the charge for the initial timestep is $1fC$, but for the next timestep, it is $1.1fC$. Thus there is a net charge of $-0.1fC$ for the timestep sequence. The solution here could be to take much smaller timesteps to reduce the accumulated error. Interestingly enough there is no way to completely remove the error unless all derivatives are known exactly, but the proof goes outside the scope of this book [4].

How can one deal with such complications? Let us examine the example by using a charge-based model instead where, for example, the charge $q(v) = C v + D v^2$. For the capacitor current, we now have $i(t) = (q(t1) - q(t0))/\Delta t$. We find

$$i(t1) = 1.05mA$$

$$\Delta q(t1) = 1.05 fC$$

$$i(t2) = -1.05mA$$

$$\Delta q(t2) = -1.05 fC$$

It is fairly obvious it will conserve charge since the algebraic expression for charge is not changing. The point here is that since charge is computed directly, it will conserve charge.

Verify how your favorite simulator conserves entities. Is it a charge conserved simulator or a voltage conserved one for the nonlinear capacitor case.

To put this more formally, we need to employ this charge formulation in the Newton-Raphson formulation. We know

$$i(v) = \frac{\partial q}{\partial t} \rightarrow i_{n+1} \approx \frac{q(u_{n+1}) - q(u_n)}{\Delta t} = g(u_{n+1}) \qquad (5.26)$$

First we need to know how the charge on the capacitor depends on voltage, $q(u) \approx C(u)u$. This we enter into the STA_nonlinear function like a subroutine call:

```
def charge(V1, V2, deltaT)
    V=V1-V2
    C=some function of V
    return C*V
```

Next we need to linearize the response and add to the Jacobian much like for the nonlinear transistor. We have by definition

$$\frac{\partial q}{\partial u} = C(u) \rightarrow \frac{\partial g}{\partial u} = \frac{C(u)}{\Delta t} \tag{5.27}$$

We then find that the current $i$ and voltage $v$ for the next Newton step $k + 1$ are

$$i = g(u_{n+1}) \approx g(u^k_{n+1}) + \frac{\partial g}{\partial u}(u - u_n) = \frac{q(u^k_{n+1}) - q(u_n)}{\Delta t} + \frac{C(u^k_{n+1})}{\Delta t}(u - u_n)$$

in the backward Euler formulation. We see that the voltages known at the previous timestep should be kept at the rhs of the STA system, and we have for the matrix itself the rest of the terms.

A python implementation will look like

```
i=(q(sol[])-q(solold[]))/deltaT
```

for the STA_nonlinear term and

```
STA_rhs=(q[sol]-q[solold])/deltaT-C(u[sol])solold[]/deltaT
Jacobian[][]=C(u[sol])/deltaT
```

for the Jacobian calculation.

We leave it to the reader to implement a Python version of this algorithm and verify charge conservation in the exercises.

## 5.4.5   Break Points

Imagine you have a circuit where the stimulus is some kind of step function or piece-wise linear function of time. Here the stimulus can be either a voltage or a current. At certain points in time, this stimulus will change, and these points need to be included in the simulation (Fig. 5.35). For example,

```
Vpwl nodea nodeb    {0ps, 0V} {100ps,1V} (120ps,1V) (150ps,0}
```

The time-voltage coordinate points correspond to specific times where the voltage has a particular value. If two consecutive points have different voltages, the intermediate voltage will be a linear ramp. The time points 0 ps, 100 ps, 120 ps, and 150 ps are referred to as break points. It simply means the simulator is forced to include these time points as it goes along solving the circuit. The netlist is searched for such stimulus and the break points included in the set of time points. In a real situation, the simulator examines if a particular timestep overreaches a break point. If it is, the timestep is adjusted to appear right at the break point. In our simple simulator, the code could look like

```
If time+deltaT > NextBreakPoint
  deltaT=NextBreakPoint-time
  BreakPointUsed
If timestepAccepted and BreakPointUsed
  NextBreakPoint=FindNextBreakPoint[NextBreakPoint]
```

We leave it to the reader to implement this function in the code examples.

Since the simulator knows when "things" happen in this case, it can take preemptive measures. Often simulators use backward Euler around these timesteps to avoid trapezoidal ringing. We noticed in Sect. 4.6 that the Euler method has a certain artificial "resistor" that dissipates energy and dampens oscillations. Therefore when such schemes are employed, there can be certain initial edge rate degradations.

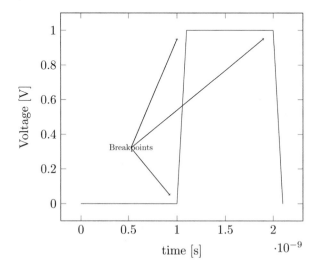

**Fig. 5.35** Picture shows the break points for a piece-wise linear function (pwl)

### 5.4.6   Transient Accuracy: Summary

The accuracy of a transient simulation is set by a number of different criteria:

- The circuit itself will amplify errors or the errors will dissipate as a function of time.
- Reltol, the relative tolerance parameter.
- $v_{nmax}$ how is it determined, globally or locally.

The actual simulator used should have plenty of information on how the accuracy is decided, and the user should spend time making sure to understand the simulator's operation.

Finally, the traditional sure-fire test is to run the simulation again with the appropriate tolerances tightened and compare the end result. The change should be less than the tolerated error.

## 5.5   Periodic Steady-State Solvers

The situation we have looked at so far involves descriptions and algorithms that were developed very early on in the SPICE development and are perhaps the most commonly used in engineering groups today. In the early 1970s, people started to consider the situation where the circuit was in a steady state where all the initial transients had died off. How can such a situation be simulated? If one thinks of a driven circuit and we have some nonlinear elements, these elements will give rise to harmonics of the fundamental tone, and in a steady state, all currents and voltages will only contain harmonics of this fundamental tone. One can then formulate the unknowns as unknown phases and magnitudes of all such harmonics. In fact the equations can be stated on a harmonic basis. This is in a nut shell how the harmonic balance method operates. We will look at a real example of this in Sect. 5.5.2. This method was studied by [3] among many others. It turns out that for some situations, a ridiculous number of harmonics need to be incorporated, particularly if one uses ideal voltage sources with fast rise-fall times. The known Gibbs phenomenon will creep up, and the inverse Fourier transform of the solution (in other words time domain) will look odd. Instead, what is called the shooting method can be used. Here an initial guess at the rate of change for each node is attempted, the time evolution is followed for one period, and then the algorithm checks if the end condition is the same as the guess of the initial condition. If so we have found a periodic solution; if not try again. We will look at this method in Sect. 5.5.1. As the wary reader no doubt recognizes, the shooting method looks cleaner but is having a much harder time converging in particular for large circuits. We will take a look at some of these interesting details.

As we did before, we will discuss a basic implementation in terms of Python codes.

## 5.5.1 Shooting Methods

There are presently two popular ways to implement steady-state solvers. We will discuss what is known as shooting methods in this section. The basic idea is that if the circuit is periodic, the response should repeat itself after a fundamental period. The other time simulators we have looked at earlier in Sect. 5.4 are known as initial value solvers where we start from an often DC-derived operating point and simulate along in time. The shooting method has different assumptions. Here we assume the signals are periodic in time with a user-defined period, $T$. At the start and end of the period, all signals have not only the same value $v_n(0) = v_n(T)$ but also the same curvatures so that $v_n(t) = v_n(t + T)$. One speaks of a boundary problem in time. The shooting method takes advantage of this in a special way we will describe here. We will not go deep into the mathematical derivations here but leave that aspect to Chap. 7 and [1, 3]. Instead we will provide a basic motivation of the theory.

We have as before the governing equation

$$f\big(v(t)\big) = i\big(v(t)\big) + \dot{q}\big(v(t)\big) + u = 0 \tag{5.28}$$

The basic idea behind the shooting method is to note that for a periodic system $f$, the system should return to the same state after time $T$. In a sense the shooting method is all about finding one correct initial condition, there can be many. Let us get a better feel for this by looking at a circuit that has no capacitors in it. We can run the amplifier in Fig. 5.1 with a fixed timestep with a netlist replacing the resistor input bias with voltage sources.

```
...
vinp inp 0 sin(0.5 0.45 1e9 0 0)
vinn inn 0 sin(0.5 0.45 1e9 0.5e-9 0)
...
```

Let us compare the initial state with the final state as shown in Table 5.19.

Clearly we do not have the correct initial state. But let us simply use the following initial state, reflecting the final state in Table 5.19. Let us simulate again and we now find in Table 5.20.

**Table 5.19** Initial and final voltage state over a harmonic period with zero initial state

| Node voltage | Initial state | Final state |
|---|---|---|
| Inp | 0 | 0.5 |
| Inn | 0 | 0.5 |
| Outp | 0 | 0.85 |
| Outn | 0 | 0.85 |
| Vs | 0 | −0.2075 |

**Table 5.20** Initial and final voltage state over a harmonic period with specific initial conditions

| Node voltage | Initial state | Final state |
|---|---|---|
| Inp | 0.5 | 0.5 |
| Inn | 0.5 | 0.5 |
| Outp | 0.85 | 0.85 |
| Outn | 0.85 | 0.85 |
| Vs | −0.2075 | −0.2075 |

**Table 5.21** Initial and final voltage state with a capacitor in the load (all initial state settings are not shown in the table)

| Node voltage | Initial state | Final state |
|---|---|---|
| Inp | 0.5 | 0.5 |
| Inn | 0.5 | 0.5 |
| Outp | 0.83008 | 0.82756 |
| Outn | 0.87354 | 0.87257 |
| Vs | −0.207503 | 0.207503 |

We now see perfect agreement. This is no surprise really since there are no capacitors or inductors in the netlist. Hence, there is no memory and each time point only depends on itself.

Let us now complicate the picture by adding capacitors to the outputs.

```
. . .
c1 vdd outp 3e-12
c2 vdd outn 3e-12
. . .
```

Going through the same iteration we find now the result in Table 5.21

We now see we are not quite at the same point in the final state as the initial state. One can say the initial transient is still not gone, and one can simulate for longer or easily iterate the procedure we outlined a few times and find the correct initial condition and a periodic solution. This type of iterations can be made systematic and we discuss this next.

Imagine we are trying to figure out what a system response should look like and we are close to finding it. We then have $f(\tilde{v}(t)) \approx f(\tilde{v}(t+T))$. Clearly the final state depends on the initial state as we just saw with the previous examples. If we now perturb the approximate solution $v(t) = v(t) + \delta v$, we can apply the familiar Newton-Raphson scheme again as long as we can figure out the transfer function from the initial state to the final one. We will here discuss specifically the case for capacitors and leave the case for inductors to the reader in the exercises.

A change in initial state turns out to be given by (see Chap. 7 for details)

$$\Delta v_i(t) = \left( I - J_{\varphi,ij}(T) \right)^{-1} \left( -v_i(t) + v_i(t+T) \right)$$

$$(5.29)$$

where

$$J_{\varphi,ij}(T) = \frac{\partial v_N}{\partial v_0} = \frac{\partial v_N}{\partial v_{N-1}} \frac{\partial v_{N-1}}{\partial v_{N-2}} \cdots \frac{\partial v_1}{\partial v_0} \frac{\partial v_0}{\partial v_0}$$

is the transfer function we were looking for. Here, the indices indicate timestep, so to get to timestep $N$, we need to see how timestep 1 depends on timestep 0 and how timestep 2 depends on timestep 1 and so on. We show in Chap. 7 for the case with capacitors in the netlist that this Jacobian can be calculated at timestep, $n$, as

$$\frac{\partial v(t_n)}{\partial v_0} = \frac{1}{\Delta t} J_f^{-1} \frac{\partial q\left(v(t_{n-1})\right)}{\partial v(t_{n-1})} \frac{\partial v(t_{n-1})}{\partial v_0}$$

where

$$J^f = \left[ \frac{\partial i\left(v(t_n)\right)}{\partial v(t_n)} + \frac{1}{\Delta t} \frac{\partial q\left(v(t_n)\right)}{\partial v(t_n)} \right]$$

is the normal network Jacobian we encountered in Sects. 5.2 and 5.4. In other words, as we step in time, we can calculate the Jacobian $J_\varphi$ as we go along.

This might seem complicated, but let us consider again the example of a circuit without memory where there are no capacitors or inductors. Now, since the signals everywhere at a given time point $t$ are only dependent on the value of the driving source at the same time point, clearly $J_{\varphi,ij} \equiv 0$, and if we solve the circuit with a DC solver at this time point, we clearly will have $v_i(t) = v_i(t+T)$; we then find $\Delta v_i(t) = 0$. This is the same conclusion we drew previously for our example circuit.

The general idea is now straightforward. We start with a certain initial state at $t = 0$ we have gotten from either a previous transient simulation or perhaps a brave guess; we step through all time steps until $t = T$ calculating the Jacobian along the way. When we get to the final state, we use Newton-Raphson with Eq. 5.29 to calculate a new initial state and start over.

Let us codify this idea. In Sect. 5.6.7, Code5.7 is an implementation of the shooting method concept with capacitors as the dynamic element.

We can apply the code to the example we used earlier in Fig. 5.1 where we have added capacitors $c_1$, $c_2$ to the input and find the netlist

```
*
vdd vdd 0 0.9
vss vss 0 0
vinp in1 0 sin(0 1.95 1e9 0 0)
vinn in2 0 sin(0 1.95 1e9 0.5e-9 0)
r1 vdd inp 100
r2 inp vss 100
r3 vdd inn 100
r4 inn vss 100
r5 vdd outp 100
r6 vdd outn 100
c1 in1 inp 1e-11
c2 in2 inn 1e-11
c3 vdd outp 1e-12
c4 vdd outn 1e-12
i1 vs vss 1e-3
m1 outn inp vs nch1
m2 outp inn vs nch1
.plot v(outp) v(outn)
```
**netlist 5.14**

and the result is shown in Fig. 5.36. We can see the algorithm converges after just a few iterations.

*Multi-tone Simulation*
The example code we just examined contained only one signal tone. It should be clear that any number of signals can be implemented in principle. One only has to keep track of the beat frequency between them and use it as a fundamental harmonic. Let us say, for example, we have a tone of 2GHz and another with frequency

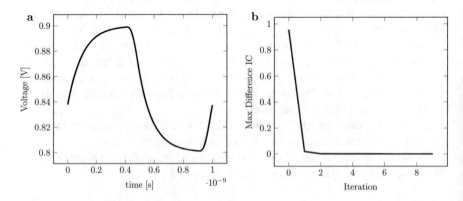

**Fig. 5.36** The output of a shooting method implementation of a steady-state simulator; (**a**) shows the output waveform of netlist, and (**b**) shows the max difference in initial state as a function of iteration. The code converges really quickly for the very simple example

3GHz. An implementation with 2GHz as a fundamental will not capture the 3GHz tone and its harmonics and likewise with using the 3GHz tone. If instead the beat frequency, 1GHz in this case, is used, then clearly all harmonics and the cross coupling of harmonics can be used. We leave to the reader to implement such a code in the exercises

### 5.5.2   Harmonic Balance Methods

Harmonic balance is a way to find the steady state of a circuit by working in frequency space instead of time. Imagine we have a circuit driven by a signal with a certain frequency. If the circuit was linear, then all nodes will contain this frequency and DC. It is simply a question of phase and magnitude. The normal topology constraints, KCL, can also be applied to frequency space. If the circuit is nonlinear, we know from elementary theory that harmonics of the fundamental frequency tone will also appear. This is the basis for harmonic balance methods, and the idea behind the various algorithms is to find the set of harmonics, both magnitude and phase, that defines all device currents and node voltages. We will here go through the basic theory without too much formalities and just hint at its inner workings. We will then move on with the actual implementation in a Python code as before.

When working in frequency space, we will define the Fourier transform as

$$H(f) = \frac{1}{T}\int_{-\infty}^{\infty} h(t)e^{-j\omega t}\, dt \tag{5.30}$$

where $\omega = 2\pi f$ and $T$ is the period over which we define the integral. Furthermore, we will work with both positive and negative frequencies. For real-valued functions $h(t)$, the following property is easy to prove

$$H(f) = H^*(-f) \tag{5.31}$$

where * denotes the complex conjugate. It is much easier to work with calculations this way. For the simulator, the discrete Fourier transform (DFT) is used

$$H(f) = \int_0^T h(t)e^{-\frac{i2\pi nt}{T}}\, dt\, \delta\left(f - \frac{n}{T}\right), \quad 0 < n < \frac{T}{2T_s}$$

where $T$ is the signal period and $T_s$ is the sampling period. Some simulators can manipulate the sampling timestep often via a parameter called sampling ratio or similar. One can use this to oversample the signal in time for a possible increase in accuracy.

We show the more formal derivation of the harmonic balance equations in Chap. 7. We here simply state the network equation, $f(v_k) = 0$, can be Fourier transformed as

$$F(V) = I(V) + \Omega Q(V) + U \tag{5.32}$$

Here the capital $V$ now represents the Fourier coefficients for the voltages at each node.

We define the Jacobian:

$$J_{ij} = \frac{\partial F_i}{\partial V_j} \tag{5.33}$$

We get

$$J(V) = \frac{\partial I(V)}{\partial V} + \Omega \frac{\partial Q(V)}{\partial V} \tag{5.34}$$

Applying the Newton-Raphson algorithm, we find the iteration:

$$V^{j+1} = V^{(j)} - J^{-1}\left(V^{(j)}\right)F\left(V^{(j)}\right) \tag{5.35}$$

The matrix $J$ is referred to as the harmonic Jacobian or conversion matrix. It tells you how a Fourier component at some node $j$ couples to another Fourier component at node $i$. Note the matrix is really a matrix within a matrix. Each circuit node contains a set of Fourier vector components that couples to Fourier components at other nodes. So now the procedure is very similar to the one we followed in the time-domain case. We set up the matrices $F$ and $J$ and iterate along. The boundary conditions set by $U$ have fixed components, and we can just easily iterate our way to the right solution. We do have many more variables; each node has k Fourier components, but once the iterations have converged, we are done. In the time domain, we simply had to continue, so the price we pay in harmonic balance is that the matrix is much bigger, but we only solve it once.

Let us now look at some examples from our device models. We will start with transistor model1 since it is fairly simple and illustrate the procedure when setting up the fundamental matrices $F,J$. We find there

$$i(t) = K\left(v_g(t) - v_s(t)\right)^2 \tag{5.36}$$

To calculate the Jacobian in frequency space, we need to first Fourier transform:

$$i(t) = \sum_{k=1-K}^{K-1} I_k e^{j2\pi k \frac{t}{T}} \tag{5.37}$$

$$I_k = \frac{1}{T}\sum_{n=0}^{N-1} i(n\tau) e^{-j2\pi k \frac{n\tau}{T}} \tag{5.38}$$

We now see

$$\frac{\partial I_k}{\partial V_l^g} = \frac{1}{T}\sum_{n=0}^{N}\frac{\partial i(t)}{\partial V_l^g}e^{-j2\pi k\frac{n\tau}{T}} = \frac{1}{T}\sum_{n=0}^{N}\frac{\partial i(t)}{\partial v_g}\frac{\partial v_g}{\partial V_l^g}e^{-i2\pi k\frac{n\tau}{T}} \tag{5.39}$$

We have

$$v_g(t) = \sum_{k=1-K}^{K-1}V_k^g e^{j2\pi k\frac{t}{T}} \rightarrow \frac{\partial v_g}{\partial V_l^g} = e^{j2\pi l\frac{t}{T}} \tag{5.40}$$

Likewise for $v_s$

$$v_s(t) = \sum_{k=1-K}^{K-1}V_k^s e^{j2\pi k\frac{t}{T}} \rightarrow \frac{\partial v_s}{\partial V_l^s} = e^{j2\pi l\frac{t}{T}} \tag{5.41}$$

The derivative term is straightforward

$$\frac{\partial i(t)}{\partial v_g} = 2K\left(v_g(t) - v_s(t)\right) \qquad \frac{\partial i(t)}{\partial v_s} = -2K\left(v_g(t) - v_s(t)\right) \tag{5.42}$$

Putting all this together, we get

$$\frac{\partial I_k}{\partial V_l^g} = \frac{1}{T}\sum_{n=0}^{N}\frac{\partial i(n\tau)}{\partial v_g}\frac{\partial v_g}{\partial V_l^g}e^{-j2\pi k\frac{n\tau}{T}} = \frac{1}{T}\sum_{n=0}^{N}\frac{\partial i(n\tau)}{\partial v_g}e^{j2\pi l\frac{n\tau}{T}}e^{-j2\pi k\frac{n\tau}{T}} =$$

$$= \frac{1}{T}\sum_{n=0}^{N}\frac{\partial i(n\tau)}{\partial v_g}e^{j2\pi(l-k)\frac{n\tau}{T}}$$

$$= \sum_{n=0}^{N}2K\left(v_g(n\tau) - v_s(n\tau)\right)e^{j2\pi(l-k)\frac{n\tau}{T}} \tag{5.43}$$

Similarly for the other voltage,

$$\frac{\partial I_k}{\partial V_l^s} = J_{l-k} = \frac{1}{T}\sum_{n=0}^{N}\frac{\partial i(n\tau)}{\partial v_s}e^{j2\pi(l-k)\frac{n\tau}{T}} \tag{5.44}$$

We see now in both these cases, and it is not hard to show this in general that the Jacobian looks like the Fourier transform of the time derivative of some function. Also note it is the difference between the indices that matters $l - k$. In matrix form, we have

$$J = \begin{pmatrix} \dfrac{\partial I_{1-K}}{\partial V^s_{1-K}} & \cdots & \dfrac{\partial I_{1-K}}{\partial V^s_{K-1}} \\ \vdots & \ddots & \vdots \\ \dfrac{\partial I_{K-1}}{\partial V^s_{1-K}} & \cdots & \dfrac{\partial I_{K-1}}{\partial V^s_{K-1}} \end{pmatrix} \tag{5.45}$$

For the case where $K = 3$, we have

$$\begin{pmatrix} \dfrac{\partial I_{-2}}{\partial V^s_{-2}} & \dfrac{\partial I_{-2}}{\partial V^s_{-1}} & \dfrac{\partial I_{-2}}{\partial V^s_{0}} & \dfrac{\partial I_{-2}}{\partial V^s_{1}} & \dfrac{\partial I_{-2}}{\partial V^s_{2}} \\[2mm] \dfrac{\partial I_{-1}}{\partial V^s_{-2}} & \dfrac{\partial I_{-1}}{\partial V^s_{-1}} & \dfrac{\partial I_{-1}}{\partial V^s_{0}} & \dfrac{\partial I_{-1}}{\partial V^s_{1}} & \dfrac{\partial I_{-1}}{\partial V^s_{2}} \\[2mm] \dfrac{\partial I_{0}}{\partial V^s_{-2}} & \dfrac{\partial I_{0}}{\partial V^s_{-1}} & \dfrac{\partial I_{0}}{\partial V^s_{0}} & \dfrac{\partial I_{0}}{\partial V^s_{1}} & \dfrac{\partial I_{0}}{\partial V^s_{2}} \\[2mm] \dfrac{\partial I_{1}}{\partial V^s_{-2}} & \dfrac{\partial I_{1}}{\partial V^s_{-1}} & \dfrac{\partial I_{1}}{\partial V^s_{0}} & \dfrac{\partial I_{1}}{\partial V^s_{1}} & \dfrac{\partial I_{1}}{\partial V^s_{2}} \\[2mm] \dfrac{\partial I_{2}}{\partial V^s_{-2}} & \dfrac{\partial I_{2}}{\partial V^s_{-1}} & \dfrac{\partial I_{2}}{\partial V^s_{0}} & \dfrac{\partial I_{2}}{\partial V^s_{1}} & \dfrac{\partial I_{2}}{\partial V^s_{2}} \end{pmatrix}$$

These terms are simply components of the Fourier transform, and as soon as we have it, we also have the Jacobian. Let us take a quick look at the integral (or discrete sum in our case of a discrete Fourier transform) again:

$$\frac{1}{T} \sum_{n=0}^{N} \frac{\partial i(n\tau)}{\partial v_s} e^{j2\pi(l-k)\frac{n\tau}{T}} \tag{5.46}$$

The sum is a sum over a product of two functions $\dfrac{\partial i(n\tau)}{\partial v_s}, e^{j2\pi(l-k)\frac{n\tau}{T}}$. This sum is always zero unless the frequency of the two functions is the same since the functions can only consist of harmonics of the fundamental, and we integrate over the fundamental harmonics period. Particularly in the case where $\dfrac{\partial i(n\tau)}{\partial v_s} = C$, a constant, the only non-zero component is the one where $l = k$ and the matrix simplifies to

$$J = \begin{pmatrix} J_0 & \cdots & 0 \\ \vdots & \ddots & \vdots \\ 0 & \cdots & J_0 \end{pmatrix} \tag{5.47}$$

We can see there is no coupling between Fourier components in that all non-diagonal elements are zero. For the case where the derivative $\dfrac{\partial i(n\tau)}{\partial v_s} \neq C$, there will be non-diagonal terms and thus coupling between harmonics. Let us assume

$$\frac{\partial i(n\tau)}{\partial v_s} = e^{-j\frac{2\pi m}{T}t} = e^{-j\omega_m t} \tag{5.48}$$

$$J_{l-k} = \frac{1}{T}\sum_{n=0}^{N} e^{-j\omega_m n\tau} e^{j2\pi(l-k)\frac{n\tau}{T}} \tag{5.49}$$

We will then only have non-zero matrix components when $l - k = \pm m$.

$$J = \begin{pmatrix} 0\cdots & J_m & \cdots & 0 \\ J_{-m} & 0 & J_m \\ 0\cdots & J_{-m} & \cdots & 0 \end{pmatrix} \tag{5.50}$$

We see we get contributions to component $1 - K$ from component $1 - K + m$ which is also obvious from the trigonometric identity:

$$\sin(l-k+m)\frac{2\pi}{T}\sin m\frac{2\pi}{T} = \frac{1}{2}\left(\cos(l-k+2m)\frac{2\pi}{T} + \cos(l-k)\frac{2\pi}{T}\right) \tag{5.51}$$

Now it should be obvious the basic approximation we are doing when we limit the harmonics from $1 - K$ to $K - 1$ is that the harmonics resulting from products going outside this range are ignored. This fact will result in convergence issues since we are only trying to solve for only a subset of all harmonics. This is why it is important to include enough harmonics to at least cover the maximum bandwidth of the system a few times over.

> It is important to include enough harmonics to at least cover the maximum bandwidth of the system a few times to minimize convergence issues.

The case for a linear resistor should now be obvious:

$$\frac{\partial i}{\partial v} = \frac{1}{R} \rightarrow J_0 = \frac{1}{T}\sum_{n=0}^{N} \frac{\partial i(n\tau)}{\partial v_s} e^{j2\pi(l-k)\frac{n\tau}{T}} = \frac{1}{T}\sum_{n=0}^{N} \frac{1}{R} = \frac{1}{R} \tag{5.52}$$

For a linear capacitor, we see

$$i = C\frac{\partial u}{\partial t} = C\sum_{k=1-K}^{K-1} j\omega_k V_k^C e^{j\omega_k t} = \sum_{k=1-K}^{K-1} I_k^C e^{j\omega_k t} \tag{5.53}$$

where

$$I_k^C = Cj\omega_k V_k^C \tag{5.54}$$

And thus

$$\frac{\partial I_k}{\partial V_l} = Cj\omega_k \delta(k-l) \tag{5.55}$$

Finally for a linear inductor, we know

$$u = L\frac{\partial i}{\partial t} = L\sum_{k=1-K}^{K-1} j\omega_k I_k^L e^{j\omega_k t} = \sum_{k=1-K}^{K-1} V_k^L e^{j\omega_k t} \tag{5.56}$$

As with the capacitor, we get now

$$V_k^L = Lj\omega_k I_k^L \tag{5.57}$$

And

$$\frac{\partial I_k}{\partial V_l} = \frac{1}{Lj\omega_k}\delta(k-l) \tag{5.58}$$

We see here there is a divergence at DC since any voltage would cause an infinite current to run through the inductor. It will need special attention; any DC voltage across an inductor is zero.

In the STA matrix description, we recall from earlier a resistor's signature looks like

$$\begin{array}{c} v_a \\ v_b \\ \vdots \\ i_R \end{array} \begin{pmatrix} & & & -1 \\ & & & +1 \\ & & & \vdots \\ -1 & +1 & \cdots & -R \end{pmatrix} \tag{5.59}$$

In the harmonic balance formulation, we simply replace the numbers with diagonal matrices corresponding to the harmonics involved, for example, the case of three harmonics results in.

$$
\begin{array}{c}
\begin{array}{c} v_a \\ v_b \\ \vdots \\ i_R \end{array}
\left(
\begin{array}{ccc}
& & \begin{pmatrix} -1 & 0 & 0 \\ 0 & -1 & 0 \\ 0 & 0 & -1 \end{pmatrix} \\
& & \begin{pmatrix} +1 & 0 & 0 \\ 0 & +1 & 0 \\ 0 & 0 & +1 \end{pmatrix} \\
& & \vdots \\
\begin{pmatrix} -1 & 0 & 0 \\ 0 & -1 & 0 \\ 0 & 0 & -1 \end{pmatrix} & \begin{pmatrix} 1 & 0 & 0 \\ 0 & 1 & 0 \\ 0 & 0 & 1 \end{pmatrix} \cdots & \begin{pmatrix} -R & 0 & 0 \\ 0 & -R & 0 \\ 0 & 0 & -R \end{pmatrix}
\end{array}
\right)
\end{array}
\tag{5.60}
$$

Since the resistor is here a linear element, each harmonic only interacts with itself, as should be clear from the matrix signature.

For an inductor, we see

$$
\begin{array}{c}
\quad\quad v_a \quad v_b \quad i_L \\
\begin{array}{c} v_a \\ v_b \\ i_L \end{array}
\left(
\begin{array}{ccc}
& & -1 \\
& & +1 \\
+1 & -1 & -L
\end{array}
\right)
\end{array}
\tag{5.61}
$$

$$
\begin{array}{c}
\begin{array}{c} v_a \\ v_b \\ \vdots \\ i_R \end{array}
\left(
\begin{array}{ccc}
& & \begin{pmatrix} -1 & 0 & 0 \\ 0 & -1 & 0 \\ 0 & 0 & -1 \end{pmatrix} \\
& & \begin{pmatrix} +1 & 0 & 0 \\ 0 & +1 & 0 \\ 0 & 0 & +1 \end{pmatrix} \\
& & \vdots \\
\begin{pmatrix} -1 & 0 & 0 \\ 0 & -1 & 0 \\ 0 & 0 & -1 \end{pmatrix} & \begin{pmatrix} 1 & 0 & 0 \\ 0 & 1 & 0 \\ 0 & 0 & 1 \end{pmatrix} \cdots & \begin{pmatrix} -j\omega_{-1}L & 0 & 0 \\ 0 & -j\omega_0 L & 0 \\ 0 & 0 & -j\omega_1 L \end{pmatrix}
\end{array}
\right)
\end{array}
\tag{5.62}
$$

A linear capacitor will look like

$$
\begin{array}{c}
\begin{array}{ccc} v_a & v_b & i_C \end{array} \\
\begin{array}{c} v_a \\ v_b \\ i_C \end{array}
\left(
\begin{array}{ccc}
 & & -1 \\
 & & +1 \\
+C & -C & -1
\end{array}
\right)
\end{array}
\tag{5.63}
$$

$$
\begin{array}{c}
v_a \\ v_b \\ \vdots \\ i_R
\end{array}
\left(
\begin{array}{cccc}
 & & & 
\begin{pmatrix} -1 & 0 & 0 \\ 0 & -1 & 0 \\ 0 & 0 & -1 \end{pmatrix} \\[6pt]
 & & & 
\begin{pmatrix} +1 & 0 & 0 \\ 0 & +1 & 0 \\ 0 & 0 & +1 \end{pmatrix} \\[6pt]
 & \phantom{(5.64)} & & \vdots \\[6pt]
\begin{pmatrix} j\omega_{-1}C & 0 & 0 \\ 0 & j\omega_0 C & 0 \\ 0 & 0 & j\omega_1 C \end{pmatrix} &
\begin{pmatrix} -j\omega_{-1}C & 0 & 0 \\ 0 & -j\omega_0 C & 0 \\ 0 & 0 & -j\omega_1 C \end{pmatrix} &
\cdots &
\begin{pmatrix} -1 & 0 & 0 \\ 0 & -1 & 0 \\ 0 & 0 & -1 \end{pmatrix}
\end{array}
\right)
\tag{5.64}
$$

Finally for a transistor modeled as a nonlinear VCCS, we need to add a nonlinear term as in the transient simulation section. Since we are working in the frequency domain, we need to do a Fourier transform of the transistor transconductance, and for our simple model, that means using

$$
I = K\left(v_g - v_s\right)^2
\tag{5.65}
$$

We find

$$
I_k = \frac{1}{T}\sum_{n=0}^{N} K\left(v_g\left(n\tau\right) - v_s\left(n\tau\right)\right)^2 e^{-i\omega_k n\tau}
\tag{5.66}
$$

For this simple case, we can of course just expand and identify terms, but let us do a proper Fourier transform, and add the resulting harmonics to the function $F(V)$. For the Jacobian (the derivative), we already derived the formula

$$
\begin{array}{c} \\ v_a \\ v_b \\ \vdots \\ i_R \end{array}
\left(
\begin{array}{ccc}
& & \begin{pmatrix} -1 & 0 & 0 \\ 0 & -1 & 0 \\ 0 & 0 & -1 \end{pmatrix} \\
& & \begin{pmatrix} +1 & 0 & 0 \\ 0 & +1 & 0 \\ 0 & 0 & +1 \end{pmatrix} \\
& & \vdots \\
\begin{pmatrix} J_0 & J_1 & J_2 \\ J_{-1} & J_0 & J_1 \\ J_{-2} & J_{-1} & J_0 \end{pmatrix} & \begin{pmatrix} -J_0 & -J_1 & -J_2 \\ -J_{-1} & -J_0 & -J_1 \\ -J_{-2} & -J_{-1} & -J_0 \end{pmatrix} & \cdots \quad \begin{pmatrix} -1 & 0 & 0 \\ 0 & -1 & 0 \\ 0 & 0 & -1 \end{pmatrix}
\end{array}
\right)
\tag{5.67}
$$

For an independent voltage source, we get likewise

$$
\begin{array}{c} \\ v_a \\ v_b \\ i_V \end{array}
\begin{array}{ccc} v_a & v_b & i_V \end{array}
\left(
\begin{array}{ccc}
& & -1 \\
& & +1 \\
+1 & -1 &
\end{array}
\right)
\begin{pmatrix} \\ \\ V \end{pmatrix}
=
\begin{pmatrix} \\ \\ \end{pmatrix}
\tag{5.68}
$$

$$
\begin{array}{c} \\ v_a \\ v_b \\ \vdots \\ i_R \end{array}
\left(
\begin{array}{ccc}
& & \begin{pmatrix} -1 & 0 & 0 \\ 0 & -1 & 0 \\ 0 & 0 & -1 \end{pmatrix} \\
& & \begin{pmatrix} +1 & 0 & 0 \\ 0 & +1 & 0 \\ 0 & 0 & +1 \end{pmatrix} \\
& & \vdots \\
\begin{pmatrix} -1 & 0 & 0 \\ 0 & -1 & 0 \\ 0 & 0 & -1 \end{pmatrix} & \begin{pmatrix} 1 & 0 & 0 \\ 0 & 1 & 0 \\ 0 & 0 & 1 \end{pmatrix} & \cdots
\end{array}
\right)
\begin{pmatrix} \\ \\ \\ \\ V_{-1} \\ V_0 \\ V_1 \end{pmatrix}
=
\begin{pmatrix} \\ \\ \\ \end{pmatrix}
\tag{5.69}
$$

We are now ready to write the code where we first assume the nonlinear transistor behavior follows model1 from Sect. 3.4. Due to the oversimplistic transistor model, it will not result in a very useful output as we will now show by reexamining the circuit in netlist 5.6. We see the harmonic balance result after an inverse Fourier transform in Fig. 5.37.

The blip in the response is due to the poor transistor model; the transistor turns on again when the gate voltage goes below the source.

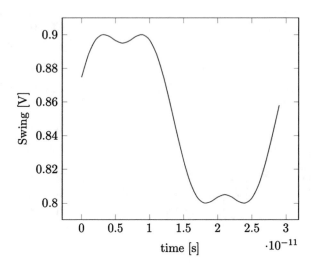

**Fig. 5.37** The result of a harmonic balance simulation of netlist 5.6 using our CMOS model1

We can take this further and implement the more realistic transistor model we discussed in Sects. 3.4.2 and 5.2.1.1. We Fourier transform the current voltage relationship just we did for the simpler model and note we need to add a derivative of the drain voltage to the Jacobian matrix. We will not go thought the details, here but they are straightforward, and we leave the mathematical derivations to the reader and present the code implementation as Code 5p8 in Sect. 5.6.8. We can simulate the same circuit one more time, **netlist 5.6**, with this improved transistor model, and we find the result in Fig. 5.38.

The improvement going from 4 harmonics to 8 is quite different from 16 to 32. The system has clearly converged.

We need to accentuate one more time the fact that the steady-state solution is what the transient would be if one waits for a long time. A properly designed steady-state simulator will always have a more accurate response than a transient. Also note the step in time is *one* period. A transient simulation ends when it ends.

*Multi-tone Implementation*
Just like we discussed for the shooting method case, a multi-tone implementation is fairly straightforward when using the beat frequency of all signal sources.

## 5.5.3   Envelope Analysis

In time-domain-based simulators, we have discussed the timesteps taken need to be smaller when higher frequencies appear in the circuitry due to LTE and GTE effects. Harmonic balance simulators do not suffer from this weakness, but they require at substantial number of harmonics, in particular if one has a combination of low-frequency modulation components riding on top of higher-frequency carriers.

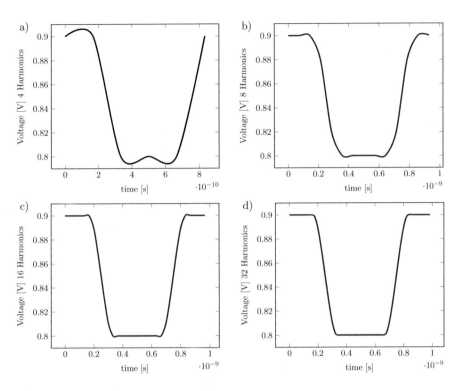

**Fig. 5.38** Harmonic balance results of using a more realistic transistor model as a function of the number of harmonics. As that number is increasing, the result improves considerably. Note: smoothing has been enabled in the rendering to make the differences stand out clearer

Envelope simulation is a combination of a steady-state simulator and a time-domain simulator. If you have ever done work in an electronics laboratory, you have perhaps had the chance to work with a spectrum analyzer. In a typical setup, such an instrument shows the spectral components of the signal at the input port. When looking at those spectral views and often seeing they vary with time, see Fig. 5.39. These time-varying components are what we are after with an envelope simulator.

Mathematically, in harmonic balance, we can write a wave form at a certain node as

$$v(t) = \sum_{k=0}^{\infty} V_k e^{-i\omega_k t} \qquad (5.70)$$

In an envelope simulation, the Fourier coefficients vary with time, so we have

$$v(t) = \sum_{k=0}^{\infty} V_k(t) e^{-i\omega_k t} \qquad (5.71)$$

just like we have seen in a spectrum analyzer.

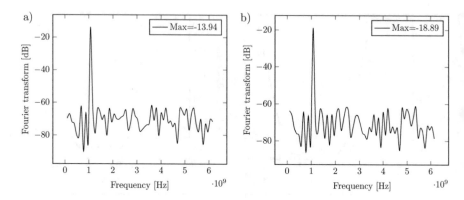

**Fig. 5.39** Typical spectrum analyzer output (**a**, **b**) shows spectrum at different times

Before the purist reading this has a fit (this type of formula is not self consistent for arbitrary waveforms), the basic assumption is that the coefficients $V_k(t)$ vary with time much slower than the carrier frequencies. In other words, as the envelope function changes, the rest of the high-frequency components have a chance to completely settle. In this sense, what we are doing in an envelope simulation is to do a steady-state analysis at a series of points in time. We start at $t = 0$ and set the modulation generators to whatever value they have at that time. The rest of the circuit is now used to solve the steady state assuming the modulation control values are constant. This will give rise to a certain set of harmonics. Having solved for the situation at $t = 0$, we can now contine to $t = t_1$ set all modulation sources to their value at that time and again solve for steady state with this new input in mind. This we continue to do until all the desired time points are solved for. Often the timestep size is fixed, and for that case the modulation bandwidth is $\pm \dfrac{1}{2T_{step}}$. Any tone in the envelope function will show up as two modulation tones around the harmonic carrier.

The advantage of envelopes simulations is that the amplitude, phase of each carrier, is available immediately. One can then use these to make models of various modulation schemes. One does not need additional calculations to convert the time-domain information to envelope information. We simply start from the steady-state solution directly, no need to wait for time-domain settling. It of course also goes the other way, and if the initial transient is of interest, the harmonic balance method does not provide a way to examine it. In short, an envelope simulator takes the carrier steady state into account through harmonic balance, and there is no need to wait for the steady state; a low-frequency modulation signal is then used in the time domain to calculate the circuit response, and there is no need to include all possible harmonics. The envelope simulator then takes advantage of what is best in both harmonic balance and time-domain simulators for certain specific applications.

### 5.5.4 Perturbation Techniques

Having established a periodic steady-state solution much can now be analyzed. Part of the analysis in the previous section showed there is a matrix called the conversion matrix, simply the harmonic Jacobian, that calculates the gain from one frequency harmonic to any other harmonic provided the signal is small enough as not to disturb the steady-state condition we have established. We can think of it as a harmonic AC analysis. A small signal riding on top of larger one will mix with such a signal via the conversion matrix and produce tones around other harmonic frequencies. This is not an uncommon situation. Think of the case of an up-conversion RF mixer where the LO signal is large and the RF signal that multiplies it is small and of baseband frequency (say 100 kHz). To resolve the gain of such, a system will require a simulation lasting some $10\,\mu s$, something that is really painful if the LO frequency is in the GHz range. Imagine our periodic steady-state system, where we can translate the gain from DC to the LO first harmonic simply with an AC like calculation; we can simulate this gain much more efficiently. The same can be said about other conversion-type systems like phase-noise translation in oscillators, etc. We will discuss details of such simulations in the last section of this chapter.

This section will discuss first what is commonly known as periodic AC simulation followed by periodic noise analysis. We will exemplify the technique in our Python setup and use it later in Sect. 5.5.7.

*Periodic AC Analysis*
Let us assume we have a solution for our circuit such that

$$f\big(v(t)\big) = i\big(v(t)\big) + \dot{q}\big(v(t)\big) + u = 0 \qquad (5.72)$$

What if we add a small signal perturbation to $u(t) \to u(t) + \tilde{u}(t)$ such that the overall solution is still approximately valid? The solution will likewise be adjusted with a small correction, $v(t) \to v(t) + \tilde{v}(t)$, and we see we seek a solution that meets

$$f\big(v(t) + \tilde{v}(t)\big) = i\big(v(t) + \tilde{v}(t)\big) + \dot{q}\big(v(t) + \tilde{v}(t)\big) + u + \tilde{u}(t) = 0 \qquad (5.73)$$

This we can expand the usual way with a Taylor series and we find

$$i\big(v(t) + \tilde{v}(t)\big) + \dot{q}\big(v(t) + \tilde{v}(t)\big) \approx i\big(v(t)\big) + \frac{\partial i(v)}{\partial v}\tilde{v} + \frac{d}{dt}\left( q\big(v(t)\big) + \frac{\partial q(v)}{\partial v}\tilde{v}\right)$$

$$(5.74)$$

We note the zeroth-order terms are simply the original solution, and they cancel out and we are left with

$$\frac{\partial i(v)}{\partial v}\tilde{v} + \frac{d}{dt}\left(\frac{\partial q(v)}{\partial v}\tilde{v}\right) + \tilde{u} = 0 \tag{5.75}$$

We can now go through similar steps we did before where we Fourier transform the variables and assuming the perturbation $u(t) = u_k e^{-i(\omega_k + \omega_p)t}$ is a single-tone perturbation of a harmonic, $k$, with an offset frequency, $\omega_p$. The Fourier transform, $\tilde{U} = U_k \delta\left(\omega - (\omega_k + \omega_p)\right) + U_{-k}\left(\delta\omega - (-\omega_k + \omega_p)\right)$.

$$\frac{\partial I(V)}{\partial V}\tilde{v} + \tilde{\Omega}\frac{\partial Q(V)}{\partial V}\tilde{v} + \tilde{u} = 0 \tag{5.76}$$

The derivative term is just the Jacobian we discussed in Sect. 5.5.2, and the equation is straightforward to solve. Note the matrix $\Omega = \Omega$ where the frequency components are offset the harmonics with the offset frequency $\omega_k \to \omega_k + \omega_p$. This simple exercise shows that we can use the Jacobian of our system to find the solution to small perturbations around the known solution. It is the same idea as perturbation calculations for other systems, like one-dimensional equations, but here we apply it to harmonic systems where the large-scale solution is known. To gain further insight, we can write the stimulus vector:

$$u(\omega) = \sum_{n}\sum_{k=-N}^{N}U_k^n\,\delta\left(\omega_k + \omega_p\right) \tag{5.77}$$

The response vector $v$ is similarly written as

$$v(\omega) = \sum_{n}\sum_{k=-N}^{N}V_k^n\,\delta\left(\omega_k + \omega_p\right) \tag{5.78}$$

Here $n$ denotes all nodes and currents defined in the nodal analysis; $k$ is the harmonic index. The coefficients $U_k$ are user-defined and depend on to what node and harmonic we want to add the stimulus. Notice the offset frequency here must be the same since we are now looking at the linearized circuit. We can now see the response $V_k$ of the circuit to a given stimulus with a specific amplitude and offset frequency to a given harmonic(s). We can, for example, sweep the offset frequency $\omega_p$ around some harmonic $\omega_k$ and see how the system responds at some other node and harmonic component. This type of analysis is often called the periodic AC analysis.

This is fairly straightforward to implement in a code. It simply involves the gains around the various harmonics.

```
val=[0 for i in range(100)]
Python code
    . # We need to recalculate the Matrix due to the frequency terms
from the inductors+capacitors
```

```
STA_rhs=[0 for i in range(MatrixSize)]
val=[[0 for i in range(100)] for j in range(4)]
for iter in range(100):
    omega=iter*1e6*2*3.14159265
    for i in range(DeviceCount):
        for row in range(TotalHarmonics):
            if DevType[i]=='capacitor':
                if DevNode1[i] != '0' :
                    Jacobian[(NumberOfNodes+i)*TotalHarmonics+
row][Nodes.index(DevNode1[i])*TotalHarmonics+row]=1j*(omegak[row]
+(numpy.sign(omegak[row])+(omegak[row]==0))*omega)*DevValue[i]
                if DevNode2[i] != '0' :
                    Jacobian[(NumberOfNodes+i)*TotalHarmonics+
row][Nodes.index(DevNode2[i])*TotalHarmonics+row]=-
1j*(omegak[row]+(numpy.sign(omegak[row])+(omegak[row]==0))*omega)
*DevValue[i]
            if DevType[i]=='inductor':
                    Jacobian[(NumberOfNodes+i)*TotalHarmonics+row]
[(NumberOfNodes+i)*TotalHarmonics+row]=-1j*(omegak[row]+(numpy.
sign(omegak[row])+(omegak[row]==0))*omega)*DevValue[i]
            if DevType[i]=='CurrentSource':
                if DevLabel[i]=='i1':
                    STA_rhs[(NumberOfNodes+i)*TotalHarmonics+ro
w]=1*(row==Jacobian_Offset)
                else:
                    STA_rhs[(NumberOfNodes+i)*TotalHarmonics+
row]=-(row==Jacobian_Offset)
    sol=numpy.matmul(numpy.linalg.inv(Jacobian),STA_rhs)
    val[0][iter]=abs(sol[6*TotalHarmonics+Jacobian_Offset])
    val[1][iter]=abs(sol[6*TotalHarmonics+Jacobian_Offset+1])
    val[2][iter]=abs(sol[6*TotalHarmonics+Jacobian_Offset+2])
    val[3][iter]=abs(sol[6*TotalHarmonics+Jacobian_Offset+3])
plt.plot(val[1])
Here we present a python code that does periodic AC simulations
.
End code
```

We will look at a few examples in Sect. 5.5.7 where we apply this code to some simple circuitry.

We have now a small-signal analysis code snippet that can look at small-signal gain riding on top of a large signal. The natural example is analog mixers where a large clock signal is used to mix down (or up) a small RF signal. In modern simulators, it is referred to as periodic ac analysis, *PAC*. Another really useful implementation is noise transfer often called periodic noise or *PNOISE*. We will look at that application next.

*Periodic Noise Analysis*

A periodic noise analysis takes advantage of the periodic AC response we discussed in the previous section to calculate the noise transfer of the system including the harmonic mixing effect. In this way, phenomena like phase-noise nonlinear transfer can be accurately simulated. The basic idea is the same as in our noise discussion in Sect. 4.4.1. The noise of a component is modeled as a current noise (or if convenient a voltage noise), and its transfer to the user-defined output is simulated. Now since the conversion matrix is known, this is fairly easy to do. A code example can look like

```
Python code
. # We need to recalculate the Matrix due to the frequency terms
from the inductors+capacitors
STA_rhs=[0 for i in range(MatrixSize)]
val=[[0 for i in range(100)] for j in range(4)]
for iter in range(100):
    omega=(iter)*1e6*2*math.pi
    for i in range(DeviceCount):
        for row in range(TotalHarmonics):
            if DevType[i]=='capacitor':
                if DevNode1[i] != '0' :
                    Jacobian[(NumberOfNodes+i)*TotalHarmonics+
row][Nodes.index(DevNode1[i])*TotalHarmonics+row]=1j*(omegak[row]
+(numpy.sign(omegak[row])+(omegak[row]==0))*omega)*DevValue[i]
    #               print("C1 adm",row,Jacobian[(NumberOfNodes
+i)*TotalHarmonics+row][Nodes.index(DevNode1[i])*TotalHarmonics+
row])
                if DevNode2[i] != '0' :
                    Jacobian[(NumberOfNodes+i)*TotalHarmonics+
row][Nodes.index(DevNode2[i])*TotalHarmonics+row]=-
1j*(omegak[row]+(numpy.sign(omegak[row])+(omegak[row]==0))*omega)
*DevValue[i]
    #               print("C2 adm",row,Jacobian[(NumberOfNodes
+i)*TotalHarmonics+row][Nodes.index(DevNode1[i])*TotalHarmonics+
row])
            if DevType[i]=='inductor':
                Jacobian[(NumberOfNodes+i)*TotalHarmonics+row]
[(NumberOfNodes+i)*TotalHarmonics+row]=-1j*(omegak[row]+(numpy.
sign(omegak[row])+(omegak[row]==0))*omega)*DevValue[i]
    #               print("L imp ",row,Jacobian[(NumberOfNodes+i)*
TotalHarmonics+row][(NumberOfNodes+i)*TotalHarmonics+row])
    #           if DevType[i]=='VoltSource':
    #               STA_rhs[(NumberOfNodes+i)*TotalHarmonics+row]=
1*((row==Jacobian_Offset+1)+(row==Jacobian_Offset-1))
```

```
                  if DevType[i]=='CurrentSource': # Adding current
source between transistor drain-source
                    STA_rhs[(NumberOfNodes+i)*TotalHarmonics+row]=1
*(row==Jacobian_Offset+1)+1*(row==Jacobian_Offset-1)
       sol=numpy.matmul(numpy.linalg.inv(Jacobian),STA_rhs)
       val[0][iter]=abs(sol[2*TotalHarmonics+Jacobian_Offset])
        val[1][iter]=20*math.log10(abs(sol[2*TotalHarmonics+Jacob
ian_Offset+1]))
       val[2][iter]=abs(sol[2*TotalHarmonics+Jacobian_Offset+2])
       val[3][iter]=math.log10(omega+1)#abs(sol[2*TotalHarmonics+J
acobian_Offset+3])**2
  plt.plot(val[1])
    Here  we  present  a  python  code  that  does  periodic  noise
simulations
    .
    End code
```

We will apply this code to an oscillator example in Sect. 5.5.7.


### 5.5.5   Periodic S-parameter, Transfer Function, Stability Analyses

The whole menagerie of analysis tools we discussed in Chap. 4 can now be applied to steady-state systems. In particular the fact we can have coupling between various harmonics can give rise to interesting phenomena. The code implementation is straightforward and involves a simple generalization of the techniques we discussed in Chap. 4. We leave the details to the reader as an exercise.


### 5.5.6   Quasi-Periodic Steady-State Analyses

We already mentioned in the discussion of shooting methods and harmonic balance methods that a multi-toned implementation is straightforward when using the beat frequency between the defined signals. We noted that using a 2GHz and 3GHz signals one can capture the needed spectrum with a 1GHz fundamental. This is fine if the tones involved are of the same order of frequencies. The situation is quite unmanageable if the tones involved have drastically different frequencies. Let us consider the situation with a 2GHz, a 3GHz, and a 1 MHz tone! The beat frequency would here be 1 MHz, and one would need thousands of harmonics to establish the steady state. This is too much for most computer systems, so instead a quasi-periodic method has been designed where only a certain number of the harmonics of each

**Fig. 5.40** Quasi-periodic illustration, where the moderate signal source has a handful of harmonics around each of the fundamental tone harmonics

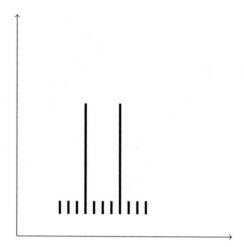

tone are considered. Often the user has to define what signals are *large* and what signals are *moderate* in size. The large tones get lots of harmonics (still user-controlled), and the moderate ones have just handful of harmonics considered. In our example, let us assume that the 2GHz and 3GHz tones are large, and the 1 MHz tone is moderate; then only the beat frequency between the 2 and 3 GHz tones will be considered, and a small set of 1 MHz tones around this fundamental 1GHz tone will be used in the analysis (Fig. 5.40). This way one can get an accurate picture of the circuitry without using all possible harmonics. The method is therefore called quasi-periodic since not all possible combinations of hamonics are used.

Such a code is simple to implement in particular with harmonic balance. We leave it as an exercise to the reader to construct such a code.

### 5.5.7   Specific Circuit Examples

With these periodic steady-state codes, let us look at a few common examples where they are extraordinarily useful. We first will discuss a phase-noise simulation using our very simple nonlinear FET model. It is a great example of how such a technique can be helpful. We follow this with a low-noise amplifier (LNA) example, where the linearity in terms of harmonics is directly given by the steady-state solution. Lastly we consider an up-conversion RF mixer where a small signal at base band is upconverted to higher frequencies. We will use the code we established with the PAC and PNOISE additions previously discussed. It should be obvious that with such a simple nonlinearity we are using, the conversion is straightforward. In a real system, there can be other much stronger nonlinear components, for example, a BJT process where the mixing into higher harmonics is a lot more complicated.

**Fig. 5.41** Simple
single-stage amplifier

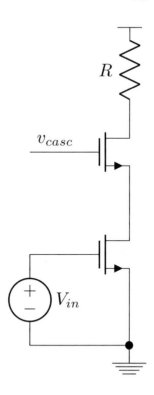

*Amplifier Linearity*
The linearity of amplifiers is a key specification in many applications. With the
steady-state simulator, we can very quickly analyze the harmonic distortion. To
illustrate we will use a very simple amplifier stage as shown in Fig. 5.41.

   This circuit has the netlist

```
*
vdd vdd 0 0.9
vss vss 0 0
vcasc casc 0 0.7
vin in 0 0.5
r1 vdd out 100
m1 out casc n1 nch1
m2 n1 in vss nch1
netlist 5.15
```

and is very simple to simulate with our steady-state simulator, **Code 5.13**
#Harmonics = 16. The resulting harmonics can be found in Table 5.22.

   We see the harmonics of the simulator appear in decreasing size as one would
expect. The higher-order terms will be reduced, thanks to the raised power terms for
those components. As an added exercise, the reader is encouraged to run the simula-

**Table 5.22** Harmonic tones
as a result of simulating the
simple amplifier in Fig. 5.41

| Harmonic | Simulated size |
|---|---|
| 0 | −1.483 |
| 1 | −45.66 |
| 2 | −67.24 |
| 3 | −85.92 |
| 4 | −102.6 |

tion with a much higher input swing (see Exercises). This will result in a typical clipping function where the output function in time is essentially a square function. The harmonic structure of such a function is well known.

*Mixer Gain (pac)*

Finally we can look at the mixer conversion gain with the help of the PAC code we developed in Sect. 5.5.4. An RF mixer is a circuit that takes a local oscillator (LO) and multiplies it with a baseband (low frequency < perhaps 1 MHz) tone. This multiplication will up-convert the baseband tone to appear around the LO harmonics. The LO swing is usually large, while the base band tone can be small. A steady simulator is ideal to investigate properties like gain for such circuits. We will here present a very simple topology to introduce the idea. In Fig. 5.42, we have a simple topology with the following netlist

```
*
vdd vdd 0 0.9
vinp in1 0 1
vinn in2 0 1
r1 vdd inp 100
r2 inp 0 100
r3 vdd inn 100
r4 inn 0 100
r5 vdd outp 100
r6 vdd outn 100
c1 in1 inp 1e-12
c2 in2 inn 1e-12
i1 vs 0 1e-3
m1 outn inp vs nch1
m2 outp inn vs nch2
i2 vs2 0 1e-3
m3 outp inp vs2 nch1
m4 outn inn vs2 nch2
l1 vdd outp 1e-9
ct1 vdd outp 25.33029e-12
l2 vdd outn 1e-9
ct2 vdd outn 25.33029e-12
```
**netlist 5.16**

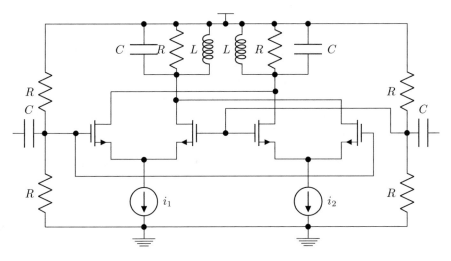

**Fig. 5.42** Simple LO mixer

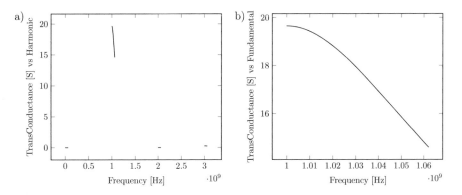

**Fig. 5.43** Mixer RF gain; (**a**) shows the response for the four lowest harmonics; (**b**) is a zoom-in of the fundamental response

For the simulation, we now simply solve for the steady state and thereafter launch a PAC simulation with the stimulus at the RF port as shown in Fig. 5.43.

The signal is suppressed around all harmonics except the first one due to the LC tank in the load.

**Verify** This can be easily verified in any simulator.

### VCO Phase Noise

The PNOISE implementation lends itself very easily to the phase-noise simulation of a voltage-controlled oscillator (VCO) circuit. A VCO is key component in many circuits where a clock is needed and that includes pretty much every chip. Sometimes the requirements are such the clock resides outside the integrated circuit in a sepa-

rate timing chip. At other times, the circuit is part of an internal phase-locked loop
or some combination of external internal circuitry.

As we already mentioned earlier, a VCO, and an LC-VCO in particular, is a great
test circuit for simulator performance. The fact there are resonances caused by various impedances cancelling each other can make it difficult to accurately predict the
oscillation amplitude, for example. Using the wrong method of integration is
another well-known source of error (see Sect. 4.6.7). Simulating oscillators with a
steady-state method also carries with it the risk of finding the DC solution instead
of the oscillating one, something we will see shortly.

Let us look at the simple oscillator circuit in Fig. 5.18 again with the netlist:

```
*
vdd vdd 0 0.9
vss vss 0 0
l1 vdd outp 1e-9
l2 vdd outn 1e-9
c1 vdd outp 1e-12
c2 vdd outn 1e-12
r1 vdd outp 1e3
r2 vdd outn 1e3
m1 outp outn vss mn1
m2 outn outp vss mn2
ins outp vss 0
netlist 5.17
```

The strategy will here be to first come up with a steady-state solution using harmonic balance. The difficulty is that there is a DC solution where the output nodes
both sit at vdd and no oscillation is taking place. The reader is encouraged to try to
get an oscillator to converge to a non-DC solution with the provided code or some
modification of it. To nudge the tool into the appropriate mode, one common strategy is to do a transient simulation in time until the oscillation starts and then launch
the steady-state simulator using the harmonics resulting from the transient simulation as a starting point. We leave this implementation to the reader as an exercise.
Another strategy is to set up the oscillator in such a way that instead of an autonomous circuit, it turns into a driven circuit [22]. Imagine we know the center frequency from a transient simulation or perhaps a simple hand calculation. We can
then drive part of the oscillator, perhaps the output node(s) with a voltage source
having the correct frequency. Imagine further the interconnect circuit between the
voltage source and the tank is such it is 0 ohm at the fundamental frequency and
infinite at any other frequency harmonic (these are wasy to do in harmonic balance).
If we now change the amplitude and phase of the voltage source such that the current through the interconnect is zero, it means the voltage phase and amplitude of
the oscillator are the identical at the connection point, and we have found the correct
steady state since the voltage source is in effect decoupled from the tank (no current

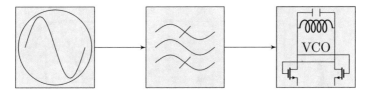

**Fig. 5.44** Using a decoupled voltage source as a way to turn an autonomous circuit into a driven one

through the interconnect). In reality since oscillator phase is a free variable, it is not bound by any physical restriction; it is usually just a matter of change the amplitude of the voltage source. For an illustration of the method, see Fig. 5.44.

A code implementation can be found in Sect. 5.6.9 **Code 5p9**.

The algorithm should now look for correct amplitude and frequency by changing the driving voltage amplitude and frequency. When the correct combination is found, the current through the filter should ideally be zero, and in that sense, the driving circuitry is not participating in the tank. This is a two-dimensional search space which is not too difficult to implement. For simplicity we here only implement the amplitude search algorithm since we know the frequency by a simple $LC$ resonance calculation: $f = 1/(2\pi\sqrt{LC}) = 5.03292 \, 10^9$ Hz.

Let us run the simulation with the following code snippet to implement search:

```
Python code
.
for AmpIndex in range(20):
        STA_rhs[(NumberOfNodes+StimulusIndex)*TotalHarmonics+Jacob
ian_Offset+1]=.235+float (AmpIndex)/10000
        STA_rhs[(NumberOfNodes+StimulusIndex)*TotalHarmonics+Jacob
ian_Offset-1]=.235+float (AmpIndex)/10000.
.
End code
```

We find the result in Fig. 5.45.

The output voltage at the minimum setting is now depicted in Fig. 5.46.

For the PNOISe simulation, let us assume that the only noise sources are noise currents sitting in parallel with the drain-source terminals. For illustrative purposes, we assume the noise amplitude is "1." We can now launch the PNOISE code, **Code 5.16**, and add one noise sources in parallel with the active devices. We plot the result in Fig. 5.47. The scaling of the response is roughly 6 dB per octave as expected.

This is a simple example of a oscillator noise simulation. A real situation has more devices and much more complicated noise contributors, but the basic idea is the same.

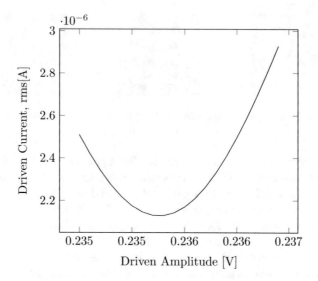

**Fig. 5.45** The residue current through the filter as a function of driving voltage amplitude

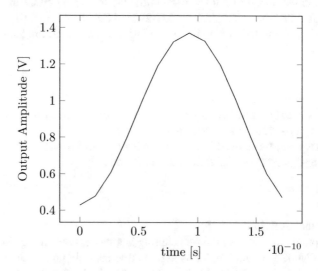

**Fig. 5.46** Steady state harmonics of simple VCO

### 5.5.8   Odd Cases: Things to Watch Out for

For the steady-state-type systems, perhaps the most obvious issue one runs into is using too few hamonics in the harmonic balance method. Oftentimes one needs to use a significantly higher number than one might initially estimate. This is most

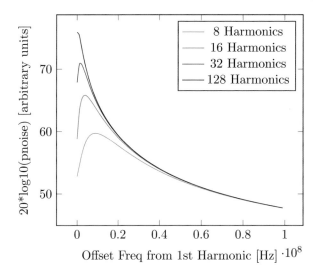

**Fig. 5.47** PNOISE output of simple VCO analysis. The number of harmonics will increase the accuracy and resolve the close to tone resonance better

often due to sharp edges in the circuitry that if not allowed for can result in the familiar Gibbs phenomenon or ringing around the edges where the high-frequency content is high.

### 5.5.9   Accuracy: How to Determine

A difficult task is to estimate the accuracy of the simulation result. Most methods have had their accuracy stepwise analyzed and is characterized as a second-order, third-order, or sometime first-order accuracy. This is referring to the local estimate at the precise timestep or frequency step or whatever. The global accuracy after many steps can be much harder to estimate, and sometimes the errors will dissipate and sometimes the errors will persist indefinitely. As we work through various systems, we will discuss the topic of accuracy when appropriate. One common and extraordinarily time consuming way to verify the accuracy is accurate is to rerun the simulation with the tolerances tightened. This cannot be done every time since the time required will be much longer to complete the task. In practice, what people do is to spot check accuracy from time to time by tightening the required tolerance.

> In practice it is good to verify the accuracy of a system by spot checking and resimulate with tighter tolerance

## *5.5.10   Simulator Options Considered*

We have considered the following options and what they mean in this chapter:

- reltol
- abstol
- vntol
- chgtol
- Trapezoidal integration method
- Euler (forward/backward) integration method
- Gear second-order integration method
- GearOnly
- TrapOnly
- Truncation errors: point local, local, global
- Source stepping
- $g_{min}$ stepping
- PTran stepping
- DPtran stepping
- Newton max iteration
- Shooting method
- Harmonic balance
- NHarmonics
- Oversampling

## *5.5.11   Summary*

We have in this chapter discussed the basic ideas behind nonlinear simulators, and we have gone from simple nonlinear models of transistors to more complicated ones. We are still far from any realistic BSIM implementation. Many of the inherent pitfalls using say nonlinear capacitors have been exemplified a few times. Finally we have constructed a simple code that can achieve a steady-state solution, and this is one of the most important tools for the modern circuit designer.

All throughout the chapter, we have built a kind of toy simulator with simple transistor modeling just to illustrate a possible way to implement the techniques. Many of the more or less well-known pitfalls with nonlinear capacitors or integration methods were highlighted with direct use of the toy simulator.

# 5.6   Codes

## 5.6.1   Code 5.1

```
#!/usr/bin/env python3
# -*- coding: utf-8 -*-
"""
Created on Thu Feb 28 22:33:04 2019

@author: mikael
"""
import numpy
import matplotlib.pyplot as plt
import math
import analogdef as ana

#
# Function definitions
#
def f_NL(STA_matrix, STA_rhs, STA_nonlinear, solution):

return numpy.matmul(STA_matrix,solution)-STA_rhs+STA_nonlinear

#
# Read netlist
#
DeviceCount=0
MaxNumberOfDevices=100
DevType=[0*i for i in range(MaxNumberOfDevices)]
DevLabel=[0*i for i in range(MaxNumberOfDevices)]
DevNode1=[0*i for i in range(MaxNumberOfDevices)]
DevNode2=[0*i for i in range(MaxNumberOfDevices)]
DevNode3=[0*i for i in range(MaxNumberOfDevices)]
DevValue=[0*i for i in range(MaxNumberOfDevices)]
DevModel=[0*i for i in range(MaxNumberOfDevices)]
Nodes=[]

modeldict=ana.readmodelfile('models.txt')
ICdict={}
Plotdict={}
Printdict={}
Optdict={}
Optdict['MaxNewtonIterations']=int(5)
```

```
#
# Read the netlist
#
DeviceCount=ana.readnetlist('netlist_dc_5p1.txt',modeldict,ICdi
ct,Plotdict,Printdict,Optdict,DevType,DevValue,DevLabel,DevNode1,
DevNode2,DevNode3,DevModel,Nodes,MaxNumberOfDevices)
#
# Create Matrix based on circuit size. We do not implement strict
Modified Nodal Analysis. We keep instead all currents
# but keep referring to the voltages as absolute voltages. We
believe this will make the operation clearer to the user.
#
NumberOfNodes=len(Nodes)
MatrixSize=DeviceCount+len(Nodes)
Jacobian=[[0 for i in range(MatrixSize)] for j in range(MatrixSize)]
Jac_inv=[[0 for i in range(MatrixSize)] for j in range(MatrixSize)]
Spare=[[0 for i in range(MatrixSize)] for j in range(MatrixSize)]
    STA_matrix=[[0   for   i   in   range(MatrixSize)]   for   j   in
range(MatrixSize)]
STA_rhs=[0 for i in range(MatrixSize)]
STA_nonlinear=[0 for i in range(MatrixSize)]
f=[0 for i in range(MatrixSize)]
#
# Create sim parameters
#
sol=[0 for i in range(MatrixSize)]
solm1=[0 for i in range(MatrixSize)]
#
# Initial conditions
#
if len(ICdict)>0:
    for i in range(len(ICdict)):
        for j in range(NumberOfNodes):
            if Nodes[j]==ICdict[i]['NodeName']:
                sol[j]=ICdict[i]['Value']
                print('Setting ',Nodes[j],' to ',sol[j])
#
# Loop through all devices and create jacobian and initial f(v)
entries according to signature
#
for i in range(DeviceCount):
    if DevType[i] != 'transistor':
        STA_matrix[NumberOfNodes+i][NumberOfNodes+i]=-DevValue[i]
        if DevNode1[i] != '0' :
        STA_matrix[NumberOfNodes+i][Nodes.index(DevNode1[i])]=1
```

```
    STA_matrix[Nodes.index(DevNode1[i])][NumberOfNodes+i]=1
  if DevNode2[i] != '0' :
    STA_matrix[NumberOfNodes+i][Nodes.index(DevNode2[i])]=-1
    STA_matrix[Nodes.index(DevNode2[i])][NumberOfNodes+i]=-1
  if DevType[i]=='capacitor':
    # Do nothing since DC sim
    STA_rhs[NumberOfNodes]=STA_rhs[NumberOfNodes]
  if DevType[i]=='inductor':
    STA_matrix[NumberOfNodes+i][NumberOfNodes+i]=0
    STA_rhs[NumberOfNodes+i]=0
  if DevType[i]=='VoltSource':
    STA_matrix[NumberOfNodes+i][NumberOfNodes+i]=0
    STA_rhs[NumberOfNodes+i]=ana.getSourceValue(DevValue[i],0)
    if DevType[i]=='CurrentSource':
        if DevNode1[i] != '0' :
    STA_matrix[NumberOfNodes+i][Nodes.index(DevNode1[i])]=0
    STA_matrix[Nodes.index(DevNode1[i])][NumberOfNodes+i]=0
        if DevNode2[i] != '0' :
    STA_matrix[NumberOfNodes+i][Nodes.index(DevNode2[i])]=0
    STA_matrix[Nodes.index(DevNode2[i])][NumberOfNodes+i]=0
        STA_matrix[NumberOfNodes+i][NumberOfNodes+i]=1
    STA_rhs[NumberOfNodes+i]=ana.getSourceValue(DevValue[i],0)
        if DevNode1[i] != '0' and DevNode2[i]!='0':
    STA_matrix[Nodes.index(DevNode1[i])][NumberOfNodes+i]=1
    STA_matrix[Nodes.index(DevNode2[i])][NumberOfNodes+i]=-1
        elif DevNode2[i] != '0' :
    STA_matrix[Nodes.index(DevNode2[i])][NumberOfNodes+i]=-1
        elif DevNode1[i] != '0' :
    STA_matrix[Nodes.index(DevNode1[i])][NumberOfNodes+i]=1
  if DevType[i]=='transistor':
    lambdaT=ana.findParameter(modeldict,DevModel[i],'lambdaT')
    VT=ana.findParameter(modeldict,DevModel[i],'VT')
    STA_matrix[NumberOfNodes+i][NumberOfNodes+i]=DevValue[i]
    STA_matrix[NumberOfNodes+i][Nodes.index(DevNode1[i])]=0
    STA_matrix[Nodes.index(DevNode1[i])][NumberOfNodes+i]=1
    STA_matrix[NumberOfNodes+i][Nodes.index(DevNode3[i])]=0
    STA_matrix[Nodes.index(DevNode3[i])][NumberOfNodes+i]=-1
    VG=sol[Nodes.index(DevNode2[i])]
    VS=sol[Nodes.index(DevNode3[i])]
    Vgs=VG-VS
    if DevModel[i][0]=='p':
        Vgs=-Vgs
    STA_nonlinear[NumberOfNodes+i]=Vgs**2
```

```
#
f=numpy.matmul(STA_matrix,sol)-STA_rhs+STA_nonlinear
#
#Loop through iteration points
#
NewIter=int(Optdict['MaxNewtonIterations'])
val=[[0 for i in range(NewIter+1)] for j in range(MatrixSize)]
for j in range(MatrixSize):
    val[j][0]=sol[j]
Iteration=[i for i in range(NewIter+1)]
for Newtoniter in range(NewIter):
    for i in range(MatrixSize):
        STA_nonlinear[i]=0
    for i in range(DeviceCount):
        if DevType[i]!='transistor':
            if DevType[i]=='capacitor':

STA_rhs[NumberOfNodes+i]=STA_rhs[NumberOfNodes+i]
            if DevType[i]=='inductor':
                STA_rhs[NumberOfNodes+i]=0
            if DevType[i]=='VoltSource':
                STA_rhs[NumberOfNodes+i]=ana.getSourceValue(Dev
Value[i],0)
            if DevType[i]=='CurrentSource':
                STA_rhs[NumberOfNodes+i]=ana.getSourceValue(Dev
Value[i],0)
        if DevType[i]=='transistor':
                VG=sol[Nodes.index(DevNode2[i])]
                VS=sol[Nodes.index(DevNode3[i])]
                Vgs=VG-VS
                if DevModel[i][0]=='p':
                    Vgs=-Vgs
                STA_nonlinear[NumberOfNodes+i]=Vgs**2
    f=numpy.matmul(STA_matrix,sol)-STA_rhs+STA_nonlinear
 #
 # Now we need the Jacobian, the transistors look like VCCS with
a specific gain = 2 K (Vg-Vs) in our case
 #
    for i in range(MatrixSize):
        for j in range(MatrixSize):
            Jacobian[i][j]=STA_matrix[i][j]
    for i in range(DeviceCount):
        if DevType[i]=='transistor':
            Jacobian[NumberOfNodes+i][NumberOfNodes+i]=DevValue[i]
# due to derfivative leading to double gain
```

```
                VG=sol[Nodes.index(DevNode2[i])]
                VS=sol[Nodes.index(DevNode3[i])]
                Vgs=VG-VS
                if DevModel[i][0]=='p':
                    PFET=-1
                    Vgs=-Vgs
                else:
                    PFET=1
                    Jacobian[NumberOfNodes+i][Nodes.index(DevNode2[
i])]]=2*PFET*Vgs
                    Jacobian[NumberOfNodes+i][Nodes.index(DevNode3[
i])]]=-2*PFET*Vgs
        sol=sol-numpy.matmul(numpy.linalg.inv(Jacobian),f)
        Jac_inv=numpy.linalg.inv(Jacobian)
        for j in range(MatrixSize):
            val[j][Newtoniter+1]=sol[j]

    ana.plotdata(Plotdict,NumberOfNodes,Iteration,val,Nodes)
    ana.printdata(Printdict,NumberOfNodes,Iteration,val,Nodes)
```

## *5.6.2   Code 5.2*

```
#!/usr/bin/env python3
# -*- coding: utf-8 -*-
"""
Created on Thu Feb 28 22:33:04 2019

@author: mikael
"""
import numpy as np
import analogdef as ana

#
# Function definitions
#
DeviceCount=0
MaxNumberOfDevices=100
DevType=[0*i for i in range(MaxNumberOfDevices)]
DevLabel=[0*i for i in range(MaxNumberOfDevices)]
DevNode1=[0*i for i in range(MaxNumberOfDevices)]
DevNode2=[0*i for i in range(MaxNumberOfDevices)]
DevNode3=[0*i for i in range(MaxNumberOfDevices)]
DevValue=[0*i for i in range(MaxNumberOfDevices)]
```

```
DevModel=[0*i for i in range(MaxNumberOfDevices)]
Nodes=[]

modeldict=ana.readmodelfile('models.txt')
ICdict={}
Plotdict={}
Printdict={}
Optdict={}
Optdict['MaxNewtonIterations']=int(5)
#
# Read the netlist
#
DeviceCount=ana.readnetlist('netlist_5p4.txt',modeldict,ICdict,
Plotdict,Printdict,Optdict,DevType,DevValue,DevLabel,DevNode1,Dev
Node2,DevNode3,DevModel,Nodes,MaxNumberOfDevices)
#
# Create Matrix based on circuit size. We do not implement strict
Modified Nodal Analysis. We keep instead all currents
# but keep referring to the voltages as absolute voltages. We
believe this will make the operation clearer to the user.
#
NumberOfNodes=len(Nodes)
MatrixSize=DeviceCount+len(Nodes)
Jacobian=[[0 for i in range(MatrixSize)] for j in range(MatrixSize)]
Jac_inv=[[0 for i in range(MatrixSize)] for j in range(MatrixSize)]
Spare=[[0 for i in range(MatrixSize)] for j in range(MatrixSize)]
STA_matrix=[[0 for i in range(MatrixSize)] for j in range(
MatrixSize)]
STA_rhs=[0 for i in range(MatrixSize)]
STA_nonlinear=[0 for i in range(MatrixSize)]
f=[0 for i in range(MatrixSize)]
#
# Create sim parameters
#
sol=[0 for i in range(MatrixSize)]
solm1=[0 for i in range(MatrixSize)]
#
# Initial conditions
#
if len(ICdict)>0:
    for i in range(len(ICdict)):
        for j in range(NumberOfNodes):
            if Nodes[j]==ICdict[i]['NodeName']:
                sol[j]=ICdict[i]['Value']
                print('Setting ',Nodes[j],' to ',sol[j])
```

```
#
# Loop through all devices and create jacobian and initial f(v)
entries according to signature
#
for i in range(DeviceCount):
    if DevType[i] != 'transistor':
        STA_matrix[NumberOfNodes+i][NumberOfNodes+i]=-DevValue[i]
        if DevNode1[i] != '0' :
            STA_matrix[NumberOfNodes+i][Nodes.index(DevNode1[i])]=1
            STA_matrix[Nodes.index(DevNode1[i])][NumberOfNodes+i]=1
        if DevNode2[i] != '0' :
            STA_matrix[NumberOfNodes+i][Nodes.index(DevNode2[i])]=-1
            STA_matrix[Nodes.index(DevNode2[i])][NumberOfNodes+i]=-1
        if DevType[i]=='capacitor':
            # Do nothing since DC sim
            STA_rhs[NumberOfNodes]=STA_rhs[NumberOfNodes]
        if DevType[i]=='inductor':
            STA_matrix[NumberOfNodes+i][NumberOfNodes+i]=0
            STA_rhs[NumberOfNodes+i]=0
        if DevType[i]=='VoltSource':
            STA_matrix[NumberOfNodes+i][NumberOfNodes+i]=0
            STA_rhs[NumberOfNodes+i]=ana.getSourceValue(DevValue[i],0)
        if DevType[i]=='CurrentSource':
            if DevNode1[i] != '0' :
                STA_matrix[NumberOfNodes+i][Nodes.index(DevNode1[i])]=0
                STA_matrix[Nodes.index(DevNode1[i])][NumberOfNodes+i]=0
            if DevNode2[i] != '0' :
                STA_matrix[NumberOfNodes+i][Nodes.index(DevNode2[i])]=0
                STA_matrix[Nodes.index(DevNode2[i])][NumberOfNodes+i]=0
            STA_matrix[NumberOfNodes+i][NumberOfNodes+i]=1
            STA_rhs[NumberOfNodes+i]=ana.getSourceValue(DevValue[i],0)
            if DevNode1[i] != '0' and DevNode2[i]!='0':
                STA_matrix[Nodes.index(DevNode1[i])][NumberOfNodes+i]=1
                STA_matrix[Nodes.index(DevNode2[i])][NumberOfNodes+i]=-1
            elif DevNode2[i] != '0' :
                STA_matrix[Nodes.index(DevNode2[i])][NumberOfNodes+i]=-1
            elif DevNode1[i] != '0' :
                STA_matrix[Nodes.index(DevNode1[i])][NumberOfNodes+i]=1
    if DevType[i]=='transistor':
        lambdaT=ana.findParameter(modeldict,DevModel[i],'lambdaT')
        VT=ana.findParameter(modeldict,DevModel[i],'VT')
        STA_matrix[NumberOfNodes+i][NumberOfNodes+i]=DevValue[i]
        STA_matrix[NumberOfNodes+i][Nodes.index(DevNode1[i])]=0
        STA_matrix[Nodes.index(DevNode1[i])][NumberOfNodes+i]=1
        STA_matrix[NumberOfNodes+i][Nodes.index(DevNode3[i])]=0
```

```
             STA_matrix[Nodes.index(DevNode3[i])][NumberOfNodes+i]=-1
             VD=sol[Nodes.index(DevNode1[i])]
             VG=sol[Nodes.index(DevNode2[i])]
             VS=sol[Nodes.index(DevNode3[i])]
             Vgs=VG-VS
             Vds=VD-VS
             if DevModel[i][0]=='p':
                 Vds=-Vds
                 Vgs=-Vgs
             if Vds < Vgs-VT :
                 STA_nonlinear[NumberOfNodes+i]=2*((Vgs-VT)*Vds-0.5*
Vds**2)
             else :
                 STA_nonlinear[NumberOfNodes+i]=(Vgs-VT)**2*(1+lambda
T*Vds)


    #
    f=np.matmul(STA_matrix,sol)-STA_rhs+STA_nonlinear
    #
    #Loop through iteration points
    #
    NewIter=int(Optdict['MaxNewtonIterations'])
    val=[[0 for i in range(NewIter+1)] for j in range(MatrixSize)]
    for j in range(MatrixSize):
        val[j][0]=sol[j]
    Iteration=[i for i in range(NewIter+1)]
    for Newtoniter in range(NewIter):
        for i in range(MatrixSize):
            STA_nonlinear[i]=0
        for i in range(DeviceCount):
            if DevType[i]!='transistor':
                if DevType[i]=='capacitor':
                    STA_rhs[NumberOfNodes+i]=STA_rhs[NumberOfNodes+i]
                if DevType[i]=='inductor':
                    STA_rhs[NumberOfNodes+i]=0
                if DevType[i]=='VoltSource':
                    STA_rhs[NumberOfNodes+i]=ana.getSourceValue(Dev
Value[i],0)
                if DevType[i]=='CurrentSource':
                    STA_rhs[NumberOfNodes+i]=ana.getSourceValue(Dev
Value[i],0)
             if DevType[i]=='transistor':
                    lambdaT=ana.findParameter(modeldict,DevModel[i],
'lambdaT')
                    VT=ana.findParameter(modeldict,DevModel[i],'VT')
```

```
                    VD=sol[Nodes.index(DevNode1[i])]
                    VG=sol[Nodes.index(DevNode2[i])]
                    VS=sol[Nodes.index(DevNode3[i])]
                    Vgs=VG-VS
                    Vds=VD-VS
                    if DevModel[i][0]=='p':
                        Vds=-Vds
                        Vgs=-Vgs
                    if Vgs<VT:
                        STA_nonlinear[NumberOfNodes+i]=1e-5
                    elif Vds < Vgs-VT:
                        STA_nonlinear[NumberOfNodes+i]=2*((Vgs-VT)*
Vds-0.5*Vds**2)
                    else :
                        STA_nonlinear[NumberOfNodes+i]=(Vgs-VT)**2*
(1+lambdaT*Vds)
        f=np.matmul(STA_matrix,sol)-STA_rhs+STA_nonlinear
    #
    # Now we need the Jacobian, the transistors look like VCCS with
a specific gain = 2 K (Vg-Vs) in our case
    #
        for i in range(MatrixSize):
            for j in range(MatrixSize):
                Jacobian[i][j]=STA_matrix[i][j]
        for i in range(DeviceCount):
            if DevType[i]=='transistor':
                lambdaT=ana.findParameter(modeldict,DevModel[i],'lam
bdaT')
                VT=ana.findParameter(modeldict,DevModel[i],'VT')
                Jacobian[NumberOfNodes+i][NumberOfNodes+i]=DevValue
[i] # due to derfivative leading to double gain
                VD=sol[Nodes.index(DevNode1[i])]
                VG=sol[Nodes.index(DevNode2[i])]
                VS=sol[Nodes.index(DevNode3[i])]
                Vgs=VG-VS
                Vds=VD-VS
                Vgd=VG-VD
                if DevModel[i][0]=='p':
                    PFET=-1
                    Vgs=-Vgs
                    Vds=-Vds
                    Vgd=-Vgd
                else:
                    PFET=1
```

```
        if Vgs<VT :
            Jacobian[NumberOfNodes+i][Nodes.index(DevNode1[i])]=
PFET*1e-1
            Jacobian[NumberOfNodes+i][Nodes.index(DevNode2[i])]=
PFET*1e-1
            Jacobian[NumberOfNodes+i][Nodes.index(DevNode3[i])]=-
PFET*1e-1
            Jacobian[Nodes.index(DevNode1[i])][NumberOfNodes+i]=1
            Jacobian[Nodes.index(DevNode3[i])][NumberOfNodes+i]=-1
        elif Vds <= Vgs-VT:
            Jacobian[NumberOfNodes+i][Nodes.index(DevNode1[i])]=
PFET*2*(Vgd-VT)
            Jacobian[NumberOfNodes+i][Nodes.index(DevNode2[i])]=
PFET*2*Vds
            Jacobian[NumberOfNodes+i][Nodes.index(DevNode3[i])]=
-PFET*2*(Vgs-VT)
            Jacobian[Nodes.index(DevNode1[i])][NumberOfNodes+i]=1
            Jacobian[Nodes.index(DevNode3[i])][NumberOfNodes+i]=-1
        else :
            Jacobian[NumberOfNodes+i][Nodes.index(DevNode1[i])]=
PFET*lambdaT*(Vgs-VT)**2
            Jacobian[NumberOfNodes+i][Nodes.index(DevNode2[i])]=
PFET*2*(Vgs-VT)*(1+lambdaT*Vds)
            Jacobian[NumberOfNodes+i][Nodes.index(DevNode3[i])]=
PFET*(-2*(Vgs-VT)*(1+lambdaT*Vds)-lambdaT*(Vgs-VT)**2)
            Jacobian[Nodes.index(DevNode1[i])][NumberOfNodes+i]=1
            Jacobian[Nodes.index(DevNode3[i])][NumberOfNodes+i]=-1
    sol=sol-np.matmul(np.linalg.inv(Jacobian),f)
    Jac_inv=np.linalg.inv(Jacobian)
    for j in range(MatrixSize):
        val[j][Newtoniter+1]=sol[j]

  ana.plotdata(Plotdict,NumberOfNodes,Iteration,val,Nodes)
  ana.printdata(Printdict,NumberOfNodes,Iteration,val,Nodes)
```

### 5.6.3   Code 5.3

```
    #!/usr/bin/env python3
# -*- coding: utf-8 -*-
"""
Created on Thu Feb 28 22:33:04 2019

@author: mikael
```

```
"""
import numpy as np
import math
import analogdef as ana

#
# Function definitions
#
DeviceCount=0
MaxNumberOfDevices=100
DevType=[0*i for i in range(MaxNumberOfDevices)]
DevLabel=[0*i for i in range(MaxNumberOfDevices)]
DevNode1=[0*i for i in range(MaxNumberOfDevices)]
DevNode2=[0*i for i in range(MaxNumberOfDevices)]
DevNode3=[0*i for i in range(MaxNumberOfDevices)]
DevValue=[0*i for i in range(MaxNumberOfDevices)]
DevModel=[0*i for i in range(MaxNumberOfDevices)]
Nodes=[]

modeldict=ana.readmodelfile('models.txt')
ICdict={}
Plotdict={}
Printdict={}
Optdict={}
Optdict['MaxNewtonIterations']=int(5)
#
# Read the netlist
#
DeviceCount=ana.readnetlist('netlist_bandgap.txt',modeldict,ICd
ict,Plotdict,Printdict,Optdict,DevType,DevValue,DevLabel,DevNode1
,DevNode2,DevNode3,DevModel,Nodes,MaxNumberOfDevices)
#
# Create Matrix based on circuit size. We do not implement strict
Modified Nodal Analysis. We keep instead all currents
# but keep referring to the voltages as absolute voltages. We
believe this will make the operation clearer to the user.
#
NumberOfNodes=len(Nodes)
MatrixSize=DeviceCount+len(Nodes)
Jacobian=[[0 for i in range(MatrixSize)] for j in range(MatrixSize)]
Jac_inv=[[0 for i in range(MatrixSize)] for j in range(MatrixSize)]
Spare=[[0 for i in range(MatrixSize)] for j in range(MatrixSize)]
STA_matrix=[[0 for i in range(MatrixSize)] for j in range(Matrix
Size)]
STA_rhs=[0 for i in range(MatrixSize)]
```

```
STA_nonlinear=[0 for i in range(MatrixSize)]
f=[0 for i in range(MatrixSize)]
#
# Create sim parameters
#
Vthermal=1.38e-23*300/1.602e-19
deltaT=1e-12
sol=[0 for i in range(MatrixSize)]
solm1=[0 for i in range(MatrixSize)]
#sol[3]=sol[4]=0.45
if len(ICdict)>0:
    for i in range(len(ICdict)):
        for j in range(NumberOfNodes):
            if Nodes[j]==ICdict[i]['NodeName']:
                sol[j]=ICdict[i]['Value']
                print('Setting ',Nodes[j],' to ',sol[j])
#sol[0]=1
#sol[1]=0.5
#sol[2]=0.5
#sol[3]=-0.3
##sol[4]=1
#
# Loop through all devices and create jacobian and initial f(v)
entries according to signature
#
for i in range(DeviceCount):
    if DevType[i] != 'transistor' and DevType[i] != 'bipolar':
        STA_matrix[NumberOfNodes+i][NumberOfNodes+i]=-DevValue[i]
        if DevNode1[i] != '0' :
        STA_matrix[NumberOfNodes+i][Nodes.index(DevNode1[i])]=1
        STA_matrix[Nodes.index(DevNode1[i])][NumberOfNodes+i]=1
        if DevNode2[i] != '0' :
        STA_matrix[NumberOfNodes+i][Nodes.index(DevNode2[i])]=-1
        STA_matrix[Nodes.index(DevNode2[i])][NumberOfNodes+i]=-1
        if DevType[i]=='capacitor':
            # Do nothing
            STA_rhs[NumberOfNodes]=STA_rhs[NumberOfNodes]
        if DevType[i]=='inductor':
            # For DC we treat this as a voltage source with V=0
            STA_matrix[NumberOfNodes+i][NumberOfNodes+i]=0
            STA_rhs[NumberOfNodes+i]=0
        if DevType[i]=='VoltSource':
            STA_matrix[NumberOfNodes+i][NumberOfNodes+i]=0
            STA_rhs[NumberOfNodes+i]=DevValue[i]
        if DevType[i]=='CurrentSource':
```

```
        if DevNode1[i] != '0' :
    STA_matrix[NumberOfNodes+i][Nodes.index(DevNode1[i])]=0
    STA_matrix[Nodes.index(DevNode1[i])][NumberOfNodes+i]=0
        if DevNode2[i] != '0' :
    STA_matrix[NumberOfNodes+i][Nodes.index(DevNode2[i])]=0
    STA_matrix[Nodes.index(DevNode2[i])][NumberOfNodes+i]=0
        STA_matrix[NumberOfNodes+i][NumberOfNodes+i]=1
        STA_rhs[NumberOfNodes+i]=DevValue[i]
        if DevNode1[i] != '0' and DevNode2[i]!='0':
    STA_matrix[Nodes.index(DevNode1[i])][NumberOfNodes+i]=1
    STA_matrix[Nodes.index(DevNode2[i])][NumberOfNodes+i]=-1
        elif DevNode2[i] != '0' :
    STA_matrix[Nodes.index(DevNode2[i])][NumberOfNodes+i]=-1
        elif DevNode1[i] != '0' :
    STA_matrix[Nodes.index(DevNode1[i])][NumberOfNodes+i]=1
    if DevType[i]=='transistor':
    lambdaT=ana.findParameter(modeldict,DevModel[i],'lambdaT')
    VT=ana.findParameter(modeldict,DevModel[i],'VT')
    STA_matrix[NumberOfNodes+i][NumberOfNodes+i]=DevValue[i]
    STA_matrix[NumberOfNodes+i][Nodes.index(DevNode1[i])]=0
    STA_matrix[Nodes.index(DevNode1[i])][NumberOfNodes+i]=1
    STA_matrix[NumberOfNodes+i][Nodes.index(DevNode3[i])]=0
    STA_matrix[Nodes.index(DevNode3[i])][NumberOfNodes+i]=-1
    VD=sol[Nodes.index(DevNode1[i])]
    VG=sol[Nodes.index(DevNode2[i])]
    VS=sol[Nodes.index(DevNode3[i])]
    Vgs=VG-VS
    Vds=VD-VS
    if DevModel[i][0]=='p':
        Vds=-Vds
        Vgs=-Vgs
    if Vds < Vgs-VT :
        STA_nonlinear[NumberOfNodes+i]=2*((Vgs-VT)*Vds-0.5*
Vds**2)
        else :
        STA_nonlinear[NumberOfNodes+i]=(Vgs-VT)**2*(1+lambda
T*Vds)
    if DevType[i]=='bipolar':
    VEarly=ana.findParameter(modeldict,DevModel[i],'Early')
    STA_matrix[NumberOfNodes+i][NumberOfNodes+i]=DevValue[i]
    STA_matrix[NumberOfNodes+i][Nodes.index(DevNode1[i])]=0
    STA_matrix[Nodes.index(DevNode1[i])][NumberOfNodes+i]=1
    STA_matrix[NumberOfNodes+i][Nodes.index(DevNode3[i])]=0
    STA_matrix[Nodes.index(DevNode3[i])][NumberOfNodes+i]=-1
    VC=sol[Nodes.index(DevNode1[i])]
```

```
        VB=sol[Nodes.index(DevNode2[i])]
        VE=sol[Nodes.index(DevNode3[i])]
        Vbe=VB-VE
        Vce=VC-VE
        if Vbe < 0 :
            STA_nonlinear[NumberOfNodes+i]=0
        else :
            STA_nonlinear[NumberOfNodes+i]=math.exp(Vbe/Vthermal)
*(1+Vce/VEarly)

    #
    f=np.matmul(STA_matrix,sol)-STA_rhs+STA_nonlinear
    #
    #Loop through iteration points
    #
    NewIter=int(Optdict['MaxNewtonIterations'])
    val=[[0 for i in range(NewIter+1)] for j in range(MatrixSize)]
    for j in range(MatrixSize):
        val[j][0]=sol[j]
    Iteration=[i for i in range(NewIter+1)]
    for Newtoniter in range(NewIter):
        for i in range(MatrixSize):
            STA_nonlinear[i]=0
        for i in range(DeviceCount):
            if DevType[i]=='capacitor':
                STA_rhs[NumberOfNodes+i]=STA_rhs[NumberOfNodes+i]
            elif DevType[i]=='inductor':
                STA_rhs[NumberOfNodes+i]=0
            elif DevType[i]=='VoltSource':
          STA_rhs[NumberOfNodes+i]=ana.getSourceValue(DevValue[i],0)
            elif DevType[i]=='CurrentSource':
          STA_rhs[NumberOfNodes+i]=ana.getSourceValue(DevValue[i],0)
            elif DevType[i]=='transistor':
                lambdaT=ana.findParameter(modeldict,DevModel[i],'lam
bdaT')
                VT=ana.findParameter(modeldict,DevModel[i],'VT')
            STA_matrix[NumberOfNodes+i][NumberOfNodes+i]=DevValue[i]
            STA_matrix[NumberOfNodes+i][Nodes.index(DevNode1[i])]=0
            STA_matrix[Nodes.index(DevNode1[i])][NumberOfNodes+i]=1
            STA_matrix[NumberOfNodes+i][Nodes.index(DevNode3[i])]=0
            STA_matrix[Nodes.index(DevNode3[i])][NumberOfNodes+i]=-1
                VD=sol[Nodes.index(DevNode1[i])]
                VG=sol[Nodes.index(DevNode2[i])]
                VS=sol[Nodes.index(DevNode3[i])]
                Vgs=VG-VS
```

```
                    Vds=VD-VS
                    if DevModel[i][0]=='p':
                        Vds=-Vds
                        Vgs=-Vgs
                    if Vds < Vgs-VT :
                        STA_nonlinear[NumberOfNodes+i]=2*((Vgs-VT)*Vds-
0.5*Vds**2)
                    else :
                        STA_nonlinear[NumberOfNodes+i]=(Vgs-VT)**2*(1+
lambdaT*Vds)
                elif DevType[i]=='bipolar':
                        VEarly=ana.findParameter(modeldict,DevModel[i],
'Early')
                STA_matrix[NumberOfNodes+i][NumberOfNodes+i]=DevValue[i]
                STA_matrix[NumberOfNodes+i][Nodes.index(DevNode1[i])]=0
                STA_matrix[Nodes.index(DevNode1[i])][NumberOfNodes+i]=1
                STA_matrix[NumberOfNodes+i][Nodes.index(DevNode3[i])]=0
                STA_matrix[Nodes.index(DevNode3[i])][NumberOfNodes+i]=-1
                    VC=sol[Nodes.index(DevNode1[i])]
                    VB=sol[Nodes.index(DevNode2[i])]
                    VE=sol[Nodes.index(DevNode3[i])]
                    Vbe=VB-VE
                    Vce=VC-VE
                    if Vbe<0:
                        STA_nonlinear[NumberOfNodes+i]=0
                    else :
                        STA_nonlinear[NumberOfNodes+i]=math.exp(Vbe/Vthe
rmal)*(1+Vce/VEarly)
        f=np.matmul(STA_matrix,sol)-STA_rhs+STA_nonlinear
    #
    # Now we need the Jacobian, the transistors look like VCCS with
a specific gain = 2 K (Vg-Vs) in our case
    #
        for i in range(MatrixSize):
            for j in range(MatrixSize):
                Jacobian[i][j]=STA_matrix[i][j]
        for i in range(DeviceCount):
            if DevType[i]=='transistor':
                lambdaT=ana.findParameter(modeldict,DevModel[i],'lam
bdaT')
                VT=ana.findParameter(modeldict,DevModel[i],'VT')
                Jacobian[NumberOfNodes+i][NumberOfNodes+i]=DevValue
[i] # due to derfivative leading to double gain
                VD=sol[Nodes.index(DevNode1[i])]
                VG=sol[Nodes.index(DevNode2[i])]
```

```
                    VS=sol[Nodes.index(DevNode3[i])]
                    Vgs=VG-VS
                    Vds=VD-VS
                    Vgd=VG-VD
                    if DevModel[i][0]=='p':
                        PFET=-1
                        Vgs=-Vgs
                        Vds=-Vds
                        Vgd=-Vgd
                    else:
                        PFET=1
                    if Vgs<VT :
                        Jacobian[NumberOfNodes+i][Nodes.index(DevNode1[
i])]=PFET*1e-1
                        Jacobian[NumberOfNodes+i][Nodes.index(DevNode2[
i])]=PFET*1e-1
                        Jacobian[NumberOfNodes+i][Nodes.index(DevNode3[
i])]=-PFET*1e-1
                    Jacobian[Nodes.index(DevNode1[i])][NumberOfNodes+i]=1
                    Jacobian[Nodes.index(DevNode3[i])][NumberOfNodes+i]=-1
                        elif Vds <= Vgs-VT:
                        Jacobian[NumberOfNodes+i][Nodes.index(DevNode1[
i])]=PFET*2*(Vgd-VT)
                        Jacobian[NumberOfNodes+i][Nodes.index(DevNode2[
i])]=PFET*2*Vds
                        Jacobian[NumberOfNodes+i][Nodes.index(DevNode3[
i])]=-PFET*2*(Vgs-VT)
                    Jacobian[Nodes.index(DevNode1[i])][NumberOfNodes+i]=1
                    Jacobian[Nodes.index(DevNode3[i])][NumberOfNodes+i]=-1
                        else :
                        Jacobian[NumberOfNodes+i][Nodes.index(DevNode1[
i])]=PFET*lambdaT*(Vgs-VT)**2
                        Jacobian[NumberOfNodes+i][Nodes.index(DevNode2[
i])]=PFET*2*(Vgs-VT)*(1+lambdaT*Vds)
                        Jacobian[NumberOfNodes+i][Nodes.index(DevNode3[
i])]=PFET*(-2*(Vgs-VT)*(1+lambdaT*Vds)-lambdaT*(Vgs-VT)**2)
                    Jacobian[Nodes.index(DevNode1[i])][NumberOfNodes+i]=1
                    Jacobian[Nodes.index(DevNode3[i])][NumberOfNodes+i]=-1
                elif DevType[i]=='bipolar':
                        VEarly=ana.findParameter(modeldict,DevModel[i],
'Early')
                        Jacobian[NumberOfNodes+i][NumberOfNodes+i]=Dev
Value[i] # due to derfivative leading to double gain
                    VC=sol[Nodes.index(DevNode1[i])]
                    VB=sol[Nodes.index(DevNode2[i])]
```

```
                    VE=sol[Nodes.index(DevNode3[i])]
                    Vbe=VB-VE
                    Vce=VC-VE
                    Vbc=VB-VC
          if Vbe<=0 :
              Jacobian[NumberOfNodes+i][Nodes.index(DevNode1[i])]=1e-5
              Jacobian[NumberOfNodes+i][Nodes.index(DevNode2[i])]=1e-5
              Jacobian[NumberOfNodes+i][Nodes.index(DevNode3[i])]=-1e-5
              Jacobian[Nodes.index(DevNode1[i])][NumberOfNodes+i]=1
              Jacobian[Nodes.index(DevNode3[i])][NumberOfNodes+i]=-1
          else :
              Jacobian[NumberOfNodes+i][Nodes.index(DevNode1[i])]=math.
exp(Vbe/Vthermal)/VEarly
              Jacobian[NumberOfNodes+i][Nodes.index(DevNode2[i])]=math.
exp(Vbe/Vthermal)*(1+Vce/VEarly)/Vthermal
              Jacobian[NumberOfNodes+i][Nodes.index(DevNode3[i])]=(-math.
exp(Vbe/Vthermal)/VEarly-math.exp(Vbe/Vthermal)*(1+Vce/VEarly)/
Vthermal)
              Jacobian[Nodes.index(DevNode1[i])][NumberOfNodes+i]=1
              Jacobian[Nodes.index(DevNode3[i])][NumberOfNodes+i]=-1
        sol=sol-np.matmul(np.linalg.inv(Jacobian),f)
        Jac_inv=np.linalg.inv(Jacobian)
        for j in range(MatrixSize):
            val[j][Newtoniter+1]=sol[j]
    ana.plotdata(Plotdict,NumberOfNodes,Iteration,val,Nodes)
    ana.printdata(Printdict,NumberOfNodes,Iteration,val,Nodes)


Code 5.3.1
#!/usr/bin/env python3
# -*- coding: utf-8 -*-
"""
Created on Thu Feb 28 22:33:04 2019

@author: mikael
"""
import numpy
import matplotlib.pyplot as plt
import math
import analogdef as ana


#
# Function definitions
#
def f_NL(STA_matrix, STA_rhs, STA_nonlinear, solution):
```

```
return numpy.matmul(STA_matrix,solution)-STA_rhs+STA_nonlinear

    #
    # Read netlist
    #
    DeviceCount=0
    MaxNumberOfDevices=100
    DevType=[0*i for i in range(MaxNumberOfDevices)]
    DevLabel=[0*i for i in range(MaxNumberOfDevices)]
    DevNode1=[0*i for i in range(MaxNumberOfDevices)]
    DevNode2=[0*i for i in range(MaxNumberOfDevices)]
    DevNode3=[0*i for i in range(MaxNumberOfDevices)]
    DevValue=[0*i for i in range(MaxNumberOfDevices)]
    DevModel=[0*i for i in range(MaxNumberOfDevices)]
    Nodes=[]

    modeldict=ana.readmodelfile('models.txt')
    ICdict={}
    Plotdict={}
    Printdict={}
    Optdict={}
    Optdict['MaxNewtonIterations']=int(5)
    Optdict['reltol']=1e-3
    Optdict['iabstol']=1e-6
    Optdict['vabstol']=1e-6
    Optdict['deltaT']=1e-12
    #
    # Read the netlist
    #
    print('This version has convergence checks')
    DeviceCount=ana.readnetlist('netlist_bandgap.txt',modeldict,ICd
ict,Plotdict,Printdict,Optdict,DevType,DevValue,DevLabel,DevNode1
,DevNode2,DevNode3,DevModel,Nodes,MaxNumberOfDevices)
    #
    # Create Matrix based on circuit size. We do not implement strict
Modified Nodal Analysis. We keep instead all currents
    # but keep referring to the voltages as absolute voltages. We
believe this will make the operation clearer to the user.
    #
    NumberOfNodes=len(Nodes)
    NumberOfCurrents=DeviceCount
    MatrixSize=DeviceCount+len(Nodes)
    Jacobian=[[0 for i in range(MatrixSize)] for j in range(MatrixSize)]
    Jac_inv=[[0 for i in range(MatrixSize)] for j in range(MatrixSize)]
    Spare=[[0 for i in range(MatrixSize)] for j in range(MatrixSize)]
```

```
STA_matrix=[[0 for i in range(MatrixSize)] for j in range(Matrix
Size)]
STA_rhs=[0 for i in range(MatrixSize)]
STA_nonlinear=[0 for i in range(MatrixSize)]
f=[0 for i in range(MatrixSize)]
#
# Create sim parameters
#
reltol=Optdict['reltol']
iabstol=Optdict['iabstol']
vabstol=Optdict['vabstol']
Vthermal=1.38e-23*300/1.602e-19
deltaT=Optdict['deltaT']
sol=[0 for i in range(MatrixSize)]
solm1=[0 for i in range(MatrixSize)]
#
if len(ICdict)>0:
    for i in range(len(ICdict)):
        for j in range(NumberOfNodes):
            if Nodes[j]==ICdict[i]['NodeName']:
                sol[j]=ICdict[i]['Value']
                print('Setting ',Nodes[j],' to ',sol[j])
#
# Loop through all devices and create jacobian and initial f(v)
entries according to signature
#
for i in range(DeviceCount):
    if DevType[i] != 'transistor' and DevType[i] != 'bipolar':
      STA_matrix[NumberOfNodes+i][NumberOfNodes+i]=-DevValue[i]
        if DevNode1[i] != '0' :
        STA_matrix[NumberOfNodes+i][Nodes.index(DevNode1[i])]=1
        STA_matrix[Nodes.index(DevNode1[i])][NumberOfNodes+i]=1
        if DevNode2[i] != '0' :
        STA_matrix[NumberOfNodes+i][Nodes.index(DevNode2[i])]=-1
        STA_matrix[Nodes.index(DevNode2[i])][NumberOfNodes+i]=-1
        if DevType[i]=='capacitor':
            # Do nothing
            STA_rhs[NumberOfNodes]=STA_rhs[NumberOfNodes]
        if DevType[i]=='inductor':
            # For DC we treat this as a voltage source with V=0
            STA_matrix[NumberOfNodes+i][NumberOfNodes+i]=0
            STA_rhs[NumberOfNodes+i]=0
        if DevType[i]=='VoltSource':
            STA_matrix[NumberOfNodes+i][NumberOfNodes+i]=0
            STA_rhs[NumberOfNodes+i]=DevValue[i]
```

```
        if DevType[i]=='CurrentSource':
            if DevNode1[i] != '0' :
        STA_matrix[NumberOfNodes+i][Nodes.index(DevNode1[i])]=0
        STA_matrix[Nodes.index(DevNode1[i])][NumberOfNodes+i]=0
            if DevNode2[i] != '0' :
        STA_matrix[NumberOfNodes+i][Nodes.index(DevNode2[i])]=0
        STA_matrix[Nodes.index(DevNode2[i])][NumberOfNodes+i]=0
            STA_matrix[NumberOfNodes+i][NumberOfNodes+i]=1
            STA_rhs[NumberOfNodes+i]=DevValue[i]
            if DevNode1[i] != '0' and DevNode2[i]!='0':
        STA_matrix[Nodes.index(DevNode1[i])][NumberOfNodes+i]=1
        STA_matrix[Nodes.index(DevNode2[i])][NumberOfNodes+i]=-1
            elif DevNode2[i] != '0' :
        STA_matrix[Nodes.index(DevNode2[i])][NumberOfNodes+i]=-1
            elif DevNode1[i] != '0' :
        STA_matrix[Nodes.index(DevNode1[i])][NumberOfNodes+i]=1
    if DevType[i]=='transistor':
        lambdaT=ana.findParameter(modeldict,DevModel[i],'lambdaT')
        VT=ana.findParameter(modeldict,DevModel[i],'VT')
        STA_matrix[NumberOfNodes+i][NumberOfNodes+i]=DevValue[i]
        STA_matrix[NumberOfNodes+i][Nodes.index(DevNode1[i])]=0
        STA_matrix[Nodes.index(DevNode1[i])][NumberOfNodes+i]=1
        STA_matrix[NumberOfNodes+i][Nodes.index(DevNode3[i])]=0
        STA_matrix[Nodes.index(DevNode3[i])][NumberOfNodes+i]=-1
        VD=sol[Nodes.index(DevNode1[i])]
        VG=sol[Nodes.index(DevNode2[i])]
        VS=sol[Nodes.index(DevNode3[i])]
        Vgs=VG-VS
        Vds=VD-VS
        if DevModel[i][0]=='p':
            Vds=-Vds
            Vgs=-Vgs
        if Vds < Vgs-VT :
            STA_nonlinear[NumberOfNodes+i]=2*((Vgs-VT)*Vds-0.5*
Vds**2)
        else :
            STA_nonlinear[NumberOfNodes+i]=(Vgs-VT)**2*(1+lambda
T*Vds)
    if DevType[i]=='bipolar':
        VEarly=ana.findParameter(modeldict,DevModel[i],'Early')
        STA_matrix[NumberOfNodes+i][NumberOfNodes+i]=DevValue[i]
        STA_matrix[NumberOfNodes+i][Nodes.index(DevNode1[i])]=0
        STA_matrix[Nodes.index(DevNode1[i])][NumberOfNodes+i]=1
        STA_matrix[NumberOfNodes+i][Nodes.index(DevNode3[i])]=0
        STA_matrix[Nodes.index(DevNode3[i])][NumberOfNodes+i]=-1
```

```
            VC=sol[Nodes.index(DevNode1[i])]
            VB=sol[Nodes.index(DevNode2[i])]
            VE=sol[Nodes.index(DevNode3[i])]
            Vbe=VB-VE
            Vce=VC-VE
            if Vbe < 0 :
                STA_nonlinear[NumberOfNodes+i]=0
            else :
                STA_nonlinear[NumberOfNodes+i]=math.exp(Vbe/Vthermal)
*(1+Vce/VEarly)

    #
    f=numpy.matmul(STA_matrix,sol)-STA_rhs+STA_nonlinear
    #
    #Loop through iteration points
    #
    NewIter=int(Optdict['MaxNewtonIterations'])
    val=[[0 for i in range(NewIter+1)] for j in range(MatrixSize)]
    for j in range(MatrixSize):
        val[j][0]=sol[j]
    Iteration=[i for i in range(NewIter+1)]
    NewtonConverged=False
    Newtoniter=0
    while not NewtonConverged and Newtoniter<NewIter:
        for i in range(MatrixSize):
            STA_nonlinear[i]=0
        for i in range(DeviceCount):
            if DevType[i]=='capacitor':
                STA_rhs[NumberOfNodes+i]=STA_rhs[NumberOfNodes+i]
            elif DevType[i]=='inductor':
                STA_rhs[NumberOfNodes+i]=0
            elif DevType[i]=='VoltSource':
          STA_rhs[NumberOfNodes+i]=ana.getSourceValue(DevValue[i],0)
            elif DevType[i]=='CurrentSource':
          STA_rhs[NumberOfNodes+i]=ana.getSourceValue(DevValue[i],0)
            elif DevType[i]=='transistor':
                lambdaT=ana.findParameter(modeldict,DevModel[i],'lam
bdaT')
                VT=ana.findParameter(modeldict,DevModel[i],'VT')
            STA_matrix[NumberOfNodes+i][NumberOfNodes+i]=DevValue[i]
            STA_matrix[NumberOfNodes+i][Nodes.index(DevNode1[i])]=0
            STA_matrix[Nodes.index(DevNode1[i])][NumberOfNodes+i]=1
            STA_matrix[NumberOfNodes+i][Nodes.index(DevNode3[i])]=0
            STA_matrix[Nodes.index(DevNode3[i])][NumberOfNodes+i]=-1
                VD=sol[Nodes.index(DevNode1[i])]
```

```
                    VG=sol[Nodes.index(DevNode2[i])]
                    VS=sol[Nodes.index(DevNode3[i])]
                    Vgs=VG-VS
                    Vds=VD-VS
                    if DevModel[i][0]=='p':
                        Vds=-Vds
                        Vgs=-Vgs
                    if Vds < Vgs-VT :
                        STA_nonlinear[NumberOfNodes+i]=2*((Vgs-VT)*Vds-
0.5*Vds**2)
                    else :
                        STA_nonlinear[NumberOfNodes+i]=(Vgs-VT)**2*(1+
lambdaT*Vds)
                elif DevType[i]=='bipolar':
                    VEarly=ana.findParameter(modeldict,DevModel[i],'Ea
rly')
                STA_matrix[NumberOfNodes+i][NumberOfNodes+i]=DevValue[i]
                STA_matrix[NumberOfNodes+i][Nodes.index(DevNode1[i])]=0
                STA_matrix[Nodes.index(DevNode1[i])][NumberOfNodes+i]=1
                STA_matrix[NumberOfNodes+i][Nodes.index(DevNode3[i])]=0
                STA_matrix[Nodes.index(DevNode3[i])][NumberOfNodes+i]=-1
                    VC=sol[Nodes.index(DevNode1[i])]
                    VB=sol[Nodes.index(DevNode2[i])]
                    VE=sol[Nodes.index(DevNode3[i])]
                    Vbe=VB-VE
                    Vce=VC-VE
                    if Vbe<0:
                        STA_nonlinear[NumberOfNodes+i]=0
                    else :
                        STA_nonlinear[NumberOfNodes+i]=math.exp(Vbe/Vther
mal)*(1+Vce/VEarly)
        f=numpy.matmul(STA_matrix,sol)-STA_rhs+STA_nonlinear
        ResidueConverged=True
        node=0
        while ResidueConverged and node<NumberOfNodes:
      # Let us find the maximum current going into node, Nodes[node]
            MaxCurrent=0
            for current in range(NumberOfCurrents):
                MaxCurrent=max(MaxCurrent,abs(STA_matrix[node][Numb
erOfNodes+current]*(sol[NumberOfNodes+current])))
            if f[node] > reltol*MaxCurrent+iabstol:
                print('f:',node,f[node],MaxCurrent)
                ResidueConverged=False
            node=node+1
    #
```

```
    # Now we need the Jacobian, the transistors look like VCCS with
a specific gain = 2 K (Vg-Vs) in our case
    #
        for i in range(MatrixSize):
            for j in range(MatrixSize):
                Jacobian[i][j]=STA_matrix[i][j]
        for i in range(DeviceCount):
            if DevType[i]=='transistor':
                lambdaT=ana.findParameter(modeldict,DevModel[i],'lam
bdaT')
                VT=ana.findParameter(modeldict,DevModel[i],'VT')
                Jacobian[NumberOfNodes+i][NumberOfNodes+i]=DevValue
[i] # due to derfivative leading to double gain
                VD=sol[Nodes.index(DevNode1[i])]
                VG=sol[Nodes.index(DevNode2[i])]
                VS=sol[Nodes.index(DevNode3[i])]
                Vgs=VG-VS
                Vds=VD-VS
                Vgd=VG-VD
                if DevModel[i][0]=='p':
                    PFET=-1
                    Vgs=-Vgs
                    Vds=-Vds
                    Vgd=-Vgd
                else:
                    PFET=1
                if Vgs<VT :
                    Jacobian[NumberOfNodes+i][Nodes.index(DevNode1[
i])]=PFET*1e-1
                    Jacobian[NumberOfNodes+i][Nodes.index(DevNode2[
i])]=PFET*1e-1
                    Jacobian[NumberOfNodes+i][Nodes.index(DevNode3[
i])]=-PFET*1e-1
                Jacobian[Nodes.index(DevNode1[i])][NumberOfNodes+i]=1
                Jacobian[Nodes.index(DevNode3[i])][NumberOfNodes+i]=-1
                    elif Vds <= Vgs-VT:
                    Jacobian[NumberOfNodes+i][Nodes.index(DevNode1[
i])]=PFET*2*(Vgd-VT)
                    Jacobian[NumberOfNodes+i][Nodes.index(DevNode2[
i])]=PFET*2*Vds
                    Jacobian[NumberOfNodes+i][Nodes.index(DevNode3[
i])]=-PFET*2*(Vgs-VT)
                Jacobian[Nodes.index(DevNode1[i])][NumberOfNodes+i]=1
                Jacobian[Nodes.index(DevNode3[i])][NumberOfNodes+i]=-1
                    else :
```

```
                        Jacobian[NumberOfNodes+i][Nodes.index(DevNode1[
i])]=PFET*lambdaT*(Vgs-VT)**2
                        Jacobian[NumberOfNodes+i][Nodes.index(DevNode2[
i])]=PFET*2*(Vgs-VT)*(1+lambdaT*Vds)
                        Jacobian[NumberOfNodes+i][Nodes.index(DevNode3[
i])]=PFET*(-2*(Vgs-VT)*(1+lambdaT*Vds)-lambdaT*(Vgs-VT)**2)
                Jacobian[Nodes.index(DevNode1[i])][NumberOfNodes+i]=1
                Jacobian[Nodes.index(DevNode3[i])][NumberOfNodes+i]=-1
            elif DevType[i]=='bipolar':
                    VEarly=ana.findParameter(modeldict,DevModel[i],
'Early')
                    Jacobian[NumberOfNodes+i][NumberOfNodes+i]=Dev
Value[i] # due to derfivative leading to double gain
                VC=sol[Nodes.index(DevNode1[i])]
                VB=sol[Nodes.index(DevNode2[i])]
                VE=sol[Nodes.index(DevNode3[i])]
                Vbe=VB-VE
                Vce=VC-VE
                Vbc=VB-VC
                if Vbe<=0 :
            Jacobian[NumberOfNodes+i][Nodes.index(DevNode1[i])]=1e-5
            Jacobian[NumberOfNodes+i][Nodes.index(DevNode2[i])]=1e-5
                Jacobian[NumberOfNodes+i][Nodes.index(DevNode3[i])]=-
1e-5
                Jacobian[Nodes.index(DevNode1[i])][NumberOfNodes+i]=1
                Jacobian[Nodes.index(DevNode3[i])][NumberOfNodes+i]=-1
                else :
                Jacobian[NumberOfNodes+i][Nodes.index(DevNode1[i])]=
math.exp(Vbe/Vthermal)/VEarly
                    Jacobian[NumberOfNodes+i][Nodes.index(DevNode2[i])]=
math.exp(Vbe/Vthermal)*(1+Vce/VEarly)/Vthermal
                    Jacobian[NumberOfNodes+i][Nodes.index(DevNode3[i])]=
(-math.exp(Vbe/Vthermal)/VEarly-math.exp(Vbe/Vthermal)*(1+Vce/
VEarly)/Vthermal)
                Jacobian[Nodes.index(DevNode1[i])][NumberOfNodes+i]=1
                Jacobian[Nodes.index(DevNode3[i])][NumberOfNodes+i]=-1
    SolutionCorrection=numpy.matmul(numpy.linalg.inv(Jacobian),f)
        UpdateConverged=True
        for node in range(NumberOfNodes) :
                vkmax=max(abs(sol[node]),abs(sol[node]-SolutionCorrect
ion[node]))
            if abs(SolutionCorrection[node])>vkmax*reltol+vabstol:
                UpdateConverged=False
        NewtonConverged=ResidueConverged and UpdateConverged
        sol=sol-SolutionCorrection
```

```
    Jac_inv=numpy.linalg.inv(Jacobian)
    for j in range(MatrixSize):
        val[j][Newtoniter+1]=sol[j]
    Newtoniter=Newtoniter+1
print('THis one seems not to work right when ic for vs is too
high?? Even vs=-0.03 is not converging netlist_dc_5p3')
ana.plotdata(Plotdict,NumberOfNodes,Iteration,val,Nodes)
ana.printdata(Printdict,NumberOfNodes,Iteration,val,Nodes)
```

## 5.6.4  Code 5.4

```
#!/usr/bin/env python3
# -*- coding: utf-8 -*-
"""
Created on Thu Feb 28 22:33:04 2019

@author: mikael
"""
import numpy as np
import analogdef as ana

#
# Function definitions
#
DeviceCount=0
MaxNumberOfDevices=100
DevType=[0*i for i in range(MaxNumberOfDevices)]
DevLabel=[0*i for i in range(MaxNumberOfDevices)]
DevNode1=[0*i for i in range(MaxNumberOfDevices)]
DevNode2=[0*i for i in range(MaxNumberOfDevices)]
DevNode3=[0*i for i in range(MaxNumberOfDevices)]
DevValue=[0*i for i in range(MaxNumberOfDevices)]
DevModel=[0*i for i in range(MaxNumberOfDevices)]
Nodes=[]

modeldict=ana.readmodelfile('models.txt')
ICdict={}
Plotdict={}
Printdict={}
Optdict={}
Optdict['MaxNewtonIterations']=int(5)
#
# Read the netlist
```

```
    #
        DeviceCount=ana.readnetlist('netlist_crossPandNinv_5p4.txt-
    ',modeldict,ICdict,Plotdict,Printdict,Optdict,DevType,DevValue,De
    vLabel,DevNode1,DevNode2,DevNode3,DevModel,Nodes,MaxNumberOfDevi
    ces)
    #
    # Create Matrix based on circuit size. We do not implement strict
    Modified Nodal Analysis. We keep instead all currents
    # but keep referring to the voltages as absolute voltages. We
    believe this will make the operation clearer to the user.
    #
    NumberOfNodes=len(Nodes)
    MatrixSize=DeviceCount+len(Nodes)
    Jacobian=[[0 for i in range(MatrixSize)] for j in range(MatrixSize)]
    Jac_inv=[[0 for i in range(MatrixSize)] for j in range(MatrixSize)]
    Spare=[[0 for i in range(MatrixSize)] for j in range(MatrixSize)]
    STA_matrix=[[0 for i in range(MatrixSize)] for j in range(Matrix
    Size)]
    STA_rhs=[0 for i in range(MatrixSize)]
    STA_nonlinear=[0 for i in range(MatrixSize)]
    f=[0 for i in range(MatrixSize)]
    #
    # Create sim parameters
    #
    deltaT=1e-12
    sol=[0 for i in range(MatrixSize)]
    solm1=[0 for i in range(MatrixSize)]
    if len(ICdict)>0:
        for i in range(len(ICdict)):
            for j in range(NumberOfNodes):
                if Nodes[j]==ICdict[i]['NodeName']:
                    sol[j]=ICdict[i]['Value']
                    print('Setting ',Nodes[j],' to ',sol[j])
    #
    # Loop through all devices and create jacobian and initial f(v)
    entries according to signature
    #

    for i in range(DeviceCount):
        if DevType[i] != 'transistor':
          STA_matrix[NumberOfNodes+i][NumberOfNodes+i]=-DevValue[i]
            if DevNode1[i] != '0' :
            STA_matrix[NumberOfNodes+i][Nodes.index(DevNode1[i])]=1
            STA_matrix[Nodes.index(DevNode1[i])][NumberOfNodes+i]=1
            if DevNode2[i] != '0' :
```

```
STA_matrix[NumberOfNodes+i][Nodes.index(DevNode2[i])]=-1
STA_matrix[Nodes.index(DevNode2[i])][NumberOfNodes+i]=-1
  if DevType[i]=='capacitor':
      # Do nothing
      STA_rhs[NumberOfNodes]=STA_rhs[NumberOfNodes]
  if DevType[i]=='inductor':
      # For DC we treat this as a voltage source with V=0
      STA_matrix[NumberOfNodes+i][NumberOfNodes+i]=0
      STA_rhs[NumberOfNodes+i]=0
  if DevType[i]=='VoltSource':
      STA_matrix[NumberOfNodes+i][NumberOfNodes+i]=0
      STA_rhs[NumberOfNodes+i]=DevValue[i]
  if DevType[i]=='CurrentSource':
      if DevNode1[i] != '0' :
  STA_matrix[NumberOfNodes+i][Nodes.index(DevNode1[i])]=0
  STA_matrix[Nodes.index(DevNode1[i])][NumberOfNodes+i]=0
      if DevNode2[i] != '0' :
  STA_matrix[NumberOfNodes+i][Nodes.index(DevNode2[i])]=0
  STA_matrix[Nodes.index(DevNode2[i])][NumberOfNodes+i]=0
      STA_matrix[NumberOfNodes+i][NumberOfNodes+i]=1
      STA_rhs[NumberOfNodes+i]=DevValue[i]
      if DevNode1[i] != '0' and DevNode2[i]!='0':
  STA_matrix[Nodes.index(DevNode1[i])][NumberOfNodes+i]=1
  STA_matrix[Nodes.index(DevNode2[i])][NumberOfNodes+i]=-1
      elif DevNode2[i] != '0' :
  STA_matrix[Nodes.index(DevNode2[i])][NumberOfNodes+i]=-1
      elif DevNode1[i] != '0' :
  STA_matrix[Nodes.index(DevNode1[i])][NumberOfNodes+i]=1
if DevType[i]=='transistor':
  lambdaT=ana.findParameter(modeldict,DevModel[i],'lambdaT')
  VT=ana.findParameter(modeldict,DevModel[i],'VT')
  STA_matrix[NumberOfNodes+i][NumberOfNodes+i]=DevValue[i]
  STA_matrix[NumberOfNodes+i][Nodes.index(DevNode1[i])]=0
  STA_matrix[Nodes.index(DevNode1[i])][NumberOfNodes+i]=1
  STA_matrix[NumberOfNodes+i][Nodes.index(DevNode3[i])]=0
  STA_matrix[Nodes.index(DevNode3[i])][NumberOfNodes+i]=-1
  VD=sol[Nodes.index(DevNode1[i])]
  VG=sol[Nodes.index(DevNode2[i])]
  VS=sol[Nodes.index(DevNode3[i])]
  Vgs=VG-VS
  Vds=VD-VS
  if DevModel[i][0]=='p':
      Vds=-Vds
      Vgs=-Vgs
  if Vds < Vgs-VT :
```

```
                        STA_nonlinear[NumberOfNodes+i]=2*((Vgs-VT)*Vds-0.5*
Vds**2)
            else :
                        STA_nonlinear[NumberOfNodes+i]=(Vgs-VT)**2*(1+lambda
T*Vds)
  #
  f=np.matmul(STA_matrix,sol)-STA_rhs+STA_nonlinear
  #
  #Loop through iteration points
  #
  NSourceSteps=100
  NewIter=int(Optdict['MaxNewtonIterations'])
  val=[[0 for i in range(NSourceSteps+1)] for j in range(MatrixSize)]
  for j in range(MatrixSize):
      val[j][0]=sol[j]
  Iteration=[i for i in range(NSourceSteps+1)]
  for step in range(NSourceSteps):
      for Newtoniter in range(NewIter):
          for i in range(MatrixSize):
              STA_nonlinear[i]=0
          for i in range(DeviceCount):
              if DevType[i]!='transistor':
                  if DevType[i]=='capacitor':

STA_rhs[NumberOfNodes+i]=STA_rhs[NumberOfNodes+i]
                  if DevType[i]=='inductor':
                      STA_rhs[NumberOfNodes+i]=0
                  if DevType[i]=='VoltSource':
                      if DevLabel[i]=='vdd':
                          if step < NSourceSteps/2:
                              STA_rhs[NumberOfNodes+i]=DevValue[i
]*step*2/NSourceSteps
              if DevType[i]=='transistor':
                  lambdaT=ana.findParameter(modeldict,DevModel[i],
'lambdaT')
                  VT=ana.findParameter(modeldict,DevModel[i],'VT')
                  VD=sol[Nodes.index(DevNode1[i])]
                  VG=sol[Nodes.index(DevNode2[i])]
                  VS=sol[Nodes.index(DevNode3[i])]
                  Vgs=VG-VS
                  Vds=VD-VS
                  if DevModel[i][0]=='p':
                      Vds=-Vds
                      Vgs=-Vgs
                  if Vgs<VT:
```

```
                    STA_nonlinear[NumberOfNodes+i]=1e-5
                elif Vds < Vgs-VT:
                    STA_nonlinear[NumberOfNodes+i]=2*((Vgs-VT)*
Vds-0.5*Vds**2)
                else :
                    STA_nonlinear[NumberOfNodes+i]=(Vgs-VT)**2*
(1+lambdaT*Vds)
        f=np.matmul(STA_matrix,sol)-STA_rhs+STA_nonlinear

    #
    # Now we need the Jacobian, the transistors look like VCCS
with a specific gain = 2 K (Vg-Vs) in our case
    #
        for i in range(MatrixSize):
            for j in range(MatrixSize):
                Jacobian[i][j]=STA_matrix[i][j]
        for i in range(DeviceCount):
            if DevType[i]=='transistor':
                lambdaT=ana.findParameter(modeldict,DevModel[i],
'lambdaT')
                VT=ana.findParameter(modeldict,DevModel[i],'VT')
                Jacobian[NumberOfNodes+i][NumberOfNodes+i]=Dev
Value[i] # due to derfivative leading to double gain
                if DevNode1[i] != '0' :
                    VD=sol[Nodes.index(DevNode1[i])]
                else:
                    VD=0
                if DevNode2[i] != '0' :
                    VG=sol[Nodes.index(DevNode2[i])]
                else:
                    VG=0
                if DevNode3[i] != '0' :
                    VS=sol[Nodes.index(DevNode3[i])]
                else:
                    VS=0
                Vgs=VG-VS
                Vds=VD-VS
                Vgd=VG-VD
                if DevModel[i][0]=='p':
                    PFET=-1
                    Vgs=-Vgs
                    Vds=-Vds
                    Vgd=-Vgd
                else:
                    PFET=1
```

```python
                    if Vgs<VT :
                        Jacobian[NumberOfNodes+i][Nodes.index(Dev
Node1[i])]=PFET*1e-10
                        Jacobian[NumberOfNodes+i][Nodes.index(Dev
Node2[i])]=PFET*1e-10
                        Jacobian[NumberOfNodes+i][Nodes.index(Dev
Node3[i])]=-PFET*1e-10
                        Jacobian[Nodes.index(DevNode1[i])][Number
OfNodes+i]=1
                        Jacobian[Nodes.index(DevNode3[i])][Number
OfNodes+i]=-1
                    elif Vds <= Vgs-VT:
                        Jacobian[NumberOfNodes+i][Nodes.index(Dev
Node1[i])]=PFET*2*(Vgd-VT)
                        Jacobian[NumberOfNodes+i][Nodes.index(Dev
Node2[i])]=PFET*2*Vds
                        Jacobian[NumberOfNodes+i][Nodes.index(Dev
Node3[i])]=-PFET*2*(Vgs-VT)
                        Jacobian[Nodes.index(DevNode1[i])][Number
OfNodes+i]=1
                        Jacobian[Nodes.index(DevNode3[i])][Number
OfNodes+i]=-1
                    else :
                        Jacobian[NumberOfNodes+i][Nodes.index(Dev
Node1[i])]=PFET*lambdaT*(Vgs-VT)**2
                        Jacobian[NumberOfNodes+i][Nodes.index(Dev
Node2[i])]=PFET*2*(Vgs-VT)*(1+lambdaT*Vds)
                        Jacobian[NumberOfNodes+i][Nodes.index(Dev
Node3[i])]=PFET*(-2*(Vgs-VT)*(1+lambdaT*Vds)-lambdaT*(Vgs-VT)**2)
                        Jacobian[Nodes.index(DevNode1[i])][Number
OfNodes+i]=1
                        Jacobian[Nodes.index(DevNode3[i])][Number
OfNodes+i]=-1
        sol=sol-np.matmul(np.linalg.inv(Jacobian),f)
        for j in range(MatrixSize):
            val[j][step+1]=sol[j]

    ana.plotdata(Plotdict,NumberOfNodes,Iteration,val,Nodes)
    ana.printdata(Printdict,NumberOfNodes,Iteration,val,Nodes)
```

## 5.6.5   Code 5.5

```
#!/usr/bin/env python3
```

```
# -*- coding: utf-8 -*-
""" """
Created on Thu Feb 28 22:33:04 2019

@author: mikael
""" """
import numpy
import matplotlib.pyplot as plt
import math
import analogdef as ana

#
# Function definitions
#
def f_NL(STA_matrix, STA_rhs, STA_nonlinear, solution):

return numpy.matmul(STA_matrix,solution)-STA_rhs+STA_nonlinear

#
# Read netlist
#
DeviceCount=0
MaxNumberOfDevices=100
DevType=[0*i for i in range(MaxNumberOfDevices)]
DevLabel=[0*i for i in range(MaxNumberOfDevices)]
DevNode1=[0*i for i in range(MaxNumberOfDevices)]
DevNode2=[0*i for i in range(MaxNumberOfDevices)]
DevNode3=[0*i for i in range(MaxNumberOfDevices)]
DevValue=[0*i for i in range(MaxNumberOfDevices)]
DevModel=[0*i for i in range(MaxNumberOfDevices)]
Nodes=[]
#
modeldict=ana.readmodelfile('models.txt')
ICdict={}
Plotdict={}
Printdict={}
Optdict={}
#
# Read the netlist
#
   DeviceCount=ana.readnetlist('netlist_gminsweep_5p5.txt',model-
dict,ICdict,Plotdict,Printdict,Optdict,DevType,DevValue,DevLabel,
DevNode1,DevNode2,DevNode3,DevModel,Nodes,MaxNumberOfDevices)
#
# Create Matrix based on circuit size. We do not implement strict
```

Modified Nodal Analysis. We keep instead all currents
    # but keep referring to the voltages as absolute voltages. We
believe this will make the operation clearer to the user.
    #
    NumberOfNodes=len(Nodes)
    MatrixSize=DeviceCount+len(Nodes)
    Jacobian=[[0 for i in range(MatrixSize)] for j in range(MatrixSize)]
    Jac_inv=[[0 for i in range(MatrixSize)] for j in range(MatrixSize)]
    Spare=[[0 for i in range(MatrixSize)] for j in range(MatrixSize)]
    STA_matrix=[[0 for i in range(MatrixSize)] for j in range(Matrix
Size)]
    STA_rhs=[0 for i in range(MatrixSize)]
    STA_nonlinear=[0 for i in range(MatrixSize)]
    f=[0 for i in range(MatrixSize)]
    #
    # Create sim parameters
    #
    deltaT=1e-12
    sol=[0 for i in range(MatrixSize)]
    solm1=[0 for i in range(MatrixSize)]
    if len(ICdict)>0:
        for i in range(len(ICdict)):
            for j in range(NumberOfNodes):
                if Nodes[j]==ICdict[i]['NodeName']:
                    sol[j]=ICdict[i]['Value']
                    print('Setting ',Nodes[j],' to ',sol[j])
    #
    # Loop through all devices and create jacobian and initial f(v)
entries according to signature
    #
    for i in range(DeviceCount):
        if DevType[i] != 'transistor':
            STA_matrix[NumberOfNodes+i][NumberOfNodes+i]=-DevValue[i]
            if DevNode1[i] != '0' :
            STA_matrix[NumberOfNodes+i][Nodes.index(DevNode1[i])]=1
            STA_matrix[Nodes.index(DevNode1[i])][NumberOfNodes+i]=1
            if DevNode2[i] != '0' :
            STA_matrix[NumberOfNodes+i][Nodes.index(DevNode2[i])]=-1
            STA_matrix[Nodes.index(DevNode2[i])][NumberOfNodes+i]=-1
            if DevType[i]=='capacitor':
                # Do nothing
                STA_rhs[NumberOfNodes]=STA_rhs[NumberOfNodes]
            if DevType[i]=='inductor':
                # For DC we treat this as a voltage source with V=0
                STA_matrix[NumberOfNodes+i][NumberOfNodes+i]=0

```
                STA_rhs[NumberOfNodes+i]=0
        if DevType[i]=='VoltSource':
                STA_matrix[NumberOfNodes+i][NumberOfNodes+i]=0
                STA_rhs[NumberOfNodes+i]=DevValue[i]
        if DevType[i]=='CurrentSource':
                if DevNode1[i] != '0' :
        STA_matrix[NumberOfNodes+i][Nodes.index(DevNode1[i])]=0
        STA_matrix[Nodes.index(DevNode1[i])][NumberOfNodes+i]=0
                if DevNode2[i] != '0' :
        STA_matrix[NumberOfNodes+i][Nodes.index(DevNode2[i])]=0
        STA_matrix[Nodes.index(DevNode2[i])][NumberOfNodes+i]=0
                STA_matrix[NumberOfNodes+i][NumberOfNodes+i]=1
                STA_rhs[NumberOfNodes+i]=DevValue[i]
                if DevNode1[i] != '0' and DevNode2[i]!='0':
        STA_matrix[Nodes.index(DevNode1[i])][NumberOfNodes+i]=1
        STA_matrix[Nodes.index(DevNode2[i])][NumberOfNodes+i]=-1
                elif DevNode2[i] != '0' :
        STA_matrix[Nodes.index(DevNode2[i])][NumberOfNodes+i]=-1
                elif DevNode1[i] != '0' :
        STA_matrix[Nodes.index(DevNode1[i])][NumberOfNodes+i]=1
    if DevType[i]=='transistor':
        if DevModel[i][0]=='p':
                PFET=1
        else:
                PFET=1
        STA_matrix[NumberOfNodes+i][NumberOfNodes+i]=DevValue[i]
        if DevNode1[i] != '0' :
        STA_matrix[NumberOfNodes+i][Nodes.index(DevNode1[i])]=0
        STA_matrix[Nodes.index(DevNode1[i])][NumberOfNodes+i]=1
        if DevNode3[i] != '0' :
        STA_matrix[NumberOfNodes+i][Nodes.index(DevNode3[i])]=0
        STA_matrix[Nodes.index(DevNode3[i])][NumberOfNodes+i]=-1
        if DevNode1[i] != '0' :
            if DevNode2[i] != '0' and DevNode3[i] != '0':
                STA_nonlinear[NumberOfNodes+i]=(sol[Nodes.index
(DevNode2[i])]-sol[Nodes.index(DevNode3[i])])**2
            if DevNode2[i] == '0':
                STA_nonlinear[Nodes.index(DevNode1[i])]=sol[Nodes.
index(DevNode3[i])]**2
                STA_nonlinear[NumberOfNodes+1]=(sol[Nodes.index
(DevNode3[i])])**2
            if DevNode3[i] == '0':
                STA_nonlinear[Nodes.index(DevNode1[i])]=sol[Nodes.
index(DevNode2[i])]**2
        if DevNode3[i] != '0' :
```

```
                if DevNode2[i] != '0':
                    STA_nonlinear[NumberOfNodes+i]=(sol[Nodes.index
(DevNode2[i])]-sol[Nodes.index(DevNode3[i])])**2
                else:
                    STA_nonlinear[Nodes.index(DevNode3[i])]=-sol[
Nodes.index(DevNode3[i])]**2
    #
    f=numpy.matmul(STA_matrix,sol)-STA_rhs+STA_nonlinear
    #
    #Loop through iteration points
    #
    SimTime=0
    NewIter=15
    val=[0 for i in range(NewIter)]
    for Newtoniter in range(NewIter):
        for i in range(MatrixSize):
            STA_nonlinear[i]=0
        for i in range(DeviceCount):
            if DevType[i]!='transistor':
                if DevType[i]=='capacitor':
                    STA_rhs[NumberOfNodes+i]=STA_rhs[NumberOfNodes+i]
                if DevType[i]=='inductor':
                    STA_rhs[NumberOfNodes+i]=0
                if DevType[i]=='VoltSource':
                    if DevLabel[i]=='vinp':
                        STA_rhs[NumberOfNodes+i]=2*math.sin(2*math.
pi*1e9*SimTime)+1
                    if DevLabel[i]=='vinn':
                        STA_rhs[NumberOfNodes+i]=-2*math.sin(2*math.
pi*1e9*SimTime)+1
            if DevType[i]=='transistor':
                if DevNode1[i] != '0' :
                    if DevNode2[i] != '0' and DevNode3[i] != '0':
                        STA_nonlinear[NumberOfNodes+i]=(sol[Nodes.
index(DevNode2[i])]-sol[Nodes.index(DevNode3[i])])**2
                    if DevNode2[i] == '0':
                        STA_nonlinear[NumberOfNodes+i]=sol[Nodes.
index(DevNode3[i])]^2
                    if DevNode3[i] == '0':
                        STA_nonlinear[NumberOfNodes+i]=sol[Nodes.
index(DevNode2[i])]^2
        f=numpy.matmul(STA_matrix,sol)-STA_rhs+STA_nonlinear
    #
    # Now we need the Jacobian, the transistors look like VCCS with
a specific gain = 2 K (Vg-Vs) in our case
```

```
#
      for i in range(MatrixSize):
          for j in range(MatrixSize):
              Jacobian[i][j]=STA_matrix[i][j]
      for i in range(DeviceCount):
          if DevType[i]=='transistor':
              Jacobian[NumberOfNodes+i][NumberOfNodes+i]=DevValue
[i] # due to derfivative leading to double gain
              if DevNode1[i] != '0' :
                  if DevNode2[i] != '0' and DevNode3[i] != '0':
                      Jacobian[NumberOfNodes+i][Nodes.index(DevNo
de2[i])]=2*PFET*(sol[Nodes.index(DevNode2[i])]-sol[Nodes.
index(DevNode3[i])])
                      Jacobian[NumberOfNodes+i][Nodes.index(Dev
Node3[i])]=-2*PFET*(sol[Nodes.index(DevNode2[i])]-sol[Nodes.
index(DevNode3[i])])
                      Jacobian[Nodes.index(DevNode1[i])][Number
OfNodes+i]=1
                      Jacobian[Nodes.index(DevNode3[i])][Number
OfNodes+i]=-1
                  elif DevNode2[i] == '0':
                      Jacobian[NumberOfNodes+i][Nodes.index(Dev
Node3[i])]=-1
                      Jacobian[Nodes.index(DevNode1[i])][Number
OfNodes+i]=1
                      Jacobian[Nodes.index(DevNode3[i])][Number
OfNodes+i]=-1
                  elif DevNode3[i] == '0':
                      Jacobian[NumberOfNodes+i][Nodes.index(Dev
Node2[i])]=1
                      Jacobian[Nodes.index(DevNode1[i])][Number
OfNodes+i]=1
                      Jacobian[Nodes.index(DevNode3[i])][Number
OfNodes+i]=-1
      sol=sol-numpy.matmul(numpy.linalg.inv(Jacobian),f)
      Jac_inv=numpy.linalg.inv(Jacobian)
      val[Newtoniter]=sol[2]#max(f)#sol[3]

  plt.plot(val)
  #f=open("../pictures/DC_PMOS_NewtonIter.csv","w+")
  #f.write("time val\n")
  #for i in range(15):
  #    f.write("%g %g\n" % (i,val[i]) )
  #f.close()
```

## 5.6.6   Code 5.6

```python
#!/usr/bin/env python3
# -*- coding: utf-8 -*-
"""
Created on Thu Feb 28 22:33:04 2019

@author: mikael
"""
import numpy as np
import scipy.linalg as slin
import matplotlib.pyplot as plt
import math
import sys
import analogdef as ana


MaxNumberOfDevices=100
DevType=[0*i for i in range(MaxNumberOfDevices)]
DevLabel=[0*i for i in range(MaxNumberOfDevices)]
DevNode1=[0*i for i in range(MaxNumberOfDevices)]
DevNode2=[0*i for i in range(MaxNumberOfDevices)]
DevNode3=[0*i for i in range(MaxNumberOfDevices)]
DevModel=[0*i for i in range(MaxNumberOfDevices)]
DevValue=[0*i for i in range(MaxNumberOfDevices)]
Nodes=[]
vkmax=0 # This is for GlobalTruncation Criterion
#
#
# Foet netlist_osc.txt: If you take too large steps initially,
the solver totally screws up and takes ridiculously small time
   # steps and massive voltage solutions!! Totally unreal!
MaxSimTime=1.001e-7 and deltaT=MaxSimTime/10000, nch1 model
#
# Read modelfile
#
modeldict=ana.readmodelfile('models.txt')
ICdict={}
Plotdict={}
Printdict={}
Optdict={}
Optdict['reltol']=1e-2
Optdict['iabstol']=1e-7
Optdict['vabstol']=1e-2
Optdict['lteratio']=2
```

```
Optdict['MaxTimeStep']=1e-11
Optdict['FixedTimeStep']='False'
Optdict['GlobalTruncation']='True'
Optdict['deltaT']=3e-13
Optdict['MaxSimulationIterations']=200000
Optdict['MaxSimTime']=1e-8
Optdict['ThreeLevelStep']='True'
Optdict['method']='trap'
#
# Read the netlist
#
  DeviceCount=ana.readnetlist('netlist_inv_string.txt',modeldic-
t,ICdict,Plotdict,Printdict,Optdict,DevType,DevValue,DevLabel,Dev
Node1,DevNode2,DevNode3,DevModel,Nodes,MaxNumberOfDevices)
#
# Create Matrix based on circuit size. We do not implement strict
Modified Nodal Analysis. We keep instead all currents
# but keep referring to the voltages as absolute voltages. We
believe this will make the operation clearer to the user.
#
reltol=Optdict['reltol']
iabstol=Optdict['iabstol']
vabstol=Optdict['vabstol']
lteratio=Optdict['lteratio']
MaxTimeStep=Optdict['MaxTimeStep']
FixedTimeStep=(Optdict['FixedTimeStep']=='True')
GlobalTruncation=(Optdict['GlobalTruncation']=='True')
deltaT=Optdict['deltaT']
MaxSimulationIterations=int(Optdict['MaxSimulationIterations'])
ThreeLevelStep=(Optdict['ThreeLevelStep']=='True')
MaxSimTime=Optdict['MaxSimTime']
PointLocal=not GlobalTruncation
method=Optdict['method']
#
# Create Matrix based on circuit size
#
NumberOfNodes=len(Nodes)
NumberOfCurrents=DeviceCount
MatrixSize=DeviceCount+len(Nodes)
Jacobian=[[0 for i in range(MatrixSize)] for j in range(MatrixSize)]
Jac_inv=[[0 for i in range(MatrixSize)] for j in range(MatrixSize)]
Spare=[[0 for i in range(MatrixSize)] for j in range(MatrixSize)]
STA_matrix=[[0 for i in range(MatrixSize)] for j in range(Matrix
Size)]
STA_rhs=[0 for i in range(MatrixSize)]
```

```
STA_nonlinear=[0 for i in range(MatrixSize)]
f=[0 for i in range(MatrixSize)]
SetupDict={}
SetupDict['NumberOfNodes']=NumberOfNodes
SetupDict['NumberOfCurrents']=NumberOfCurrents
SetupDict['DeviceCount']=DeviceCount
SetupDict['Nodes']=Nodes
SetupDict['DevNode1']=DevNode1
SetupDict['DevNode2']=DevNode2
SetupDict['DevNode3']=DevNode3
SetupDict['DevValue']=DevValue
SetupDict['DevType']=DevType
SetupDict['DevModel']=DevModel
SetupDict['MatrixSize']=MatrixSize
SetupDict['Jacobian']=Jacobian
SetupDict['STA_matrix']=STA_matrix
SetupDict['STA_rhs']=STA_rhs
SetupDict['STA_nonlinear']=STA_nonlinear
SetupDict['FixedTimeStep']=FixedTimeStep
SetupDict['method']=method
SetupDict['GlobalTruncation']=GlobalTruncation
SetupDict['PointLocal']=PointLocal
SetupDict['vkmax']=vkmax
SetupDict['Vthermal']=1.38e-23*300/1.602e-19
SetupDict['reltol']=reltol
SetupDict['iabstol']=iabstol
SetupDict['vabstol']=vabstol
SetupDict['lteratio']=lteratio
SetupDict['MaxTimeStep']=MaxTimeStep
#
# Create sim environment
#
sol=[0 for i in range(MatrixSize)]
solm1=[0 for i in range(MatrixSize)]
solm2=[0 for i in range(MatrixSize)]
soltemp=[0 for i in range(MatrixSize)]
SimDict={}
SimDict['deltaT']=deltaT
SimDict['ThreeLevelStep']=ThreeLevelStep
SimDict['sol']=sol
SimDict['solm1']=solm1
SimDict['solm2']=solm2
SimDict['soltemp']=soltemp
SimDict['f']=f
#
```

```
# Initial conditions
#
if len(ICdict)>0:
    for i in range(len(ICdict)):
        for j in range(NumberOfNodes):
            if Nodes[j]==ICdict[i]['NodeName']:
                sol[j]=ICdict[i]['Value']
                solm1[j]=ICdict[i]['Value']
                solm2[j]=ICdict[i]['Value']
                print('Setting ',Nodes[j],' to ',sol[j])
#
# Loop through all devices and create jacobian and initial f(v)
entries according to signature
#
ana.build_SysEqns(SetupDict, SimDict, modeldict)
#
f=np.matmul(STA_matrix,sol)-STA_rhs+STA_nonlinear
#
# Initialize Variables
#
val=[[0 for i in range(MaxSimulationIterations)] for j in range
(MatrixSize)]
vin=[0 for i in range(MaxSimulationIterations)]
timeVector=[0 for i in range(MaxSimulationIterations)]
TotalIterations=0
iteration=0
SimTime=0
NewtonIter=0
NewtonConverged=False
LTEIter=0
Converged=False
for i in range(MatrixSize):
    soltemp[i]=sol[i]
#
#Loop through time points
#
while SimTime<MaxSimTime and iteration<MaxSimulationIterations:
    if iteration%100==0:
        print("Iter=",iteration,NewtonIter,LTEIter,deltaT,vkmax
,SimTime)
    SimTime=SimTime+deltaT
    NewtonIter=0
    LTEIter=0
    ResidueConverged=False
    UpdateConverged=False
```

```
        NewtonConverged=False
        Converged=False
        while (not NewtonConverged) and NewtonIter<50:
            NewtonIter=NewtonIter+1
            for i in range(MatrixSize):
                STA_nonlinear[i]=0
                    ana.update_SysEqns(SimTime,  SetupDict,  SimDict,
modeldict)
                    SimDict['f']=f=np.matmul(STA_matrix,soltemp)-STA_
rhs+STA_nonlinear
                    ResidueConverged=ana.DidResidueConverge(SetupDict,
SimDict)
  #
  # Now we need the Jacobian
  #
            for i in range(MatrixSize):
                for j in range(MatrixSize):
                    Jacobian[i][j]=STA_matrix[i][j]
                ana.build_Jacobian(SetupDict, SimDict, modeldict)
            UpdateConverged=ana.DidUpdateConverge(SetupDict, SimDict)
                SolutionCorrection=np.matmul(np.linalg.inv(Jacobian),f)
                for i in range(MatrixSize):
                    soltemp[i]=soltemp[i]-SolutionCorrection[i]
  #
            NewtonConverged=ResidueConverged and UpdateConverged
  #
  # Verify if LTE is within set accuracy
  #
            LTEConverged,  MaxLTERatio=ana.DidLTEConverge(SetupDict,
SimDict, iteration, LTEIter, NewtonConverged, timeVector, SimTime,
SolutionCorrection)
        if not LTEConverged:
  #
  #      if LTE did not converge we need to reset the Newton
counter and the temp solution back to the previously accepted time
step
  #
            NewtonIter=0
            for i in range(MatrixSize):
                soltemp[i]=sol[i]
  #
  # Update Time Step
  #
                    deltaT,   iteration,   SimTime,   Converged=ana.
UpdateTimeStep(SetupDict, SimDict, LTEConverged, NewtonConverged,
```

```
val, iteration, NewtonIter, MaxLTERatio, timeVector, SimTime)
    TotalIterations=TotalIterations+NewtonIter

    SimDict['deltaT']=deltaT
    if Converged:
        for i in range(MatrixSize):
            sol[i]=soltemp[i]
        for node in range(NumberOfNodes):
            vkmax=max(vkmax,abs(sol[node]))
            SetupDict['vkmax']=vkmax
    if deltaT<1e-15:
        print('Warning: Timestep too short: ',deltaT)
        sys.exit(0)

reval=[[0 for i in range(iteration)] for j in range(MatrixSize)]
retime=[0 for i in range(iteration)]
logvalue=[0 for i in range(iteration)]
for i in range(iteration):
    for j in range(MatrixSize):
        reval[j][i]=val[j][i]
    retime[i]=timeVector[i]

ana.plotdata(Plotdict,NumberOfNodes,retime,reval,Nodes)
if len(Printdict)> 0:
    ana.printdata(Printdict,NumberOfNodes,retime,reval,Nodes)
print('TotalIterations ',TotalIterations)
```

### 5.6.7   Code 5.7

```python
#!/usr/bin/env python3
# -*- coding: utf-8 -*-
"""
Created on Thu Feb 28 22:33:04 2019

@author: mikael
"""
import numpy as np
import sys
import analogdef as ana

#
# Initial definitions
#
```

```
MaxNumberOfDevices=100
DevType=[0*i for i in range(MaxNumberOfDevices)]
DevLabel=[0*i for i in range(MaxNumberOfDevices)]
DevNode1=[0*i for i in range(MaxNumberOfDevices)]
DevNode2=[0*i for i in range(MaxNumberOfDevices)]
DevNode3=[0*i for i in range(MaxNumberOfDevices)]
DevModel=[0*i for i in range(MaxNumberOfDevices)]
DevValue=[0*i for i in range(MaxNumberOfDevices)]
Nodes=[]
vkmax=0

#
# Read modelfile
#
modeldict=ana.readmodelfile('models.txt')
ICdict={}
Plotdict={}
Printdict={}
Optionsdict={}
Optionsdict['reltol']=1e-2
Optionsdict['iabstol']=1e-7
Optionsdict['vabstol']=1e-2
Optionsdict['lteratio']=2
Optionsdict['deltaT']=1e-12
Optionsdict['NIterations']=200
Optionsdict['GlobalTruncation']=True
Optionsdict['method']='be'
#
# Read the netlist
#
DeviceCount=ana.readnetlist('netlist_tran_5p11.txt',modeldict,I
Cdict,Plotdict,Printdict,Optionsdict,DevType,DevValue,DevLabel,De
vNode1,DevNode2,DevNode3,DevModel,Nodes,MaxNumberOfDevices)
#
# Create Matrix based on circuit size. We do not implement strict
Modified Nodal Analysis. We keep instead all currents
# but keep referring to the voltages as absolute voltages. We
believe this will make the operation clearer to the user.
#
#
# Create Matrix based on circuit size
#
MatrixSize=DeviceCount+len(Nodes)
NumberOfNodes=len(Nodes)
NumberOfCurrents=DeviceCount
```

```
Jacobian=[[0 for i in range(MatrixSize)] for j in range(MatrixSize)]
Jac_inv=[[0 for i in range(MatrixSize)] for j in range(MatrixSize)]
Spare=[[0 for i in range(MatrixSize)] for j in range(MatrixSize)]
    STA_matrix=[[0  for  i  in  range(MatrixSize)]  for  j  in
range(MatrixSize)]
STA_rhs=[0 for i in range(MatrixSize)]
STA_nonlinear=[0 for i in range(MatrixSize)]
CapMatrix=[[0 for i in range(MatrixSize)] for j in range(MatrixSize)]
    FrachetMatrix=[[0  for  i  in  range(MatrixSize)]  for  j  in
range(MatrixSize)]
for i in range(MatrixSize):
    FrachetMatrix[i][i]=1
IdentityMatrix=[[0 for i in range(MatrixSize)] for j in range
(MatrixSize)]
for i in range(MatrixSize):
    IdentityMatrix[i][i]=1
f=[0 for i in range(MatrixSize)]
maxSol=[0 for i in range(10)]
#
deltaT=Optionsdict['deltaT']
NIterations=int(Optionsdict['NIterations'])
GlobalTruncation=Optionsdict['GlobalTruncation']
PointLocal=not GlobalTruncation
reltol=Optionsdict['reltol']
iabstol=Optionsdict['iabstol']
vabstol=Optionsdict['vabstol']
lteratio=Optionsdict['lteratio']
method=Optionsdict['method']
#
SetupDict={}
SetupDict['NumberOfNodes']=NumberOfNodes
SetupDict['NumberOfCurrents']=NumberOfCurrents
SetupDict['DeviceCount']=DeviceCount
SetupDict['Nodes']=Nodes
SetupDict['DevNode1']=DevNode1
SetupDict['DevNode2']=DevNode2
SetupDict['DevNode3']=DevNode3
SetupDict['DevValue']=DevValue
SetupDict['DevType']=DevType
SetupDict['DevModel']=DevModel
SetupDict['MatrixSize']=MatrixSize
SetupDict['Jacobian']=Jacobian
SetupDict['STA_matrix']=STA_matrix
SetupDict['STA_rhs']=STA_rhs
SetupDict['STA_nonlinear']=STA_nonlinear
```

```
SetupDict['method']=method
SetupDict['reltol']=reltol
SetupDict['iabstol']=iabstol
SetupDict['vabstol']=vabstol
SetupDict['lteratio']=lteratio
SetupDict['GlobalTruncation']=GlobalTruncation
SetupDict['PointLocal']=PointLocal
SetupDict['vkmax']=vkmax
#
# Create sim parameters
#
deltaT=Optionsdict['deltaT']
sol=np.zeros(MatrixSize)#[0.0 for i in range(MatrixSize)]
solm1=np.zeros(MatrixSize)#[0.0 for i in range(MatrixSize)]
soltemp=np.zeros(MatrixSize)#[0.0 for i in range(MatrixSize)]
solInit=np.zeros(MatrixSize)#[0.0 for i in range(MatrixSize)]
SimDict={}
SimDict['deltaT']=deltaT
SimDict['sol']=sol
SimDict['solm1']=solm1
SimDict['solInit']=solInit
SimDict['soltemp']=soltemp
SimDict['f']=f
#
# Loop through all devices and create jacobian and initial f(v)
entries according to signature
#
ana.build_SysEqns(SetupDict, SimDict, modeldict)
f=np.matmul(STA_matrix,sol)-STA_rhs+STA_nonlinear
#
#Loop through frequency points
#
 # We first calculated the STA matrix STA_rhs + STA_nonlinear
element
Npnts=int(Period/deltaT)
val=[[0 for i in range(Npnts)] for j in range(MatrixSize)]
vin=[0 for i in range(20)]
for ShootingIter in range(10):
    for i in range(MatrixSize):
        sol[i]=solInit[i]
    timepnts=[i*deltaT for i in range(Npnts)]
    for iter in range(Npnts):
        if iter%100==0:
            print("Iter=",iter)
        SimTime=iter*deltaT
```

```
              for i in range(MatrixSize):
                  soltemp[i]=sol[i]
              NewIter=500
              ResidueConverged=False
              UpdateConverged=False
              NewtonConverged=False
              NewtonIter=0
    #
    #     for Newtoniter in range(NewIter):
    #
              while not NewtonConverged and NewtonIter<50:
                  NewtonIter=NewtonIter+1

                  for i in range(MatrixSize):
                      STA_nonlinear[i]=0
                      ana.update_SysEqns(SimTime, SetupDict, SimDict,
    modeldict)
                          SimDict['f']=f=np.matmul(STA_matrix,soltemp)-STA_
    rhs+STA_nonlinear

                      ResidueConverged=ana.DidResidueConverge(SetupDict,
    SimDict)
        #
        # Now we need the Jacobian
        #
                      for i in range(MatrixSize):
                          for j in range(MatrixSize):
                              Jacobian[i][j]=STA_matrix[i][j]
                      SetupDict['Jacobian']=Jacobian
                      ana.build_Jacobian(SetupDict, SimDict, modeldict)
                          UpdateConverged=ana.DidUpdateConverge(SetupDict,
    SimDict)
                      Jacobian=SetupDict['Jacobian']
                  SolutionCorrection=np.matmul(np.linalg.inv(Jacobian),f)
                      for i in range(MatrixSize):
                          soltemp[i]=soltemp[i]-SolutionCorrection[i]

                  NewtonConverged=ResidueConverged and UpdateConverged
                  if NewtonConverged:
                      for i in range(MatrixSize):
                          solm1[i]=sol[i]
                      for node in range(NumberOfNodes):
                          vkmax=max(vkmax,abs(sol[node]))
                      SetupDict['vkmax']=vkmax
```

```
f=np.matmul(STA_matrix,soltemp)-STA_rhs+STA_nonlinear
                for i in range(MatrixSize):
                    sol[i]=soltemp[i]
                    val[i][iter]=sol[i]
                #
                #Calculate new FrachetMatrix
                #
                for i in range(DeviceCount):
                    if DevType[i]=='capacitor':
                        CapMatrix[Nodes.index(DevNode1[i])][Nodes.
index(DevNode2[i])]=DevValue[i]
                        CapMatrix[Nodes.index(DevNode2[i])][Nodes.
index(DevNode1[i])]=DevValue[i]
                    elif DevType[i]=='inductor':
                        print('Error: this shooting code only imple-
ments capacitors as dynamic elements\n')
                        sys.exit(1)
                    FrachetMatrix=np.matmul(np.linalg.inv(Jacobian),np.
matmul(CapMatrix,FrachetMatrix))/deltaT
            else:
                print('Newtoniteration did not converge.\nexiting
...\n')
                sys.exit(0)
        #
        # Find out maximum difference in initial value
        #
        maxSol[ShootingIter]=max(abs(sol-solInit))
        #
        # Calculate new initial value based on previous step result
        #
                        solInit=solInit+np.matmul(np.linalg.
inv(IdentityMatrix-FrachetMatrix),(sol-solInit))
    ana.plotdata(Plotdict,NumberOfNodes,timepnts,val,Nodes)
    #
    if len(Printdict)> 0:
        ana.printdata(Printdict,NumberOfNodes,timepnts,val,Nodes)
    #
```

## 5.6.8   Code 5.8

```
#!/usr/bin/env python3
# -*- coding: utf-8 -*-
"""
```

```
Created on Thu Feb 28 22:33:04 2019

@author: mikael
"""
import numpy as np
from scipy.fftpack import fft, ifft
import matplotlib.pyplot as plt
import math
import analogdef as ana

#
# Function definitions
#
MaxNumberOfDevices=100
DevType=[0*i for i in range(MaxNumberOfDevices)]
DevLabel=[0*i for i in range(MaxNumberOfDevices)]
DevNode1=[0*i for i in range(MaxNumberOfDevices)]
DevNode2=[0*i for i in range(MaxNumberOfDevices)]
DevNode3=[0*i for i in range(MaxNumberOfDevices)]
DevModel=[0*i for i in range(MaxNumberOfDevices)]
DevValue=[0*i for i in range(MaxNumberOfDevices)]
Nodes=[]
#
# Read modelfile
#
modeldict=ana.readmodelfile('models.txt')
ICdict={}
Plotdict={}
Printdict={}
Optionsdict={}
SetupDict={}
Optionsdict['NHarmonics']=32
Optionsdict['Period']=1e-9
Optionsdict['PAC']='False'
Optionsdict['MaxNewtonIterations']=15
Optionsdict['iabstol']=1e-7
#
# Read the netlist
#netlist_mixer_hb_5p14.txt
    DeviceCount=ana.readnetlist('netlist_tran_5p6.txt',modeldict,IC
dict,Plotdict,Printdict,Optionsdict,DevType,DevValue,DevLabel,Dev
Node1,DevNode2,DevNode3,DevModel,Nodes,MaxNumberOfDevices)
    #
    # Create Matrix based on circuit size. We do not implement strict
Modified Nodal Analysis. We keep instead all currents
```

```
  # but keep referring to the voltages as absolute voltages. We
believe this will make the operation clearer to the user.
  #
  #
  # Create Matrix based on circuit size
  #
  NHarmonics=Optionsdict['NHarmonics']
  Period=Optionsdict['Period']
  run_PAC=(Optionsdict['PAC']=='True')
  NSamples=2*(NHarmonics-1)
  TotalHarmonics=2*NHarmonics-1
  NumberOfNodes=len(Nodes)
  MatrixSize=(DeviceCount+len(Nodes))*TotalHarmonics
 Jacobian=[[0 for i in range(MatrixSize)] for j in range(MatrixSize)]
 Jac_inv=[[0 for i in range(MatrixSize)] for j in range(MatrixSize)]
 Spare=[[0 for i in range(MatrixSize)] for j in range(MatrixSize)]
 STA_matrix=[[0 for i in range(MatrixSize)] for j in range(Matrix
Size)]
  STA_rhs=[0 for i in range(MatrixSize)]
  STA_nonlinear=[0 for i in range(MatrixSize)]
  f=[0 for i in range(MatrixSize)]
    Template=[[0  for  i  in  range(TotalHarmonics)]  for  j  in
range(TotalHarmonics)]

  Jacobian_Offset=int(TotalHarmonics/2)
  omegak=[0 for i in range(TotalHarmonics)]
  HarmonicsList=[0 for i in range(TotalHarmonics)]
  Samples=[i*Period/NSamples for i in range(NSamples)]
  #
  # Create sim parameters
  #
  SetupDict['NumberOfNodes']=NumberOfNodes
  SetupDict['DeviceCount']=DeviceCount
  SetupDict['DevValue']=DevValue
  SetupDict['TotalHarmonics']=TotalHarmonics
  SetupDict['Jacobian_Offset']=Jacobian_Offset
  SetupDict['NSamples']=NSamples
  SetupDict['DevNode1']=DevNode1
  SetupDict['DevNode2']=DevNode2
  SetupDict['DevNode3']=DevNode3
  SetupDict['DevType']=DevType
  SetupDict['DevLabel']=DevLabel
  SetupDict['DevModel']=DevModel
  SetupDict['Nodes']=Nodes
  SetupDict['STA_matrix']=STA_matrix
```

```
SetupDict['Jacobian']=Jacobian
SetupDict['STA_rhs']=STA_rhs
SetupDict['STA_nonlinear']=STA_nonlinear
SetupDict['omegak']=omegak
iabstol=Optionsdict['iabstol']

sol=np.zeros(MatrixSize)+1j*np.zeros(MatrixSize)
SolutionCorrection=np.zeros(MatrixSize)+1j*np.zeros(MatrixSize)
TransistorOutputTime=[0 for i in range(NSamples)]
TransistorOutputTimeDerivative=[0 for i in range(TotalHarmonics)]
TransistorOutputFreq=[0 for i in range(TotalHarmonics)]
Jlkm=[0 for i in range(TotalHarmonics)]
Jlko=[0 for i in range(TotalHarmonics)]
Vg=[0 for i in range(TotalHarmonics)]
Vs=[0 for i in range(TotalHarmonics)]
Vd=[0 for i in range(TotalHarmonics)]
VgTime=[0 for i in range(NSamples)]
VsTime=[0 for i in range(NSamples)]
VdTime=[0 for i in range(NSamples)]
gm=[0 for i in range(TotalHarmonics)]
Vp=[0 for i in range(TotalHarmonics)]
Vn=[0 for i in range(TotalHarmonics)]
VpTime=[0 for i in range(NSamples)]
VnTime=[0 for i in range(NSamples)]
IOscFilterSpec=[0 for i in range(TotalHarmonics)]
IOscFilter=[0 for i in range(NSamples)]
#
SimDict={}
SimDict['sol']=sol
#
# Loop through all devices and create jacobian and initial f(v)
entries according to signature
#
for i in range(2*NHarmonics-1):
    HarmonicsList[i]=i
for i in range(TotalHarmonics):
    omegak[i]=(1-NHarmonics+i)*2*math.pi/Period

ana.build_SysEqns_HB(SetupDict, SimDict, modeldict)


f=np.matmul(STA_matrix,sol)-STA_rhs+STA_nonlinear
#
#Loop through frequency points
#
```

```
    # We first calculated the STA matrix STA_rhs + STA_nonlinear
element
    TimePnts=[i*Period/NSamples for i in range(NSamples)]
    NewIter=int(Optionsdict['MaxNewtonIterations'])
    Newtoniter=0
    while Newtoniter < NewIter and abs(max(f)) > iabstol:
        print('NewtonIteration :',Newtoniter,abs(max(f)))
        #
        # Update the nonlinear term column matrix
        #
        for i in range(MatrixSize):
            STA_nonlinear[i]=0
        ana.update_SysEqns_HB(SetupDict, SimDict, modeldict)
        f=np.matmul(STA_matrix,sol)-STA_rhs+STA_nonlinear

    #
    # Now we need the Jacobian
    #
        for i in range(MatrixSize):
            for j in range(MatrixSize):
                Jacobian[i][j]=STA_matrix[i][j]
        ana.build_Jacobian_HB(SetupDict, SimDict, modeldict)

        SolutionCorrection=np.matmul(np.linalg.inv(Jacobian),f)
        for i in range(MatrixSize):
            sol[i]=sol[i]-SolutionCorrection[i]
        Newtoniter=Newtoniter+1

    for j in range(TotalHarmonics):
        Vn[j]=sol[Nodes.index('outp')*TotalHarmonics+j]
    #    Vp[j]=sol[Nodes.index('n1')*TotalHarmonics+j]
    VnTime=ana.idft(Vn,TotalHarmonics)
    #VpTime=ana.idft(Vp,TotalHarmonics)
    #plt.plot(VpTime)
    plt.plot(TimePnts,VnTime)

    if run_PAC:
        STA_rhs=[0 for i in range(MatrixSize)]
        val=[[0 for i in range(100)] for j in range(4)]
        for iter in range(100):
            omega=iter*1e6*2*3.14159265
            print('Frequency sweep:',iter*1e6)
            for i in range(DeviceCount):
                for row in range(TotalHarmonics):
                    if DevType[i]=='capacitor':
```

```
                    if DevNode1[i] != '0' :
                        Jacobian[(NumberOfNodes+i)*TotalHarmoni
cs+row][Nodes.index(DevNode1[i])*TotalHarmonics+row]=1j*(omegak[r
ow]+(np.sign(omegak[row])+(omegak[row]==0))*omega)*DevValue[i]
                    if DevNode2[i] != '0' :
                        Jacobian[(NumberOfNodes+i)*TotalHarmoni
cs+row][Nodes.index(DevNode2[i])*TotalHarmonics+row]=-
1j*(omegak[row]+(np.sign(omegak[row])+(omegak[row]==0))*omega)*De
vValue[i]
                    if DevType[i]=='inductor':
                        Jacobian[(NumberOfNodes+i)*TotalHarmonics+
row][(NumberOfNodes+i)*TotalHarmonics+row]=-1j*(omegak[row]+(np.
sign(omegak[row])+(omegak[row]==0))*omega)*DevValue[i]
                    if DevType[i]=='CurrentSource':
                        if DevLabel[i]=='i1':
                            STA_rhs[(NumberOfNodes+i)*TotalHarmonic
s+row]=1*(row==Jacobian_Offset)
                        else:
                            STA_rhs[(NumberOfNodes+i)*TotalHarmonic
s+row]=-(row==Jacobian_Offset)
                sol=np.matmul(np.linalg.inv(Jacobian),STA_rhs)
                val[0][iter]=abs(sol[6*TotalHarmonics+Jacobian_Offset])
                val[1][iter]=20*math.log10(abs(sol[6*TotalHarmonics+Jac
obian_Offset+1]))

val[2][iter]=abs(sol[6*TotalHarmonics+Jacobian_Offset+2])

val[3][iter]=abs(sol[6*TotalHarmonics+Jacobian_Offset+3])
        plt.plot(val[1])
```

## 5.6.9 Code 5.9

```
#!/usr/bin/env python3
# -*- coding: utf-8 -*-
"""
Created on Thu Feb 28 22:33:04 2019

@author: mikael
"""
import numpy as np
from scipy.fftpack import fft, ifft
import matplotlib.pyplot as plt
import math
```

```
import analogdef as ana

#
# Function definitions
#
MaxNumberOfDevices=100
DevType=[0*i for i in range(MaxNumberOfDevices)]
DevLabel=[0*i for i in range(MaxNumberOfDevices)]
DevNode1=[0*i for i in range(MaxNumberOfDevices)]
DevNode2=[0*i for i in range(MaxNumberOfDevices)]
DevNode3=[0*i for i in range(MaxNumberOfDevices)]
DevModel=[0*i for i in range(MaxNumberOfDevices)]
DevValue=[0*i for i in range(MaxNumberOfDevices)]
Nodes=[]
#
# Read modelfile
#
modeldict=ana.readmodelfile('models.txt')
ICdict={}
Plotdict={}
Printdict={}
Optionsdict={}
SetupDict={}
Optionsdict['NHarmonics']=8
Optionsdict['Period']=1/5032661878.243104
Optionsdict['PNOISE']='False'
Optionsdict['iabstol']=1e-11
#
# Read the netlist
#
DeviceCount=ana.readnetlist('netlist_osc_hb_pn_5p16.txt',modeld
ict,ICdict,Plotdict,Printdict,Optionsdict,DevType,DevValue,DevLab
el,DevNode1,DevNode2,DevNode3,DevModel,Nodes,MaxNumberOfDevices)
#
# Create Matrix based on circuit size. We do not implement strict
Modified Nodal Analysis. We keep instead all currents
# but keep referring to the voltages as absolute voltages. We
believe this will make the operation clearer to the user.
#
#
# Create Matrix based on circuit size
#
NHarmonics=int(Optionsdict['NHarmonics'])
Period=Optionsdict['Period']
run_PNOISE=(Optionsdict['PNOISE']=='True')
```

```
NSamples=2*(NHarmonics-1)
TotalHarmonics=2*NHarmonics-1
NumberOfNodes=len(Nodes)
MatrixSize=(DeviceCount+len(Nodes))*TotalHarmonics
Jacobian=[[0 for i in range(MatrixSize)] for j in range(MatrixSize)]
Jac_inv=[[0 for i in range(MatrixSize)] for j in range(MatrixSize)]
Spare=[[0 for i in range(MatrixSize)] for j in range(MatrixSize)]
   STA_matrix=[[0   for   i   in   range(MatrixSize)]   for   j   in
range(MatrixSize)]
STA_rhs=[0 for i in range(MatrixSize)]
STA_nonlinear=[0 for i in range(MatrixSize)]
f=np.zeros(MatrixSize)+1j*np.zeros(MatrixSize)
   Template=[[0   for   i   in   range(TotalHarmonics)]   for   j   in
range(TotalHarmonics)]

Jacobian_Offset=int(TotalHarmonics/2)
omegak=[0 for i in range(TotalHarmonics)]
HarmonicsList=[0 for i in range(TotalHarmonics)]
Samples=[i*Period/NSamples for i in range(NSamples)]
#
# Create sim parameters
#
SetupDict['NumberOfNodes']=NumberOfNodes
SetupDict['DeviceCount']=DeviceCount
SetupDict['DevValue']=DevValue
SetupDict['TotalHarmonics']=TotalHarmonics
SetupDict['Jacobian_Offset']=Jacobian_Offset
SetupDict['NSamples']=NSamples
SetupDict['DevNode1']=DevNode1
SetupDict['DevNode2']=DevNode2
SetupDict['DevNode3']=DevNode3
SetupDict['DevType']=DevType
SetupDict['DevLabel']=DevLabel
SetupDict['DevModel']=DevModel
SetupDict['Nodes']=Nodes
SetupDict['STA_matrix']=STA_matrix
SetupDict['Jacobian']=Jacobian
SetupDict['STA_rhs']=STA_rhs
SetupDict['STA_nonlinear']=STA_nonlinear
SetupDict['omegak']=omegak
iabstol=Optionsdict['iabstol']
#
sol=np.zeros(MatrixSize)+1j*np.zeros(MatrixSize)
SolutionCorrection=np.zeros(MatrixSize)+1j*np.zeros(MatrixSize)
TransistorOutputTime=[0 for i in range(NSamples)]
```

```
TransistorOutputTimeDerivative=[0 for i in range(TotalHarmonics)]
TransistorOutputFreq=[0 for i in range(TotalHarmonics)]
Jlkm=[0 for i in range(TotalHarmonics)]
Jlko=[0 for i in range(TotalHarmonics)]
Vg=[0 for i in range(TotalHarmonics)]
Vs=[0 for i in range(TotalHarmonics)]
Vd=[0 for i in range(TotalHarmonics)]
VgTime=[0 for i in range(NSamples)]
VsTime=[0 for i in range(NSamples)]
VdTime=[0 for i in range(NSamples)]
gm=[0 for i in range(TotalHarmonics)]
Vp=[0 for i in range(TotalHarmonics)]
Vn=[0 for i in range(TotalHarmonics)]
VpTime=[0 for i in range(NSamples)]
VnTime=[0 for i in range(NSamples)]
IOscFilterSpec=[0 for i in range(TotalHarmonics)]
IOscFilter=[0 for i in range(NSamples)]
#
SimDict={}
SimDict['sol']=sol
#
# Loop through all devices and create jacobian and initial f(v)
entries according to signature
#
for i in range(2*NHarmonics-1):
    HarmonicsList[i]=i
for i in range(TotalHarmonics):
    omegak[i]=(1-NHarmonics+i)*2*math.pi/Period

for i in range(DeviceCount):
    if DevType[i] == 'oscfilter':
        OscFilterIndex=i
    if(DevLabel[i] == 'vinp'):
        StimulusIndex=i
ana.build_SysEqns_HB(SetupDict, SimDict, modeldict)

#f=np.matmul(STA_matrix,sol)-STA_rhs+STA_nonlinear
#
#Loop through frequency points
#
# We first calculated the STA matrix STA_rhs + STA_nonlinear
element
for AmpIndex in range(1):
    STA_rhs[(NumberOfNodes+StimulusIndex)*TotalHarmonics+Jacob
ian_Offset+1]=.23577+float (AmpIndex)/100000
```

```
        STA_rhs[(NumberOfNodes+StimulusIndex)*TotalHarmonics+Jacob
ian_Offset-1]=.23577+float (AmpIndex)/100000
      NewIter=15
      Newtoniter=0
      f[0]=1j
      while Newtoniter < NewIter and max(abs(f)) > iabstol:
          print('NewtonIteration :',Newtoniter,max(abs(f)))
          #
          # Update the nonlinear term column matrix
          #
          for i in range(MatrixSize):
              STA_nonlinear[i]=0
          ana.update_SysEqns_HB(SetupDict, SimDict, modeldict)
          f=np.matmul(STA_matrix,sol)-STA_rhs+STA_nonlinear
      #
      # Now we need the Jacobian
      #
          for i in range(MatrixSize):
              for j in range(MatrixSize):
                  Jacobian[i][j]=STA_matrix[i][j]
          ana.build_Jacobian_HB(SetupDict, SimDict, modeldict)
          SolutionCorrection=np.matmul(np.linalg.inv(Jacobian),f)
          for i in range(MatrixSize):
              sol[i]=sol[i]-SolutionCorrection[i]

          Newtoniter=Newtoniter+1


      f=np.matmul(STA_matrix,sol)-STA_rhs+STA_nonlinear
      for j in range(TotalHarmonics):
          Vp[j]=sol[3*TotalHarmonics+j]
          Vn[j]=sol[4*TotalHarmonics+j]
          IOscFilterSpec[j]=sol[(NumberOfNodes+OscFilterIndex)*To
talHarmonics+j]
      VnTime=ana.idft(Vn,TotalHarmonics)
      VpTime=ana.idft(Vp,TotalHarmonics)
      IOscFilter=ana.idft(IOscFilterSpec,TotalHarmonics)
       print('rms ',np.real(np.sqrt(np.mean(IOscFilter**2)))),STA_
rhs[(NumberOfNodes+StimulusIndex)*TotalHarmonics+Jacob
ian_Offset+1])
   #plt.plot(IOscFilter)


   if run_PNOISE:
      #
      # Decouple the driving circuit from the system
```

```
        #
        Jacobian=SetupDict['Jacobian']
        Jacobian[(NumberOfNodes+OscFilterIndex)*TotalHarmonics+Jaco
bian_Offset-1][(NumberOfNodes+OscFilterIndex)*TotalHarmonics+Jacob
ian_Offset-1]=-1e18
        Jacobian[(NumberOfNodes+OscFilterIndex)*TotalHarmonics+Jaco
bian_Offset+1][(NumberOfNodes+OscFilterIndex)*TotalHarmonics+Jacob
ian_Offset+1]=-1e18
        #
        # We need to recalculate the Matrix due to the frequency terms
from the inductors+capacitors
        #
        STA_rhs=[0 for i in range(MatrixSize)]
        val=[[0 for i in range(100)] for j in range(4)]
        for iter in range(100):
            omega=(iter)*1e6*2*math.pi
            for i in range(DeviceCount):
                for row in range(TotalHarmonics):
                    if DevType[i]=='capacitor':
                        if DevNode1[i] != '0' :
                            Jacobian[(NumberOfNodes+i)*TotalHarmoni
cs+row][Nodes.index(DevNode1[i])*TotalHarmonics+row]=1j*(omegak[r
ow]+(np.sign(omegak[row])+(omegak[row]==0))*omega)*DevValue[i]
                        if DevNode2[i] != '0' :
                            Jacobian[(NumberOfNodes+i)*TotalHarmoni
cs+row][Nodes.index(DevNode2[i])*TotalHarmonics+row]=-
1j*(omegak[row]+(np.sign(omegak[row])+(omegak[row]==0))*omega)*De
vValue[i]
                    if DevType[i]=='inductor':
                        Jacobian[(NumberOfNodes+i)*TotalHarmonics+
row][(NumberOfNodes+i)*TotalHarmonics+row]=-1j*(omegak[row]+(np.
sign(omegak[row])+(omegak[row]==0))*omega)*DevValue[i]
                    if DevType[i]=='CurrentSource': # Adding current
source between transistor drain-source
                        STA_rhs[(NumberOfNodes+i)*TotalHarmonics+ro
w]=.5*(row==Jacobian_Offset+1)+.5*(row==Jacobian_Offset-1)
            sol=np.matmul(np.linalg.inv(Jacobian),STA_rhs)
            val[0][iter]=abs(sol[3*TotalHarmonics+Jacobian_Offset])

val[1][iter]=abs(sol[3*TotalHarmonics+Jacobian_Offset+1])

val[2][iter]=abs(sol[3*TotalHarmonics+Jacobian_Offset+2])
            val[3][iter]=20*math.log10(val[1][iter]) #abs(sol[2*Tota
lHarmonics+Jacobian_Offset+3])**2
        plt.plot(val[3])
```

## 5.7 Exercises

1. Play with codes and the netlists. When does it break? Why? If anything looks off, please go to www.fastictechniques.com for an updated code version and an interesting discussion.
2. Do the details of the mathematical implementation of a PMOS, and compare to the code implementation in Appendix A. Is the code done properly? Improve it!
3. Implement a charge conserverd nonlinear capacitor following the outline in Sect. 5.4.4. Verify by applying a voltage up and down ramp, and compute the charge both at the peak and at the end. Use a capacitor that varies like

$$C(u) = \frac{C_0}{1 + \dfrac{U}{U_0}}$$

4. A consequence of the fact that a capacitor's current is calculated through the numerical approximation of a time derivative is that accuracy of such wave forms is less. Any rapid rise in voltage across a capacitor can result in an erroneously large error in current. Imagine we have a nonlinear capacitor with voltage ramp. Simulate this with backward Euler, trap, and Gear2, and explain the difference in response.
5. Simulate a CMOS inverter stage with capacitors both to ground and across gate-drain, and based on hand calculations of the delay, show the simulator gives the appropriate response.
6. Simulate the five-transistor circuit in Fig. 5.48. Does the response look reasonable? What can be improved?
7. Redesign the bandgap in Fig. 5.8 to make its temperature flat at 27C.
8. Build a simulator combining an initial transient with a subsequent steady-state simulation.
9. Simulate the amplifier in Sect. 5.5.7 with an amplitude $A = 1V$, and examine the output harmonics. Compare to the harmonics of a square wave.
10. Implement a multitone Python code using both the shooting method and the harmonic balance method.
11. Implement an envelope analysis simulation algorithm by wrapping the harmonic balance routine developed in Sect. 5.5.2 inside a time step loop where the circuitry controlling the modulation is changing slowly. Try it on a simple netlist.
12. Implement a periodic steady-state solver where inductors are also included.
13. The periodic S-parameter, transfer function, and stability analyses are simple generalizations of the algorithms we discussed in Chap. 4. Implement them in Python, and compare to simple situations to ensure they work properly.
14. Implement a quasi-periodic Python code that can generate harmonics around certain main harmonic frequencies.

**Fig. 5.48** Simple five
transistor gain stage with

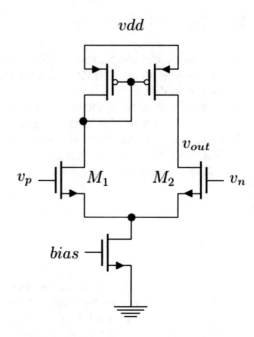

# References

1. Pedro, J., Root, D., Xu, J., & Nunes, L. (2018). *Nonlinear circuit simulation and modeling: Fundamentals for microwave design* (The Cambridge RF and Microwave Engineering Series). Cambridge: Cambridge University Press. https://doi.org/10.1017/9781316492963
2. Antognetti, P., & Massobrio, G. (2010). *Semiconductor device modeling with spice* (2nd ed.). India: McGraw Hill Education.
3. Kundert, K., White, J., & Sangiovanni-Vicentelli. (1990). *Steady-state methods for simulating analog and microwave circuits*. Norwell, MA: Kluwer Academic Publications.
4. Kundert, K. (1995). *The designers guide to spice and spectre*. Norwell, MA: Kluwer Academic Press.
5. Najm, F. N. (2010). *Circuit simulation*. Hobroken, NJ: Wiley.
6. Berry, R. D. (1971). An optimal ordering of electronic circuit equations for a sparse matrix solution. *IEEE Transactions on Circuit Theory, 18*, 40–50.
7. Calahan, D. A. (1972). *Computer-Aided Network Design* (revised ed.). New York, NY: McGraw-Hill.
8. Chua, L. O., & Lin, P.-M. (1975). *Computer-aided analysis of electronic circuits*. New York, NY: McGraw-Hill.
9. Gear, C. W. (1971). *Numerical initial value problems in ordinary differential equations*. Englewood Cliffs, NJ: Prentice-Hall.
10. Hachtel, G. D., Brayton, R. K., & Gustavson, F. G. (1971). The sparse tableau approach to network analysis and design. *IEEE Transactions on Circuit Theory, 18*, 101–113.
11. Ho, C.-W., Zein, A., Ruehli, A. E., & Brennan, P. A. (1975). The modified nodal approach to network analysis. *IEEE Transactions on Circuits and Systems, 22*, 504–509.
12. Milne, W. E. (1949). A note on the numerical integration of differential equations. *Journal of Researchof the National Bureau of Standards, 43*, 537–542.

13. Nagel L. W. (1975). *SPICE2: A computer program to simulate Semiconductor Circuits*. PhD thesis, University of Caliornia, Berkeley. Memorandum No ERL-M520.
14. Nagel, L. W., & Pederson, D. O. (1973). Simulation program with integrated circuit emphasis. In *Proceedings of the sixteenth midwest symposium on circuit theory*. Canada: Waterloo.
15. Ogrodzki, J. (1994). *Circuit simulation methods and algorithms*. Boca Raton, FL: CRC Press.
16. Vlach, J., & Singhai, K. (1994). *Computer methods for circuit analysis and design* (New York, NY, 2nd ed.). Van Nostrand Reinhold Co..
17. McCalla, W. J. (1988). *Fundamentals of computer-aided circuit simulation*. Norwell, MA: Kluwer Academic Publishers.
18. Ho, C. W., Zein, D. A., Ruehli, A. E., & Brennan, P. A. (1977). An algorithm for DC solutions in an experimental general purpose interactive circuit design program. *IEEE Transactions on Circuits and Systems, 24, 416, –422.*
19. Ruehli A. E., (Eds.). (1986). *Circuit analysis, simulation and design – Part I*, North-Holland, Amsterdam published as Volume 3 of *Advances in CAD for VLSI*.
20. Ruehli A. E., (Eds.). (1987). *Circuit analysis, simulation and design – Part 2*, North-Holland, Amsterdam published as Volume 3 of *Advances in CAD for VLSI*.
21. Vladimirescu, A. (1994). *The spice book*. New York, NY: Wiley.
22. Suarez, A. (2009). *Analysis and design of autonomous microwave circuits*. Hobroken, NJ: Wiley-IEEE Press.
23. Sahrling, M. (2019). *Fast techniques for integrated circuit design*. Cambridge: Cambridge University Press.
24. Gustafsson, K. (1988). *Stepsize control in ODE-solvers: Analysis and synthesis*, Thesis, Lund: Lund University Publications.

# Chapter 6
# Epilogue: Simulators in Practice

**Abstract** This book has so far looked at the basic principles behind circuit simulators. With the help of the provided codes, the reader has hopefully gained a better
understanding of what simulators do well and what they perhaps do less well.
Armed with this knowledge, this chapter will discuss best practices when using
simulators and working on design projects. It is by necessity a subjective viewpoint,
but the hope is the reader will learn something helpful along the way. Different
developers will have different experiences and as such slightly different approaches
to design work. All in all, we will stay away from controversial and very subjective
approaches and stay true to what a designer is most likely to encounter in the
industry.

Fundamentally, the task for an industry professional is to build circuitry with a
reasonable yield, so it is economically feasible to produce them. This important
aspect will be the undertone of the chapter. The work for an industry professional is
thus somewhat different than for an academic professional. In academia, inventing
new ways to use devices or break new grounds in unexplored adaptations of a technology is paramount. The industry professional is also very keen on inventing new
uses of devices but with the goal of making an end product that can be mass produced. These slightly different approaches will make the work a little bit different
in the two settings, and this chapter will showcase more of the industry professional's approach.

We will first discuss good practices when encountering a new process technology in Sect. 6.1. It can be a new "fresh off the press" high-end technology just
released from a foundry, or it might simply be new to the user. The worst thing to do
is to just start simulating along without any idea of how the foundry modeling team
has built their models. The model switches might not be tuned properly to a particular application, and if one does not find this out first thing, one could be in for a
surprise in the laboratory evaluating the circuit. Section 6.2 discusses small block
simulation strategies, and we wrap up the chapter with a presentation of a typical
simulation flow when designing large blocks.

© Springer Nature Switzerland AG 2021
M. Sahrling, *Analog Circuit Simulators for Integrated Circuit Designers*,
https://doi.org/10.1007/978-3-030-64206-8_6

## 6.1   Model Verification Strategies with a New Process Technology

Let us imagine we are starting a design project with a new process technology. It can be new to the user, or it might be an actual brand new process that comes online. In today's environment, new processes come online at a fairly fast pace, and if one is working on projects where the latest technologies are of interest, this is a common experience to run into. The economics of semiconductor industry dictates that cost both in terms of power and money is pushed lower and lower, and since both are tied to the transistor size, there is a big incentive to reduce the scale of the transistors. The first thing to do is to look at the device models, and we discuss that next.

Often when one encounters a new technology, one is given a set of model files, and depending on the maturity of the process, these model files will reflect the devices more or less well. A mature process such as the classic TSMC 180 nm has been in active design for decades and can be trusted much more than the latest ultra-small geometry CMOS process that just came online. However, even for TSMC 180 nm, a good understanding of what models do well and not so well is important. For new processes, the difficulty for the designer is to be able to gauge the quality of the models. A modern foundry has professional staff that characterizes the active devices with great care, but if the process is new, all kinds of time pressures can cause the models to be off, not least because the process might be in flux! It is up to the designer to judge the maturity of the model files, and if oddities are found, a heads up to the foundry is often much appreciated.

What kind of characterization should one do? The answer tends to be application specific in the details, but we will provide a general set of simulations that are most of the time very useful to have on hand when proceeding with design tasks:

- Simulate the transistors using $I_d$ vs $V_{gs}$, $V_{ds}$ for a CMOS process ($I_c$ vs $V_{be}$, $V_{ce}$ for a bipolar one). Vary channel length and width.
- Use these plots to extract the threshold voltage, $V_{th}$.
- Simulate the transition frequency $f_t$ vs channel length and width.
- Simulate the gate-source, gate-drain capacitance vs length and width.

In this section, we will discuss the details behind these three sets of simulations and what one should take away from them in Sects. 6.1.1, 6.1.2, and 6.1.3. There are of course other sets of simulations that could be useful, like noise and $1/f$-corner, and these three sets are just a good starting point and are not intended to be exhaustive to any degree.

We will discuss briefly what these properties might mean in this chapter. For example, it can be quite frustrating to extract a transistor and found huge discrepancies compared to schematic simulations. The contact and drain/source resistance of ultra-small channel length transistors can be very detrimental to performance, and if one has to do an extraction to find out how bad it is, it is good to know this early on so one can find ways to expedite the design time.

In this section, we will present some model verification strategies that have proven helpful over many years of design work. The benefit is if discrepancies are found, one can take preventive measures and so improve the likelihood of first-time success with a particular design. We first discuss the basic characterizations, and then we follow with other spot checks like drain/source resistance modeling, $g_m$ versus channel length, NQS switch $g_{m,max}$ method to extract $V_t$, and varactor's capacitance vs voltage (some models only characterize the capacitance at some average voltage and can thus be off). We do not intend this list to be exhaustive but rather reflect what the author encountered over the years of active design work and has found to be useful hints at the workings of the process devices.

## 6.1.1  DC Response Curves

The first thing to do when a new process is to be evaluated is to simulate the basic drain current versus gate-source voltage and drain-source voltage. The result will look something like Figs. 6.1 and 6.2.

Note in particular what happens at high $V_{ds}$ levels. There are effects like DIBL and ISCE that will increase the drain current significantly at higher drain voltages, and it is a good idea to verify the result is reasonable. Of course, also look at the first derivatives of these functions to make sure they are continuous (Figs.6.3 and 6.4) and their first derivatives (Fig. 6.5)

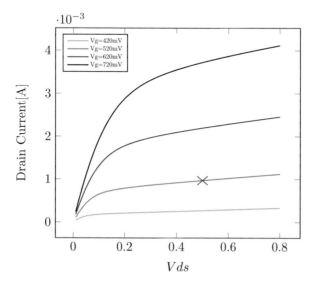

**Fig. 6.1** $I_d$ vs $V_{ds}$, $V_{gs}$. The cross marks a possible biasing point where the gate overdrive voltage is about 150 mV and the drain current is 1 mA for a $V_{ds} = 500$ mV

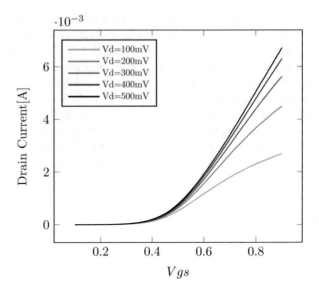

**Fig. 6.2** $I_d$ vs $V_{gs}$

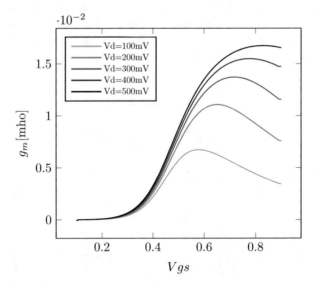

**Fig. 6.3** g_m vs $V_{gs}$

At times going even further can be useful also. We will stick to these plots here, and if in need of further data, we will simply mention that in the text.

- Do the response curves look physical?
- How does the model scale with channel length? Device width?
- How are parasitic resistors handled? Really important for ultra-short channel length processes.

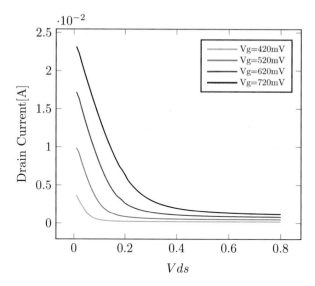

**Fig. 6.4** g_o vs $V_{ds}$

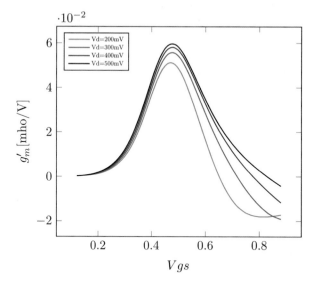

**Fig. 6.5** gm' vs $V_{gs}$

## 6.1.2   $V_{th}$ Extraction

The threshold voltage, $V_{th}$, is not a physical quantity like the thermal voltage, $V_t = kT/q$; instead it is a common way to parameterize the transition from non-inverted channel to an inverted channel where minority carriers become present in the region just below the gate dielectric (see Chap. 3). The precise value of the $V_{th}$ voltage can thus be debated. One common methodology is called the gm-max method where the transistor is biased in the linear region, and $V_{ds} < V_{gs} - V_{th}$ is fixed, the gate voltage is swept, and the peak value of $g_m$ is located. The point where the tangent to the $I_d$ vs $V_{gs}$ at gm-max crosses the $I_d = 0$ axis is the threshold voltage. The reason we look at the point where $g_m$ is maximum is we need to make sure we are out of the subthreshold regime but not that deep in the saturation region where the mobility is affected. The max $g_m$ point is thus where the following formula is approximately valid (linear region from Chap. 3):

$$I_{ds} = \mu_{eff} C_{ox} \frac{W}{L} \left( \left( V_{gs} - V_T \right) V_{ds} - \frac{m}{2} V_{ds}^{\,2} \right) \tag{6.1}$$

The expression for $I_d = 0$ when

$$\left( V_{gs} - V_T \right) V_{ds} - \frac{1}{2} V_{ds}^2 = 0 \rightarrow V_{gs} = V_T + \frac{1}{2} V_{ds} \tag{6.2}$$

We simply subtract $1/2 V_{ds}$ from $V_{gs}$ at the intercept point to get an estimate of $V_T$.

Simulate $V_T$ as a function of gate length and gate width. An example of such a simulation can be found in Figs. 6.1 and 6.3. The max $g_m$ is found around $V_{gs} = 750$ mV, and drawing a tangent from that point on Fig. 6.1 results in Fig. 6.6. The intersection point is at $V_{int} = V_{gs} = 0.45$ V resulting in a $V_{th} = V_{int} - V_{ds}/2 = 400$ mV. After simulating all these characteristics, we ask questions like:

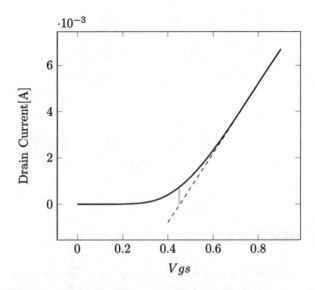

**Fig. 6.6**  $V_{th}$ extraction procedure

- Is the response reasonable in terms of
  (a)   scaling?
  (b)   physics, in terms of say output conductance for the CS stage and $V_T$ vs gate length?

- Are the parasitics, like gate-source/gate-drain resistance and sidewall capacitance, included? Do they make sense? The gate-source capacitance should be larger than the gate-drain capacitance even in the presence of sidewall capacitance. If the parasitics are not included properly when the model file switches are set correctly, we have a different kind of problem.

When the DC characterization curves and the threshold voltage has been simulated, think of a "unit" transistor in your design flow. If you do low-power design, it might be a 10 uA drain current transistor, or if you are doing high-speed designs, it might be 1 mA transistor or even higher. Here we will assume a "unit" transistor is biased at 1 mA with a 150 mV overdrive ($V_{gs} - V_{th}$) and a $V_{ds} = 500$ mV with a minimum channel length. These factors will determine the transistor width. With the characterization plots we just did, we can now find out what the trans-conductance $g_m$ is for such a "unit" transistor and of course $g_{ds}$. If we look through Figs. 6.1, 6.2, 6.3, and 6.4 again, we find a threshold voltage of 370 mV, and with a $V_{gs} = 520$ mV, we find a drain current of 1 mA from Fig. 6.1. We have a $g_m = 0.012$ and $g_{ds} = 0.75 \ 10^{-3}$ mmhos from Figs. 6.3 and 6.4. Do these numbers correspond to the model extracted values? Often a modern schematic tool will back-annotate these numbers for a given transistor, and it is surprising how often these numbers do not correspond to the simulated ones. Users beware! We summarize the DC characteristics of this transistor in Table 6.1.

In the next few sections, we will continue to add various characteristics to this "unit" transistor.

### 6.1.3   $f_t$ Characterization

The transition frequency, $f_t$, is a classic parameter used to estimate the speed of a transistor. The advanced CMOS nodes can have $f_t > 400$ GHz for certain bias settings which is quite impressive. The transition frequency is defined as the frequency where the drain current equals gate current (see Fig. 6.7). If one simulates a transistor in the common gate configuration, one can see the transition frequency is simply

**Table 6.1**  DC characteristics of a "unit" transistor

| Parameter | Value | Unit |
| --- | --- | --- |
| Width | 10 | μm |
| $g_m$ | 11 | mmho |
| $g_{ds}$ | 0.5 | mmho |
| $V_{th}$ | 370 | mV |

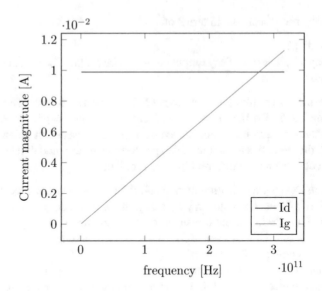

**Fig. 6.7** Bias of a "unit" transistor in saturation. The transistor is biased such that the drain current is 1 mA. Depending on the application, other transistor bias points might be more suitable

**Fig. 6.8** Common gate
stage

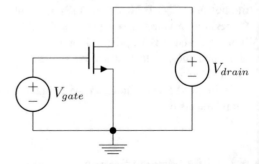

the point where the input current at the source splits equally between the drain and the gate ports. In some sense, the gate current is lost through the gate capacitance so it is a measure of the speed the transistor can handle.

The simplest way to set up a transistor for $f_t$ verification can be found in Fig. 6.8. The drain-source voltage is often kept at some voltage below maximum allowed, while the DC gate voltage is being swept. The AC currents in both drain and gate are compared, and the transition frequency is now the point where the two currents have the same magnitude.

At this point, there are a number of appropriate questions to ask:

- What is the transition frequency $f_t$? Does it match the advertised number?
- How does this point depend on the biasing? (Dependence on gate voltage can prove educational.)

**Table 6.2** Transition
frequency for our "unit"
transistor

| Parameter | Value | Unit |
|-----------|-------|------|
| $f_t$ | 280 | GHz |

- Does it make physical sense? To first order, the transition frequency should scale linearly with $g_m$ followed by a roll-off at higher bias points.

Often there is a bit back and forth between the foundry contact person and occasionally the modeling team to straighten out these kinds of questions (Table 6.2).

## 6.1.4 Gate-Source and Gate-Drain Capacitance Characterization

One convenient property to characterize the transistors is their gate capacitance both to source and drain, in particular when in saturation. A good approach is to take a "unit" transistor biased in saturation with an overdrive voltage >100 mV and a drain current of roughly 1 mA. Or pick a transistor that is typical for the particular application one is working on. Then run an AC simulation where you look at the amount of *imaginary* current going through the drain and source nodes, respectively (see Fig. 6.9). This will quickly give you the capacitances for a typical (or unit) transistor, since the imaginary current slope is equal to $\omega C$. For the case of Fig. 6.9, we find (Table 6.3).

By keeping these numbers in mind, one can estimate a transistor's loading on a circuit when looking at schematics [1]. This can be very efficient when designing circuits. Furthermore, do the drain source capacitances scale properly? With no parasitic resistors in the source and drain, one would imagine the gate-source capacitance to be larger since the sidewall capacitance should be similar to the drain side, but the channel charge in saturation should only contribute to the source capacitance. Sometimes one needs to explicitly turn off the parasitic resistors through a specific model family choice. These parasitic resistors can confuse the initial intuition one has about the relative size of the capacitors. A drain resistor can increase the effective gate-drain capacitance through the Miller effect, and a source resistor can reduce the effective gate-source capacitance through negative feedback in the source (the internal source voltage is in phase with the gate voltage).

We can calculate the resulting $f_t$ from these capacitances

$$f_t = \frac{g_m}{2\pi\left(C_{gs} + C_{gd}\right)} \approx \frac{10^{-2}}{36\cdot10^{-15}} \approx 280 \text{ GHz}$$

which confirms our direct simulation of the transition frequency.

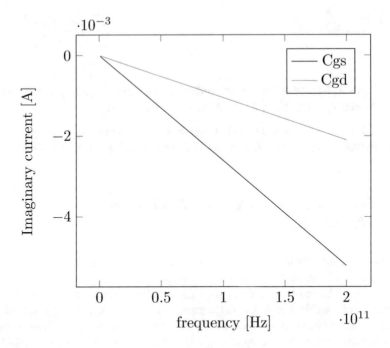

**Fig. 6.9** Simulation result of a "unit" transistor's capacitances

**Table 6.3** Gate-source/gate-drain capacitances calculated from the simulation in Fig. 6.9

| Capacitance | Value | Unit | Comment |
|---|---|---|---|
| Gate-source | 2.6e-3/6.28e11 = 4.3 | fF | Saturation |
| Gate-drain | 1.1e-3/6.28e11 = 1.6 | fF | Saturation |

### 6.1.5 Summary

We have looked at a "unit" transistor in our fictitious technology. We found it has the following characteristics as shown in Table 6.4.

Armed with this little table, we can now get an idea of the value of these parameters for any size transistor by simply multiplying by a scale factor. The further we are from this "unit" transistor size, the greater the error, but it will still provide a good sense of what, say, the capacitive load is with a particular transistor. Also things like transconductance and output conductance can be estimated really quickly without need for simulations. It is a handy little table one can use throughout the design process.

In other applications, other parameters like the on-resistance, $r_{on}$, might be valuable to add to this short list of characterization parameters. This whole idea is not new by any means (see, e.g., [1–3] for more much more details).

**Table 6.4** Final table of "unit" transistor common parameters

| Parameter | Value | Unit |
|-----------|-------|------|
| Width     | 10    | μm   |
| $g_m$     | 11    | mmho |
| $g_{ds}$  | 0.5   | mmho |
| $V_{th}$  | 370   | mV   |
| $f_t$     | 280   | GHz  |
| $C_{gs}$  | 4.3   | fF   |
| $C_{ds}$  | 1.6   | fF   |

## 6.1.6  Example of Faulty Model Behavior

To drive home the point of examining the model behavior closely, let us look at the buffer in Sect. 5.2 but use the unrealistic transistor model 1 instead and take a look at the output node. We did this already in Chap. 5 (Fig. 5.37). The unphysical blip in the middle of the up and down plateaus comes from the fact that the model is a simple square function, and if $V_{gs} < 0$, the transistor turns on again something that is unphysical. It is very unlikely the reader will run into some model that is this off, but hopefully the benefit of examining transistor behavior carefully is appreciated.

## 6.1.7  Corner Simulation Strategy

A foundry will not release only one set of models. Often they will include many different model sets often referred to as process corner models, and as a new engineer, it is easy to be convinced the word "corner" refers to some extreme case that will "never happen." This is not the case. A single wafer or die for that matter contains a certain process variation. The average over the wafer can very well be lining up with one process skew corner, but around this average, there will be a certain variation that depends on the foundry and technology used. Occasionally one finds silicon coming back from the factory and behaving much worse than any model simulation could ever accomplish. This is often due to some weakness in the design itself that shows up in a simulation corner that "could never happen." Instead of looking at corners at some kind of extreme case, it is better to view them as a stress test on your circuit. If any of the simulation results changes significantly (and "caveat lector" be careful with the meaning of the word significant) in one corner, your circuit will simply not work properly. This odd behavior will be a part, to a lesser or higher degree, of all your dice no matter what supposed process skew corner wafer they came from.

Sometimes the skew process files have all elements lumped together in the corner files. This rarely makes sense since different device types have different process dependencies. If the foundry is not providing a separate corner file for capacitors, NMOS, PMOS, resistors, diodes, etc., it often makes sense to create your own skew model files. Depending on the type of project you are getting into and its application, it might be best to add in more corners than just the standard CMOS transistor that comes directly from the foundry.

Traditionally there are two other corner simulation parameters that are needed: the supply voltage corner and the temperature corner case. We have the fearsome three, namely, process, voltage, and temperature (PVT), and they are often listed like this in descending order of importance.

Once the needed corners have been established, it is time to go to town and simulate. With modern simulator environments, it is not difficult to set up large simulation sweeps and verify compliance with simple scripting languages. Run these over the weekends or overnight, and pay careful attention to small oddities in the result (Fig. 6.10). Is there a trend over corners? Sometimes it is so by necessity, but say with a bandgap, you would expect to see a flat response around the target temperature. In this evaluation phase, it is a good thing to be a bit "paranoid," and often one has to fight one's tendency to move on.

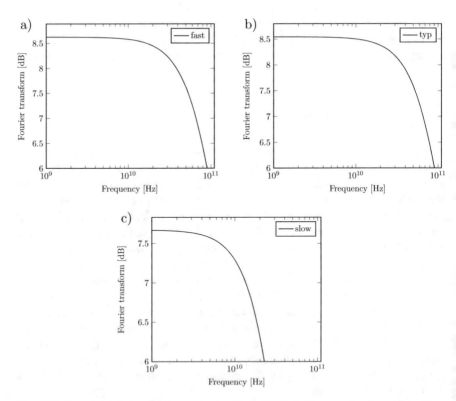

**Fig. 6.10** Corner simulation warning examples. Do not be fooled because the slow corner seems to misbehave that since it is just in one corner that can "never happen," the circuit is ok. Such a case indicates that something is off with the circuit or the simulation setup, and on silicon this circuit will not work properly

## 6.1.8   Monte Carlo Simulations

Monte Carlo simulations are another classic stress test of circuitry. The idea is that in manufacturing, there is variability when deploying the various doping and injection steps where the device characteristics will vary from device to device. This is often characterized by the foundry one way or other. This characterization is often done with devices lying right next to each other and can be a too good situation compared to devices being far apart. There are some key steps to think about:

- How is the mismatch characterized by the foundry? Are the devices assumed to lie next to each other?
- Run ≫ 100 simulations, with C-only extraction models and the RC-only extraction models.
- Study the resulting standard deviation. How close in number of sigma are we from the specifications?

## 6.2   Small Block Simulations

In this section, we will discuss general strategies concerning small block simulations. We start with simple model estimation calculations and continue with appropriate simulation strategies to characterize the schematic.

Before any circuit simulation is performed, think the circuit through in terms of simple modeling (see [1] for many examples on how to do this for a broad range of applications). The crucial idea is the simulator should only confirm what you already know will happen. If one lets the simulator tell what is going on without asking oneself why, one can easily be led astray. It can be the simulator responding incorrectly to some stimulus along the lines we outlined in Chaps. 4 and 5, or it can be a result of incorrect stimulus. Both of these problems must be caught early to avoid lengthy debugging later. With experience many of these calculations can be done in one's head.

First describe the circuit you are to design in terms of a simple model.

## 6.2.1   Analog Circuit Simulation Strategy

The simulation strategy will depend strongly on what type of circuit you are designing. What we outline here is intended for use in continuous time applications like amplifiers. Always start your design work by making a simple model of your design [1].

Once the simple model is in place, it is always best to first verify the DC bias with a DC simulation. Ask yourself:

- Is everything biased properly?
- Are all relevant transistors in saturation (for CMOS)?
- Is the current draw as expected?

Once the DC bias condition is satisfactory, one should proceed doing an AC simulation and look whether:

- AC gain is as expected from the simple model.
- AC bandwidth matches the simple model.

When the small signal response is as expected, one should run transient simulations and periodic steady-state simulations and look at things like:

- Is distortion as expected (see [1] for fast ways to estimate harmonics)?
- Do the CMOS transistors get out of the saturation condition? (This can cause distortion and change in loading.)

The flow is illustrated in Fig. 6.11.

One of the most common mistakes made by beginners is to use ideal sources everywhere. Often it is alright to do at a very initial stage, the first few days. Ideal sources are really *ideal*, and a voltage source can deliver any current with no limitation. Think, for example, of the case where we have some RF input of 50 ohms and one uses an ideal voltage source to model this. It will have infinite bandwidth! It is really disappointing at the end of a long design effort to realize the bandwidth is actually down the tubes, and it was not caught due to some idealization. But it is not just true for signal sources. Voltage supply sources are equally bad to substitute with ideal ones. The packaging around the die will often show significant inductance in the supply nodes, and to counter this huge numbers of decoupling, capacitors need to be used on chip to lower the impedance for fast-switching signals (see Fig. 6.12).

Likewise for current sources, sometimes a design needs to proceed before a full biasing block is ready. In this case, mirror an ideal current properly to set the bias up. Do not use an ideal current directly at the common source point of a differential pair. Instead use a mirror circuit to set up the bias as in Fig. 6.13. An ideal current source should be used at the point where the biasing current will end up eventually.

Always use realistic voltage and current sources.

And perhaps most important for small blocks in particular, use realistic source impedance and load impedance.

Use realistic source and load impedances from the get-go.

**Fig. 6.11** A typical design
flow diagram

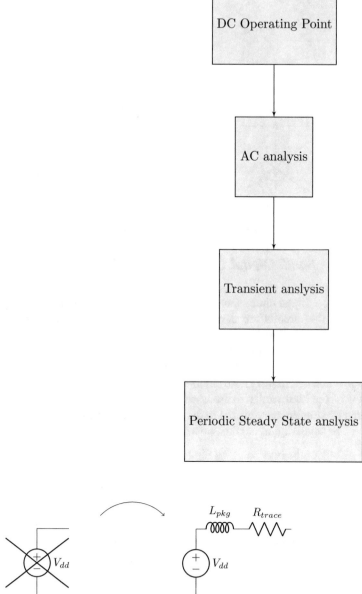

**Fig. 6.12** Good practice for voltage supplies

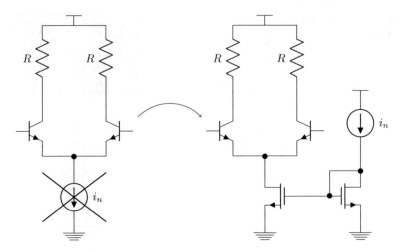

**Fig. 6.13** Good practice when using ideal current sources

## 6.2.2 Small Digital Circuit Simulation Strategy

For large digital circuits, one needs to follow the register transfer logic (RTL) flow where the circuits are described in a high-level hardware description language (RTL) like Verilog. These are simulated with event-based simulators, in contrast to the continuous time simulators we have described in this book. However, occasionally one needs relatively small digital blocks and state machines to control some analog circuit. These circuits can be simulated with a continuous time simulator of the kind we have discussed here. For these circuits, the simulation strategy can be simplified compared to the analog continuous time case.

For small digital circuits, do a DC simulation.

- Look at the bias point, all digital gate outputs should be either at ground or vdd; if not, some connection might be floating unintentionally.

There is no point running AC simulations for pure digital circuits; instead one should proceed to the transient simulation, and one should look at timing:

- Is the timing correct between the signals? If not, the sizing of some buffers might be wrong and unable to drive the load properly.
- Is the edge rate healthy? This is another indicator the load is inappropriate. What healthy means is directly dependent on the application.

The steady simulators usually have a hard time with circuits with a lot of fast-switching edges. The shooting method is in particular infamous for having difficulties because of the inherent assumption that the final state depends weakly on changes in the initial state (Sect. 5.5), and for a fast-changing edge, this is not true, and Newton-Raphson might not find an appropriate solution. Furthermore, there might not be a simple way to see periodicity in the circuit, or it might be really long, so the fundamental assumption of a steady state is not helpful. Most of the time, the designer is then limited to transient simulations.

## 6.3 Large Block Simulations

For larger blocks, the most difficult problem to deal with is the size of the circuit. As we discussed in Chaps. 2, 4, and 5, modern simulators have some really nifty features, among them really efficient matrix solvers, but for modern SOC, the device count for the analog section can easily be in the tens of millions, and even such powerful tools can start to choke. It is naturally not surprising that this happens since the one gives rise to the other with more powerful computers being able to generate even more powerful chips one can put in to make the computers more powerful that makes even more powerful chips and so on. It is an exponential development effort!

A good way to deal with these situations is to use mixed-mode simulators, where you can, for example, mix Verilog event-type simulators with differential equation solvers like SPICE. If there is no such tool available, one is often helped by constructing simplified Verilog-A models for the blocks that one can switch in and out between schematic views and Verilog-A views. These work well in SPICE simulators and can be a much faster way to move forward. For Verilog-A views, keep in mind to include appropriate driving and load impedances.

The key thing for large block simulations is to make sure the signal path is fully simulated with appropriate load and source impedances. Include also the clock distribution path if there is such.

- Does the bandgap, bias circuit come up properly after being enabled?
- Does the right block get the right bias current?
- Can the registers be programmed properly?
- Are all the blocks being enabled correctly?

Pay particular attention to the clocks. If the clocks are not coming up correctly, *nothing* can be tested in the digital domain and few analog functions. This is such an important function that there are often clock overrides that can enable the digital core with a low-frequency external clock that goes directly to the core, bypassing all analog circuitry.

### 6.3.1 Analog-Digital Co-simulation Strategies

There are in principal at least three different ways analog blocks can co-simulate with digital blocks: (1) both using SPICE-like tools, (2) analog → spice and digital → Verilog, and (3) both using Verilog tools. The first option quickly becomes impossible due to sheer circuit size; the other two are more common, the second option being the most common of all. Option 3 is occasionally used when simple signal path analysis might be warranted. It obviously cannot handle analog signals unless one uses some sort of digitization.

A good design methodology is to make sure all analog blocks are turned off when the supply voltages are applied. This means the basic biasing in terms of bandgap and associated currents are down. Then, in bring up, one analog block after another can be turned on and functionality established. Having everything turned on immediately can cause strange behavior since the state of the circuit can be unknown, and large unintended currents can start destroying the interconnect metals.

For modern analog blocks, there are normally many digital bits needed to be able to control the biasing in detail and adjust for things like temperature, voltage drift, and aging to name a few. There can easily be many thousand bits to control the analog circuitry. Naturally, there is no way the digital registers can be programmed in the analog simulator; the time scale for a register programming sequence can be several microseconds, whereas the analog circuits can have a timescale of picoseconds. The proper functionality of the registers can only be verified with a Verilog simulator, and since they control the analog block, doing it in a mixed simulation mode environment is ideal.

In circuits like data converters or RF mixers, there is usually a fast clock somewhere that sets the time for the chip functions. To generate the clock internally, it is often required to have an accurate input clock of up to a few GHz perhaps (the higher the frequency, the more expensive the solution gets on the printed circuit board design). This clock needs to be turned off during simulation of register programming due to the different timescales involved. This is usually easy to do with a delay statement in the clock input circuitry.

A chip usually generates its own biasing where internal bandgap, fixed voltage, and currents are generated. One useful approach is to program this bandgap and current generator to come up with the help of register programming. To keep a bias block from generating too much noise, there is usually a large capacitor with a large time constant present in the block. It can take several 10 s of microseconds for such a block to come up properly, depending on the circuit, and during this time, all faster time constants, like clocks we just mentioned, should be turned off so the bandgap's behavior can be observed.

When the biasing has been proven to work and come up properly and bias the correct circuit block, with 100 s–1000 s of currents needed, it is time to turn on the clocks and input signals so the detailed analog behavior can be studied.

## 6.3.2   Summary

We have in this chapter touched briefly on good practices when designing blocks using simulators and modern foundry models. We mentioned it is really important to verify the device modeling, in particular for technology nodes that are coming online. The modeling teams working for modern foundries are very professional and knowledgeable, but characterizing modern transistors with BSIM is a very measurement-intensive and time-consuming task, and it is easy to have shortcomings, in particular the early releases. Even with more established technologies, it is worthwhile to study the device models. It is then less likely one finds shortcomings in the model and much more likely one finds shortcomings in one's own understanding. In either case, you'll be thanking yourself later for doing the hard work upfront. It makes the subsequent design work much smoother.

We also discussed strategies for designing small analog circuit blocks followed by a discussion of efficient strategies for large circuit simulation verifications.

The chapter has been by its nature a subjective one, and many experienced design engineers will likely have a different viewpoint, but yours truly have followed these principles for many years with some success and think they are worth considering when building integrated circuits.

## 6.4  Exercises

1. Calculate the third harmonic response of a resistively degenerated differential pair.

## References

1. Sahrling, M. (2019). *Fast techniques for integrated circuit design*. Cambridge, UK: Cambridge University Press.
2. Jespers, P. G. A., & Murmann, B. (2017). *Systematic design of analog CMOS circuits*. Cambridge, UK: Cambridge University Press.
3. Jespers, P. G. A. (2010). *The gm/Id methodology, a sizing tool for low-voltage analog CMOS circuits*. New York: Springer.

# Chapter 7
# Mathematics Behind Simulators

**Abstract** This appendix describes some of the mathematical details behind the circuit analysis in a more formal way than elsewhere in the book. We will first discuss an electrical network in terms of directed graphs where we define basic entities relevant for circuit theory and their properties. This will lead into nodal analysis and finally modified nodal analysis. The following section will discuss the solution of differential equations in terms of difference equations. This is a huge subject, and we will only have space to highlight some of the important aspects of this subject. For the interested reader, the references will contain many pointers for future study.

## 7.1 Network Theory

An electrical network consisting of various kinds of active and passive devices can be thought of as directed graph where the nodes (sometimes called vertices) represent a voltage and the directed edges represent the current flow between the nodes [1–3, 5, 6]. This current flows through the various elements in the network as shown in Fig. 7.1.

Following the notation in [2, 3] let us define the node voltages as

$$v_j, j \in \{1,\ldots,n-1\} \tag{7.1}$$

These voltages represent the difference in electrical potential between node $j$ and a reference node, commonly referred to as node "0." The electrical currents are represented by the edges, and they have magnitude and direction. We will denote the currents by

$$i_k, k \in \{0,1,\ldots,m-1\} \tag{7.2}$$

and the edges by

$$e_k, k \in \{0,1,\ldots,m-1\} \tag{7.3}$$

With these definitions, we can now define an incidence matrix $M_{jk}$ as follows. Let each node be represented by a row in $M_{jk}$, and let each edge be represented by a

© Springer Nature Switzerland AG 2021
M. Sahrling, *Analog Circuit Simulators for Integrated Circuit Designers*,
https://doi.org/10.1007/978-3-030-64206-8_7

**Fig. 7.1** Graph
representation of a
directed network

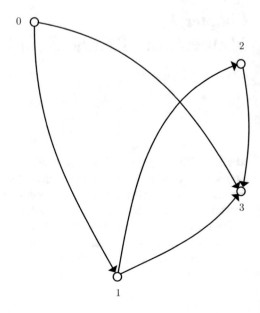

column in $M_{jk}$ such that if the current *leaves* the node, the entry is $+1$ and if the current *enters* the node, the entry is $-1$. All other entries are 0. Clearly each column has one $+1$ and one $-1$ since each current only leaves and enters one node. With the example in Fig. 7.1, we find

$$M_{jk} = \begin{cases} 0 & 0 & 1 & 1 & 0 \\ 1 & 0 & -1 & 0 & 1 \\ -1 & -1 & 0 & -1 & 0 \\ 0 & 1 & 0 & 0 & -1 \end{cases} \tag{7.4}$$

It is here evident that the rows are linearly dependent since if we treat them as vectors, $s_j$, and add them together,

$$\sum_{j=0}^{n} s_j \equiv 0 \rightarrow s_l = -\sum_{\substack{j=0 \\ j \neq l}}^{n} s_j \tag{7.5}$$

In other words, there is superfluous information in the matrix, and we can remove one row. This new matrix, $A$, is commonly referred to as the reduced incidence matrix. In our example, we get by removing the first row of $M_{jk}$:

$$A = \begin{cases} 1 & 0 & -1 & 0 & 1 \\ -1 & -1 & 0 & -1 & 0 \\ 0 & 1 & 0 & 0 & -1 \end{cases} \tag{7.6}$$

This is really useful in that we can now define the voltage across any edge as

$$u_k = v\left(e_{k,tail}\right) - v\left(e_{k,head}\right) \qquad (7.7)$$

In matrix form, we can compactly write this as

$$u = A^T v \qquad (7.8)$$

where the superscript $T$ refers to matrix transpose. We recognize this equation as Kirchhoff's voltage law (KVL), and we found a compact way to describe it.

Kirchhoff's current law can be similarly described as

$$\sum_{e_l \in E_i} i_{kl} - \sum_{e_l \in E_o} i_{kl} = 0 \qquad (7.9)$$

The sum of all incoming and outgoing currents for each node (or row in $A$) is zero. In our matrix formulation, this becomes

$$Ai = 0 \qquad (7.10)$$

Notice that we have removed one row, conveniently chosen as node "0" which in circuit theory we refer to as ground. We will not keep track of the currents going in and out of this reference node since it will be given by the sum of all currents in all other nodes. This is a convenience we have already taken advantage of in the main text.

The two laws of Kirchhoff can now be combined into one matrix equation:

$$\begin{bmatrix} A & 0 & 0 \\ 0 & I & -A^T \end{bmatrix} \begin{bmatrix} i \\ u \\ v \end{bmatrix} = \begin{bmatrix} 0 \\ 0 \end{bmatrix} \qquad (7.11)$$

This equation is often called the topological constraints since it only concerns itself with the network topology, the details of the circuit elements, and their response to stimuli is not taken into account. This equation contains $2m + n - 1$ unknowns and $m + n - 1$ unknowns. We are then short with $m$ equations to be able to have a chance of solving it. These $m$ equations called the branch equations come from the circuit elements. With these included, we then have a full system we can attempt to solve. Another advantage of these set of equations is the matrix on the left-hand side is sparse in that most entries are zero. Such systems are much easier to solve than more general matrix systems.

### 7.1.1   Sparse Tableau Analysis

Let us first look at linear circuit elements and see how the branch equations arise. Such a circuit element can be characterized as either a current response to a voltage stimulus

$$i = yv \tag{7.12}$$

or vice versa

$$v = zi \tag{7.13}$$

As the reader no doubt knows, $y$ is the admittance and $z$ is the impedance of the element. In general for a set of $m$ branch equations, we find

$$Zi - Yu = s \tag{7.14}$$

where $Z$, $Y$ are sparse $m \times m$ matrices and $s$ is known $1 \times m$ vector (often zero). We can now combine these set of branch equations with the topological constraints to find

$$\begin{bmatrix} A & 0 & 0 \\ 0 & I & -A^T \\ Z & Y & 0 \end{bmatrix} \begin{bmatrix} i \\ u \\ v \end{bmatrix} = \begin{bmatrix} 0 \\ 0 \\ s \end{bmatrix} \tag{7.15}$$

This is known as the sparse tableau analysis (STA) formulation (see [1]). This formulation is general and sparse making it fast to solve for. However, there is some redundancy, and in modern circuit simulators where often memory space is a concern due to the large circuitry often under study, there are other more compact formulations. Let us first substitute KVL ($u = A^T v$) into the branch equations. We find

$$Zi + Yu = Zi + YA^T v = s \tag{7.16}$$

The network matrix equation now becomes

$$\begin{bmatrix} A & 0 \\ Z & YA^T \end{bmatrix} \begin{bmatrix} i \\ v \end{bmatrix} = \begin{bmatrix} 0 \\ s \end{bmatrix} \tag{7.17}$$

This is sometimes called the reduced tableau form.

### 7.1.2   Nodal Analysis

Let us now make the observation that if we can write $i$ as a function of the admittance $Y$ and $s$

$$i = Yu + s \tag{7.18}$$

and KVL, we get a much reduced matrix system that do not depend on the branch currents at all:

$$AYA^T v = -As \tag{7.19}$$

The formulation is called the nodal analysis (NA) form of the network equations. It is much smaller $(n - 1) \times (n - 1)$ matrix with only the nodes as unknowns. It turns out the requirement in the branch currents excludes ideal voltage sources since the equation for an ideal independent voltage source does not contain its current:

$$v(a) - v(b) = V \tag{7.20}$$

The node voltage between two nodes is fixed independent on current (put differently, the ideal independent voltage source has 0 ohm output resistance). This nice property of the network equation where the currents are eliminated with the unfortunate requirement on the absence of an independent voltage source is quite attractive, and we will discuss the common way to work around this problem next.

### 7.1.3   Modified Nodal Analysis

The modified nodal analysis (MNA) was first proposed by [2]. The starting point is the nodal analysis we just discussed with the modification that elements that need to keep their currents are simply added to the system of branch equations. One talks about elements belonging to one of two *groups* see for example [2, 3]:

**Definition of Element Groups**
Elements whose currents are to be eliminated belong to Group 1 and all other belong to Group 2.

The branch currents and voltages can now be separated depending on the elements' group membership. We have

$$i = \begin{bmatrix} i_1 \\ i_2 \end{bmatrix}, u = \begin{bmatrix} u_1 \\ u_2 \end{bmatrix} \tag{7.21}$$

where currents in the Group 1 elements current are denoted by $i_1$, voltages by $u_2$, and so on. The branch equations now become for elements in Group 1

$$i_1 + Z_{12}i_2 = Y_{11}u_1 + Y_{12}u_2 + s_1 \tag{7.22}$$

and for elements in Group 2

$$Z_{22}i_2 = Y_{21}u_1 + Y_{22}u_2 + s_2 \tag{7.23}$$

In matrix form

$$\begin{bmatrix} I & Z_{12} \\ 0 & Z_{22} \end{bmatrix} \begin{bmatrix} i_1 \\ i_2 \end{bmatrix} - \begin{bmatrix} Y_{11} & Y_{12} \\ Y_{21} & Y_{22} \end{bmatrix} \begin{bmatrix} u_1 \\ u_2 \end{bmatrix} = \begin{bmatrix} s_1 \\ s_2 \end{bmatrix} \tag{7.24}$$

Kirchhoff laws are also written as a function of the group memberships:

$$A_1 i_1 + A_2 i_2 = 0 \tag{7.25}$$

$$u_1 = A_1^T v \qquad u_2 = A_2^T v \tag{7.26}$$

A direct substitution of the branch currents $i_1$ into KCL and using the KVL equations for $u_1$ and $u_2$, we find

$$A_1 \left( -Z_{12}i_2 + Y_{11}A_1^T v + Y_{12}A_2^T v + s_1 \right) + A_2 i_2 = 0 \tag{7.27}$$

After recasting in terms of the unknowns and moving the knowns to the right-hand side, we end up with

$$A_1 \left( Y_{11}A_1^T + Y_{12}A_2^T \right) v + \left( A_2 - A_1 Z_{12} \right) i_2 = -A_1 s_1 \tag{7.28}$$

To get the full modified nodal set of equations, we now need to simply add the branch equation for the elements in Group 2, and we find upon using the KVL equations again

$$\begin{bmatrix} A_2 - A_1 Z_{12} & A_1 \left( Y_{11}A_1^T + Y_{12}A_2^T \right) \\ Z_{22} & -Y_{21}A_1^T - Y_{22}A_2^T \end{bmatrix} \begin{bmatrix} i_2 \\ v \end{bmatrix} = \begin{bmatrix} -A_1 s_1 \\ s_2 \end{bmatrix} \tag{7.29}$$

This is the general form of the MNA system of equations. One can make further simplifications in certain situations. For example, assume there are no controlled sources. This means the matrices $Z_{12}$, $Y_{12}$, $Y_{21} = 0$ and $Y_{11}$ are diagonal. We have

$$\begin{bmatrix} A_2 & A_1 Y_{11}A_1^T \\ Z_{22} & -Y_{22}A_2^T \end{bmatrix} \begin{bmatrix} i_2 \\ v \end{bmatrix} = \begin{bmatrix} -A_1 s_1 \\ s_2 \end{bmatrix} \tag{7.30}$$

If in addition there are no current sources in the network, $Y_{22} = I$, and we find the common formulation of MNA in the literature as

$$\begin{bmatrix} A_2 & A_1 Y_{11} A_1^T \\ Z_{22} & -A_2^T \end{bmatrix} \begin{bmatrix} i_2 \\ v \end{bmatrix} = \begin{bmatrix} -A_1 s_1 \\ s_2 \end{bmatrix} \tag{7.31}$$

Note here that the first set of equations is simply a result of KCL formulated at all the nodes and the second set comes from the branch equations. We can now follow the following recipe when constructing the network system of equations:

1. Read an element from the circuit netlist.
2. If the element belongs to Group 1, eliminate its current by using its branch equation, and replace the branch voltages using KVL.
   If the element belongs to Group 2, write its branch equation including its current, and eliminate all branch voltages using KVL.

This recipe leads naturally to a specific signature for each element in a circuit netlist like what was described in the main text.

The situation is a little bit different when discussing dynamic elements like capacitors and inductors where the derivative of currents and branch voltages come into play. We have then in general for the branch equations

$$Zi + L\frac{di}{dt} - Yu - C\frac{du}{dt} = s \tag{7.32}$$

or in matrix form

$$\begin{bmatrix} I & Z_{12} \\ 0 & Z_{22} \end{bmatrix} \begin{bmatrix} i_1 \\ i_2 \end{bmatrix} + \begin{bmatrix} 0 & 0 \\ 0 & L_{22} \end{bmatrix} \begin{bmatrix} di_1/dt \\ di_2/dt \end{bmatrix} - \begin{bmatrix} Y_{11} & Y_{12} \\ Y_{21} & Y_{22} \end{bmatrix} \begin{bmatrix} u_1 \\ u_2 \end{bmatrix}$$
$$- \begin{bmatrix} C_{11} & 0 \\ 0 & C_{22} \end{bmatrix} \begin{bmatrix} du_1/dt \\ du_2/dt \end{bmatrix} = \begin{bmatrix} s_1 \\ s_2 \end{bmatrix} \tag{7.33}$$

Note the inductor is part of Group 2. This is due to its zero impedance at DC; if one can guarantee DC never happens in your analysis, one can move it to Group 1, but in general, it should be a Group 2 element. We can now proceed along the same lines we did earlier and arrive at the MNA with dynamic elements:

$$\begin{bmatrix} A_2 - A_1 Z_{12} & A_1\left(Y_{11}A_1^T + Y_{12}A_2^T\right) \\ Z_{22} & -Y_{21}A_1^T - Y_{22}A_2^T \end{bmatrix} \begin{bmatrix} i_2 \\ v \end{bmatrix} + \begin{bmatrix} 0 & A_1 C_{11} A_1^T \\ L_{22} & -C_{22} A_2^T \end{bmatrix} \begin{bmatrix} di_2/dt \\ dv/dt \end{bmatrix} = \begin{bmatrix} -A_1 s_1 \\ s_2 \end{bmatrix} \tag{7.34}$$

## 7.2   Numerical Solution Techniques of Differential Equations

We have included the preceding derivation of the standard circuit matrix system for the sake of completeness. The standard references contain much more details [1–5]. We will in this section offer a brief discussion on solvability of these equations. We know from matrix theory that in order for a matrix equation to be solvable, the inverse of the matrix needs to exist in general. Obviously, if the rhs is zero, the trivial solution (zero) will always be possible (if there's nothing, you get nothing). So a study of solvability of the MNA set of equations is a study of the matrix itself and what can be said of its inverse. It is clear that if the matrix is diagonal, the solution will exist, so in the weaker case where the matrix is strongly diagonal, one can make the case that it probably exists. We will spend a few pages here discussing these various interesting cases.

It is clear in general that a solution to the MNA equations do not exist. We need to put additional constraints on the elements and the network to increase the probability of a solution. Imagine, for instance, the case where there are two independent voltage sources across the same branch with different voltages. This will result in an impossible situation with a resulting infinite current. Sometimes when setting up simulations in the schematic environment, one can make a mistake resulting in precisely this situation. A modern simulator will always complain about illegal branch topologies and exit with an error. There is also the complimentary problem of two independent current sources in series with different current values. This will also often result in an impossible situation. These types of pathological situations must not be allowed. They are often called *consistency requirements*.

The equations we have discussed so far are most generally known as differential-algebraic equations (DAE). They are more difficult to solve than the ordinary differential equations (ODE) we often think about regarding network systems. An ODE has a long history of study and many methods are known. The difficulty with network equations comes when the dynamic element matrix $\begin{bmatrix} 0 & A_1 C_{11} A_1^T \\ L_{22} & -C_{22} A_2^T \end{bmatrix}$ is nonlinear or singular. Nonlinear capacitance is very common in circuitry where, for example, reverse biased junction diodes and effects like channel capacitance variation with gate voltage in MOSFETs can significantly complicate the problem. In this book, we have mostly concerned ourselves with the simpler problem of constant $L$, $C$ which makes traditional ODE methods working well. With this cautionary tale of the difficulty surrounding the solution of general network equations, we will focus the remainder of this section on ODEs.

### 7.2.1   ODE Solution Methods

This section will discuss the basic mathematics and theorems surrounding ODEs. As in the previous section, we will present the material for the purpose of self-containment. We rely on references [3, 4] for much of the presentation.

### 7.2.1.1   Initial Value Problems

An initial value problem (IVP) is one where the solution is known at some time, let us call it time $t = 0$, and the system is governed by a differential equation that drives it forward in time to some other final time $t_f$. Formally we have

$$\frac{dx(t)}{dt} = f\big(x(t),t\big) \qquad x(0) = x_0 \tag{7.35}$$

where $x$ is an $m$-dimensional real vector and the vector $x_0$ is the known state at time $t = 0$.

Not all IVPs have a solution obviously, and sometimes there can be several solutions. We call these existence and uniqueness properties of a particular equation. Before we go into these properties, we need to be more specific about what kind of solutions we are discussing and over what domain (region in space and time) they are defined. Let us first define the domain $\mathcal{D}$. We are interested in solutions defined on $\mathbb{R}^m \times \mathbb{R}$ such that

$$\mathcal{D} = \big\{ (x,t) \,|\, x \in \mathbb{R}^m \text{ and } 0 \le t \le t_f \big\} \tag{7.36}$$

A problem like the one in Eq. 7.35 needs to be well behaved under small perturbations (remember things like numerical round-off errors) and not run haywire. Such systems are defined as well-posed.

---

**Definition 7.1**

Let $(\delta(t), \delta_0)$ and $(\tilde{\delta}(t), \tilde{\delta}_0)$ be two perturbations to the equation $\dot{x}(t) = f(x,t), x(t_0) = x_0$, and let $\hat{x}(t)$ and $\tilde{x}(t)$ be the resulting perturbed solutions. If there exist an $S > 0$ such that, for all $t \in [0, t_f]$, $\epsilon > 0$, we have

$$\| \hat{x}(t) - \tilde{x}(t) \| \le S\epsilon, \text{ whenever } \| \delta(t) - \tilde{\delta}(t) \| \le \epsilon \text{ and } \| \delta_0 - \tilde{\delta}_0 \| \le \epsilon$$

then the equation is said to be totally stable, or well-posed.

---

The idea is that if the solution does not differ too much with a small perturbation, the solution is stable. Now, think about a digital CMOS gate whose input changes slightly when it sits around the trigger point. Is such a system well-posed? Given that gain of the gate, the output will change drastically. However, it is still possible to find an $S > 0$ that matches the condition in the definition, so such a system is still technically well-posed. That does not mean it is a helpful thing when looking for a solution, and sure enough some analysis methods have a really hard time around trip points (see, e.g., the shooting method in Chap. 4).

We also need to define what is known as the Lipschitz condition on the function $f$ in Eq. 7.35.

**Definition 7.2**
Lipschitz continuity. Let us assume $f$ is such that there exists a constant $L$ such that

$$\left\| f\left(x,t\right)-f(x',t)\right\| \le \left\| Lx-x'\right\| \tag{7.37}$$

holds for every $\left(x,t\right),\left(x',t\right) \in \mathcal{D}$. $f$ is then called Lipschitz continuous.

It is a little more than a continuity condition but not quite as strong as differentiability [4].

The following standard theorem now guarantees the existence of a unique solution:

**Theorem 7.1**
Let $f(x,t)$ be continuous in $t$ on $\mathcal{D}$ and Lipschitz continuous in $x$ on $\mathcal{D}$. Then, for any $x_0 \in \mathrm{R}^m$, there exists a unique solution $x(t)$ of the equation $\dot{x}(t) = f\left(x,t\right), x\left(t_0\right) = x_0$ where $x(t)$ is continuous and differentiable on $\mathcal{D}$. In addition the equation is well-posed.

Finally, let us wrap up the mathematical definitions by the following definition of a linear system. We will assume the system under study in this chapter is such that Theorem 7.1 is met.

**Definition 7.3 Linear System**
An equation $\dot{x}\left(t\right) = f\left(x,t\right)$ is considered linear if $f(x,t)$ can be described as

$$f\left(x,t\right) = A\left(t\right)x + b\left(t\right)$$

where $A(t)$ is in general an $m \times m$ matrix and $b$ is a column vector of size $m$. If $A(t) = A$, in other words, constant independent of time, we say the system is linear with constant coefficients:

$$\dot{x}\left(t\right) = Ax + b\left(t\right)$$

If $b = 0$ we have a homogeneous system:

$$\dot{x}\left(t\right) = Ax$$

Imagine a general solution $\hat{x}(t)$ to the homogeneous system and $\tilde{x}(t)$ is a particular solution to constant coefficient equation, one can show that
$x(t) = \hat{x}(t) + \tilde{x}(t)$ is the general solution to the constant coefficient equation [3, 4] . Since $A$ is a matrix, one can solve for its eigenvalues and eigenvectors by solving

$$Aq_i = \lambda_i q_i, i \in \{1,2,,,m\} \tag{7.38}$$

One can now show that

$$\hat{x}(t) = \sum_{i=1}^{m} c_i e^{\lambda_i t} q_i \tag{7.39}$$

is the general solution to the homogeneous system where the constants $c_i$ are arbitrary [3, 4]. As a result, the solution to the constant coefficient equation is

$$x(t) = \bar{x}(t) + \sum_{i=1}^{m} c_i e^{\lambda_i t} q_i \tag{7.40}$$

We will need these solutions and properties of linear equations later when we discuss stability of ODEs.

### 7.2.1.2   Classification of Linear Multistep Methods

The most common solution methods are called linear multistep methods. They rely on more than one previous solution in time to calculate the new solution at the new timestep. Formally we have from [4] originally but also [3]

$$\sum_{j=-1}^{k-1} \alpha_j x_{n-j} = h \sum_{j=-1}^{k-1} \beta_j f\left(x_{n-j}, t_{n-j}\right), k \geq 1, \alpha_{-1} \equiv 1 \tag{7.41}$$

where $h$ is the timestep which is assumed uniform for now. If $\beta_{-1} \neq 0$, we say the method is implicit; else we say it is explicit.

For any solution method, there are a few important properties that are needed for them to be useful. First, the method should converge and we will define that properly shortly. Second, the method should be stable. These two properties will be discussed in the next few sections.

### Convergence

We can think of convergence as the way a numerical method approaches the exact solution as the timestep, $h \to 0$ (we leave round-off errors out of the discussion for now). The following definition is then fairly intuitive:

**Definition 7.4**

A numerical method is said to be convergent if, for all IVPs satisfying the hypotheses in Theorem 7.1, we have that

$$\lim_{h \to 0} \left( \max_{t_n \in [t_0, t_f]} x(t_n) - x_n \right) = 0$$

Notice we evaluate the exact solution $x$ at the discrete time points, $t_n$. Basically it means the numerical solution approaches the exact one as the timestep goes to zero. It is clearly necessary to have a method that approaches the exact solution in this limit.

**Definition 7.5**

A numerical method is said to be consistent if, for all IVPs satisfying the hypotheses in Theorem 7.1, we have that

$$\lim_{h \to 0} \left( \max_{t_n \in [t_0, t_f]} \frac{1}{h} R_{n+1} \right) = 0$$

Here $R_{n+1}$ is the residual defined as

$$R_{n+1} = \sum_{j=-1}^{k-1} \alpha_j x(t_{n-j}) - h \sum_{j=-1}^{k-1} \beta_j f\left( x(t_{n-j}), t_{n-j} \right) \tag{7.42}$$

It is essentially equal to the local truncation error (LTE) we discussed in Chap. 4. If $R_{n+1}$ is zero, it means the numerical solution follows the exact solution precisely. We now come to the important definition of the first characteristic polynomial. We will use this later in this section when discussing stability.

**Definition 7.6a**

For the linear multistep method, we define the first characteristic polynomial in $z \in C$ as

$$\rho(z) = \sum_{j=-1}^{k-1} \alpha_j z^{k-j-1} = \alpha_{-1} z^k + \alpha_0 z^{k-1} + \ldots + \alpha_{k-1}$$

As the reader sees, it is defined using only the coefficients on the left-hand side of the LMS method. This is due to fact that as $h \to 0$, it is only the terms on the left-hand side that determines the dynamics of the algorithm. As such it should be an important entity to study. We have also the second characteristic polynomial.

**Definition 7.6b**

For the linear multistep method, we define the second characteristic polynomial in $z \in C$ as

$$\sigma(z) = \sum_{j=-1}^{k-1} \beta_j z^{k-j-1} = \beta_{-1} z^k + \beta_0 z^{k-1} + \ldots + \beta_{k-1}$$

In this section, we have concerned ourselves with the implementation of a numerical method to solve the original IVP. To have a chance at a solution, the IVP had to meet certain well-posedness qualities as we discussed in the previous section. For the numerical method itself, there are similar considerations, and we will discuss first what is called zero-stable difference systems.

**Definition 7.7**

Let $\{\delta_n\}$ and $\{\hat{\delta}_n\}$ be any two perturbations of the difference system, and let $\{x_n\}$ and $\{\hat{x}_n\}$ be the resulting perturbed solutions. If there exists constants $S$ and $h_0$ such that, for all $0 < h \le h_0$, we have

$$x_n - \hat{x}_n \le S\epsilon, \forall n \quad \text{whenever} \, \delta_n - \hat{\delta}_n \le \epsilon, \forall n$$

we say the difference system is zero-stable.

Notice the similarity to the continuous time definition of well-posedness for the differential equation.

As we hinted earlier, the first characteristic polynomial is important to study zero stability. In fact we have the following theorem:

**Theorem 7.2**

A difference system is zero-stable if and only if it satisfies what is called the root condition: if every root of the characteristic polynomial $\rho(z)$ is either inside the unit circle (in the complex plane) or alone on the unit circle.

With these definitions and theorems in hand, we can now state the fundamental theorem for the study of IVPs as

**Theorem 7.3**

A difference system is convergent if and only if it is both consistent and zero-stable.

We will go through the various methods we have defined in the book and see how these conditions line up.

**Order of Difference Equation**

The concept of order of a difference equation will be defined shortly, but first we need to define what is called a difference operator.

The linear difference operator of a LMS method, denoted by $D$, is an operator which applied to a smooth time function $s(t)$ produces another time function:

$$D\big[s(t);h\big] \equiv \sum_{j=-1}^{k-1} \alpha_j s(t-jh) - h \sum_{j=-1}^{k-1} \beta_j s'(t-jh) \tag{7.43}$$

Notice that when $D$ is applied to the exact solution, it becomes the residual. With this definition, one can define the order of an LMS method easily. A Taylor expansion is most often used to describe the order of a function, and we start here with the expansion of $s(\tau)$ around $\tau = t$:

$$s(\tau) = s(t) + \sum_{q=1}^{\infty} \frac{1}{q!} \frac{d^q s(t)}{dt^q} (\tau - t)^q$$

The derivative of $s(\tau)$ with respect to $\tau$ gives us

$$\frac{ds(\tau)}{d\tau} = \sum_{q=1}^{\infty} \frac{1}{(q-1)!} \frac{d^q s(t)}{dt^q} (\tau - t)^{q-1}$$

After evaluating these two expressions at $\tau = t - jh$, we have

$$s(t-jh) = s(t) + \sum_{q=1}^{\infty} \frac{1}{q!} \frac{d^q s(t)}{dt^q} (-jh)^q$$

$$\frac{ds(t-jh)}{d\tau} = \sum_{q=1}^{\infty} \frac{1}{(q-1)!} \frac{d^q s(t)}{dt^q} (-jh)^{q-1}$$

We can now put these two expressions into the definition of $D$ and collect terms of the same kind

$$D\big[s(t);h\big] = C_0 s(t) + C_1 h \frac{ds(t)}{dt} + \ldots + C_q h^q \frac{d^q s(t)}{dt^q} + \ldots$$

with constants

$$C_0 = \sum_{j=-1}^{k-1} \alpha_j$$

$$C_1 = \sum_{j=-1}^{k-1} j\alpha_j - \sum_{j=-1}^{k-1} \beta_j$$

$$C_q = \frac{(-1)^q}{q!} \sum_{j=-1}^{k-1} j^q \alpha_j - \frac{(-1)^{q-1}}{(q-1)!} \sum_{j=-1}^{k-1} j^{q-1} \beta_j$$

**Definition 7.8**
An LMS method is said to be of order p if $C_0 = \ldots = C_p = 0$, but $C_{p+1} \neq 0$ and $C_{p+1}$ is called the error constant of the LMS method.

One can show that this order is a well-defined intrinsic property of an LMS method [3, 4]. The residual of an LMS method of order $p$ becomes

$$R_{n+1} = C_{p+1} h^{p+1} \frac{d^{p+1} x(t_n)}{dt^{p+1}} + \ldots h^{p+2}$$

This will be convenient later when discussing the local truncation error. Armed with these definitions and theorems, we can now make some interesting observations:

$$C_0 = \rho(1) \quad \text{and} \quad C_1 = \frac{d\rho(1)}{dz} - \sigma(1) - (k-1)\rho(1)$$

We now have an LMS method is consistent if and only if

$$\lim_{h \to 0} \frac{1}{h} \left( C_0 x(t_n) + C_1 h x^{\cdot(t_n)} + C_2 h^2 x^{\cdot\cdot(t_n)} + \ldots \right) = 0$$

Evidently for the method to be consistent, $C_0 = C_1 = 0$. In other words, the order has to be $\geq 1$. It is also clear to be consistent:

$$\rho(1) = 0 \quad \text{and} \quad \frac{d\rho(1)}{dz} = \sigma(1)$$

Interestingly, if $\sigma(1) = 0 \rightarrow \rho(1) = \rho'(1) = 0$, meaning $z = +1$ is a double root of $\rho(z)$, and the method cannot be zero-stable. Basically we have shown that consistent and zero-stable methods must have

$$\sigma(1) \neq 0$$

After all these definitions, it should come as no surprise that the location truncation error (LTE) is defined as

$$\text{LTE} = C_{p+1} h^{p+1} \frac{d^{p+1} x(t_n)}{dt^{p+1}} + \dots h^{p+2}$$

The first term above is often referred to as the principal local truncation error (PLTE).

With all this prep work, we can now take a look at our integration methods and see their characteristics.

**Forward Euler**

Forward Euler has the following sequence: $x_{n+1} = x_n + h f_n$ can be written as

$$x_{n+1} - x_n = h f_n$$

From a Taylor expansion, we have

$$x(t_{n+1}) = x(t_n) + h x'(t_n) + \frac{1}{2} h^2 x''(t_n) + \dots h^3$$

The LTE order of FE is thus 2.

**Backward Euler**

Similarly for BE we have $x_{n+1} - x_n = h f_{n+1}$, we have $\alpha_{-1} = 1, \alpha_0 = -1, \beta_{-1} = 1$ giving

$$C_0 = 0, \quad C_1 = 0, \quad C_2 = -\frac{1}{2}$$

Meaning BE is of order 1, whereas its LTE is of second order given by

$$\tau_{n+1}(h) = -\frac{1}{2} h^2 x''(t_n) + \dots h^3$$

**Trapezoidal Method**

The trapezoidal method has

$$x_{n+1} - x_n = \frac{h}{2}(f_{n+1} + f_n)$$

We can identify the coefficients as $\alpha_{-1} = 1, \alpha_0 = -1, \beta_{-1} = \beta_0 = 1/2$ giving the

$$C_0 = 0, \quad C_1 = 0, \quad C_2 = 0, \quad C_3 = -\frac{1}{12}$$

So that trap is of order 2 with LTE of order 3

$$\tau_{n+1}(h) = -\frac{1}{12}h^3 x'''(t_n) + \dots h^4$$

**Second-Order Gear Method**
We leave this to the reader to explore further.

### 7.2.1.3  Stability of Linear Multistep Methods

In the last section, we defined what is called zero stability. Essentially we looked at an LMS method to see if it approached the exact solution as the timestep $h \to 0$. We noted this is obviously a desirable property of a numerical solution scheme. The similarity between this zero stability property and the concept of well-posedness for the differential equation was also quite a natural consequence of the numerical method approaches of the continuous time formulation when $h \to 0$. Now, in a real implementation, we cannot put $h = 0$, and we instead need an idea of stability for finite $h$. In reality errors can accumulate as we step along the solution, and such errors can completely blow up the solution. Chapter 3 demonstrates such behavior for the FE method. The concept of stability has obviously been studied to great depths over the years due to its importance. Even before the arrival of the electronic computer, such studies were performed. It is a widely researched topic, and we will here limit ourselves to the presentation of linear stability theory. There are also intense research going on in the field of nonlinear stability, but we will not touch upon that here. Linear stability concerns the examination of a test system. In continuous time, we have

$$\frac{dx(t)}{dt} = Ax(t) \tag{7.44}$$

It is common to assume the solution to such systems die down with time, but strictly speaking, there are situations one can imagine where this is not true, for example, oscillators. We will not overly complicate the discussion but assume here the system solution is such $x(t) \to 0$, $t \to \infty$. A numerical approximation to this system can then be expected to also die down for large times, $x_n \to 0$, $n \to \infty$. This is what we mean when talking about stability of the numerical system. How can we convince ourselves our numerical method behaves this way? As we saw earlier, such a linear system has the solution

$$x(t) = \sum_{i=1}^{m} c_i e^{\lambda_i t} q_i$$

where $\lambda$ is the complex eigenvalue and $q_i$ is the eigenvector of the matrix $A$. For simplicity we will assume that all eigenvalues are distinct. For this solution to die down with time, we then simply require

$$\text{Re}(\lambda_i) < 0, \forall i$$

If we apply an LMS method to the system, we end up with the following sequence:

$$\sum_{j=-1}^{k-1} \alpha_j x_{n-j} - h \sum_{j=-1}^{k-1} \beta_j A x_{n-j} = \sum_{j=-1}^{k-1} (\alpha_j I - h\beta_j A) x_{n-j} = 0$$

We now need to understand under what conditions the solutions die down with time:

$$x_n \to 0, n \to \infty$$

From matrix theory, we know we can find a non-singular matrix $Q$ that transforms $A$ into a diagonal matrix $\Lambda = Q^{-1}AQ$ whose diagonal elements are equal to the eigenvalues of $A$, $\lambda_i$. Let us multiply the sequence by $Q^{-1}$ from the left and insert $I = QQ^{-1}$ next to $A$.

$$\sum_{j=-1}^{k-1} (\alpha_j Q^{-1} - h\beta_j Q^{-1} A Q Q^{-1}) x_{n-j} = \sum_{j=-1}^{k-1} (\alpha_j Q^{-1} - h\beta_j \Lambda Q^{-1}) x_{n-j} = 0$$

We then define $x_n = Q y_n$, and we get

$$\sum_{j=-1}^{k-1} (\alpha_j Q^{-1} - h\beta_j \Lambda Q^{-1}) Q y_{n-j} = \sum_{j=-1}^{k-1} (\alpha_j I - h\beta_j \Lambda) y_{n-j} = 0$$

The sequence $y_n$ can in general be complex. We see these define a sequence of $m$ single-variable equations (one for each eigenvalue), and we need to look for $y_n \to 0$, $n \to \infty$ which is simply a condition that each of the sequences

$$\lim_{n \to \infty} y_n = 0$$

We can then simply restate the problem in terms of single-variable sequence

$$\sum_{j=-1}^{k-1} (\alpha_j - h\beta_j \lambda_i) y_{n-j} = \sum_{j=-1}^{k-1} \gamma_j y_{n-j} = 0$$

and ask what conditions must exist so the solution satisfies the limit. Since all the $m$ equations are the same, we only need to look at one of them to find the right conditions. We have here assumed the coefficients are constant, and for such sequences, there are solutions one can find as follows. Let us write $y_n$ in terms of a candidate solution, $y_n = r_i^n$, where $r_i \in \mathbb{C}$ & $r_i \neq 0$:

$$\sum_{j=-1}^{k-1} \gamma_j y_{n-j} = \sum_{j=-1}^{k-1} \gamma_j r_i^{n-j} = r_i^{n-k+1}\left(\gamma_{-1} r_i^k + \gamma_0 r_i^{k-1} + \ldots + \gamma_{k-1}\right)$$

One can now see the sequence is a solution if $r_i$ is a root to the so-called characteristic polynomial:

$$\sum_{j=-1}^{k-1} \gamma_j r_i^{k-j-1} = \gamma_{-1} r_i^k + \gamma_0 r_i^{k-1} + \ldots + \gamma_{k-1}$$

In fact one can show that if $r_i$ is a solution to the characteristic polynomial, then the following sequences are also solutions:

$$\left\{r_i^n\right\}, \left\{n r_i^n\right\}, \left\{n^2 r_i^n\right\}, \ldots, \left\{n^{\mu_i - 1} r_i^n\right\}$$

One can further show the general solution is simple a linear combination of these sequences. With this in mind, it should be clear that for the solution to die down with time, we must have $|r_i| < 1$ for all roots of the characteristic polynomial. We can write the characteristic polynomial in terms of the first and second polynomials for the LMS method:

$$\pi\left(r, \hat{h}\right) = \sum_{j=-1}^{k-1}\left(\alpha_j - \hat{h}\beta_j\right)r^{k-j-1} = \rho(r) - \hat{h}\sigma(r), \hat{h} = h\lambda_i$$

This polynomial is often called the stability polynomial in the literature. We can now define

**Definition 7.9**
A LMS method is said to be absolutely stable for a given $\hat{h}$ if, for that $\hat{h}$, all the roots of $\pi\left(r, \hat{h}\right)$ are strictly inside the unit circle of the complex plane. Otherwise it is said to be absolutely unstable.

We can see that if $\rho(r)$ and $\sigma(r)$ have the same root, the polynomial has the same root. In general this is not true, so the root, $r_0$, can be found for a given $\hat{h}$ by

$$\hat{h}_0 = \frac{\rho(r_0)}{\sigma(r_0)}$$

Provided $\sigma(r_0) \neq 0$. In fact one can see the roots as being parameterized by $h$. The roots will change as $\hat{h}$ changes, so for same values of $\hat{h}$, the system is stable. This leads us to the following definition:

**Definition 7.10**
An LMS method is said to have a region of absolute stability $\mathcal{R}_A$ if it is absolutely stable for all $\hat{h} \in \mathcal{R}_A$.

This means one can use the timestep as a means to remedy stability. This will be in addition to using timestep as an accuracy control. It will restrict the timesteps available and can seriously prolong any simulation. The preference is then to use methods that do not depend on the timestep for their stability.

We can now study our integration methods to find out their region of stability. We will only look closely at Euler's methods.

**Forward Euler**
Forward Euler has the following sequence $x_{n+1} = x_n + hf_n$ we can write as

$$x_{n+1} - x_n = hf_n$$

We can easily identify the coefficients $\alpha_{-1} = 1$, $\alpha_0 = -1$, $\beta_0 = 1$. The characteristic polynomials are

$$\rho(z) = z - 1 \quad \text{and} \quad \sigma(z) = 1$$

The stability polynomial is now

$$\pi(r, \hat{h}) = \rho(r) - \hat{h}\sigma(r) = r - \hat{h} - 1$$

leading to the single root

$$r = \hat{h} + 1$$

For this sequence to be stable, we must then require $\hat{h}$ to be inside the unit circle centered at $-1$. We see here one can always bring this method into stability by choosing a small enough $h$.

**Backward Euler**
Backward Euler reads $x_{n+1} - x_n = hf_{n+1}$. We find $\alpha_{-1} = 1$, $\alpha_0 = -1$, $\beta_{-1} = 1$ so here

$$\rho(z) = z - 1 \quad \text{and} \quad \sigma(z) = z$$

With the stability polynomial

$$\pi\left(r,\hat{h}\right) = \rho\left(r\right) - \hat{h}\sigma\left(r\right) = r - 1 - \hat{h}r$$

which has a root

$$r = \frac{1}{1-\hat{h}}$$

In contrast to Forward Euler, we here need to keep $\hat{h}$ outside the circle of radius 1 centered at 1. $\mathcal{R}_A$ contains the whole left half plane, and the method is then absolutely stable for all timesteps $h$! One peculiar thing with the method is that for cases where the real solution is constant and does not die down with time, like an oscillator, the method still produces a sequence that approaches zero. It is, in a sense, overly stable.

**Trapezoidal Method, Second-Order Gear Method**
These trap and Gear2 methods can be analyzed similarly, but most of the time the roots are found numerically. We will not proceed any further here.

### 7.2.1.4   Timestep Control Algorithms

The method used to control the timestep when using variable timesteps is obvious of some import. Basically one can surmise that if a timestep was successful, one can increase the next timestep. The purpose would be to reach the final goal in as short a physical time as possible within the required accuracy. How much should one increase the step? If a timestep was not successful in reaching the desired accuracy, how much should one decrease it? The actual choice in professional circuit simulators is not well known, but the literature on numerical solution of differential equations has quite a few suggestions [3, 4]. The basic idea is to measure somehow how close to failing the error requirements a given timestep was, and if you are close to failing, you take somewhat smaller timesteps. If you are far away from failing, you take longer timesteps. If you are in the middle, you do not change the timestep. The problem then becomes finding out how far we are from failing the timestep. Assume we have such a norm, let us call it $\rho$. We have then

$$\text{if} \begin{cases} \rho \geq 0.9, \text{decrease timestep by } x\% \\ 0.2 < \rho < 0.9, \text{leave timestep unchanged} \\ \rho \leq 0.2 \text{ increase time step by } y\% \end{cases}$$

How can we find such a norm, and what are the appropriate changes to the timestep when required? The timestep itself is most governed by the LTE requirement. We can then define a norm as

$$\rho = \frac{\left| v_n(t) - v_{npred}(t) \right|}{\left| \alpha \left( \text{reltol } v_{nmax} + \text{vabstol} \right) \right|}$$

Naturally one can tier the timestep change more if desired and the reader is encouraged to do so. In our implementation, we decrease the timestep by 10% when needed and increase it by 0.1%. The reader is strongly encouraged to explore more.

## 7.3   Newton-Raphson Theory

We will in this section go through more of the details of the derivation of the Newton-Raphson algorithm and point out some of the potential difficulties one can run into when deploying it. As we already discussed in Chap. 2, the Newton-Raphson is a true workhorse in solving nonlinear equations. It is employed almost everywhere such solvers are needed and performs generally really well, provided the function it tracks is reasonably smooth. We will discuss briefly such shortcomings in this section.

### 7.3.1   Basic Derivation for Arbitrary Dimensions

Let us look at the following n-dimensional equation

$$f(x) = 0, f = f_n, n > 1 \tag{7.45}$$

where $f$ is a vector with components $f_n$ and we need to solve for $x$, having components $x_n$. Let us do a Taylor expansion around a point $x_0$ which does not solve the equation but is close. We find

$$f_n(x = x_0 + \Delta x) = f_n(x_0) + \frac{\partial f_i}{\partial x_j}(x_0)\Delta x = 0$$

The entity $\partial f_i/\partial x_j$ is a matrix commonly referred to as the Jacobian. By finding the inverse to the Jacobian, we can then solve for the error $\Delta x$ which can be written as

$$\Delta x = -\left[ \frac{\partial f_i}{\partial x_j} \right]^{-1} f(x_0) \tag{7.46}$$

Clearly if the Jacobian is non-singular, the procedure is very similar to the one-dimensional case.

**Convergence Rate**

The Newton-Raphson method converges with a rate proportional to $\Delta x^2$. This follows from the error term in the Taylor expansion.

### 7.3.2 Common Difficulties and Workarounds

The main difficulty with Newton-Raphson lies in the behavior of the Jacobian. If it is singular, the method obviously fails, but also if it is close to zero, the solution can jump far away from the true solution even if it is close to it at the beginning of the iteration. A non-singular Jacobian is usually indicative of a circuit problem most likely in the branch equations where poor device models give rise to poorly defined Jacobians.

In addition, there are also situations where there is a local minimum to the function $f$, and if during the course of iteration the solution gets close to such a point, it can iterate around this local minima endlessly [5]. The Newton-Raphson algorithm is guaranteed to converge if the starting point is close "enough" to the solution. Luckily in most circuit situations, the starting point is the previous timestep, and if the timestep itself is small enough, the Newton-Raphson method usually finds its way.

## 7.4 Shooting Method Theory

We have as before the governing equation [6]

$$f\big(v(t)\big) = i\big(v(t)\big) + \dot{q}\big(v(t)\big) + u = 0 \tag{7.47}$$

Let us spell out the full terms for all elements. For resistors

$$i(v) = \frac{1}{R} v$$

For inductors

$$i(v) = \int \frac{1}{L} v(t)\, dt$$

The capacitors appear naturally in the second term in Eq. 7.47

$$q(v) = C v$$

or

$$i(v) = \dot{q}(v) = C \frac{dv}{dt}$$

The transistors finally are the well-known

$$i(v) = g_m v$$

for the linear case.

We denote by $v(t)$ the state of the circuit voltages at a specific time $t$ similar to [5]. We are looking for solutions that have the property:

$$v(t) = v(t + T)$$

Let us define a function that is the difference in state between these two times:

$$h_i \left( v_i(t), v_i(t+T) \right) = v_i(t) - v_i(t+T)$$

We know this function is supposed to be zero when we have found the proper solution. For the intermediate iterations, we can do a Taylor expansion of $h$:

$$h_{i+1} = h_i + \frac{\partial h_i}{\partial v_i} \Delta v_i = v_i(t) - v_i(t+T) + \frac{\partial \left( v_i(t) - v_i(t+T) \right)}{\partial v_i} \Delta v_i(t)$$

$$\approx v_i(t) - v_i(t+T) + \left( I - \frac{\Delta v_i(t+T)}{\Delta v_i(t)} \right) \Delta v_i(t) = v_i(t) - v_i(t+T)$$

$$+ \left( I - J_{\varphi,ij}(T) \right) \Delta v_i(t) = 0$$

Here the Jacobian $J_{\varphi,\,ij}(T)$ represents the sensitivity of the final state due to changes in the initial state. The equation can now be rearranged to yield

$$\Delta v_i(t) = \left( I - J_{\varphi,ij}(T) \right)^{-1} \left( -v_i(t) + v_i(t+T) \right)$$

Following Newton-Raphson's idea, we see that a change in initial state can be calculated from the Jacobian of the previous iteration.

We can write the Jacobian using the chain rule as

$$J_{\varphi,ij}(T) = \frac{\partial v_N}{\partial v_{N-1}} \frac{\partial v_{N-1}}{\partial v_{N-2}} \cdots \frac{\partial v_1}{\partial v_0} \frac{\partial v_0}{\partial v_0}$$

It is in other words a product of the state sensitivities between all the timesteps where the last factor on the right is the identity matrix $\boldsymbol{I}$. This sensitivity of the state to the previous one can be derived by taking the derivative of Eq. 7.47 with respect to $s_0$:

$$
\frac{\partial f\left(v\left((t_n)\right)\right)}{\partial v_0} = \frac{\partial \left(i\left(v\left(t_n\right)\right)+\dot{q}\left(v\left(t_n\right)\right)+u\right)}{\partial v_0}
$$

$$
= \frac{\partial i\left(v\left(t_n\right)\right)}{\partial v_0} + \frac{\partial \left(q\left(v\left(t_n\right)\right)-q\left(v\left(t_{n-1}\right)\right)\right)}{\Delta t \partial v_0} = 0
$$

After applying the chain rule, we find

$$
\frac{\partial i\left(v\left(t_n\right)\right)}{\partial v\left(t_n\right)} \frac{\partial v\left(t_n\right)}{\partial v_0} + \frac{1}{\Delta t}\left[\frac{\partial q\left(v\left(t_n\right)\right)}{\partial v\left(t_n\right)} \frac{\partial v\left(t_n\right)}{\partial v_0} - \frac{\partial q\left(v\left(t_{n-1}\right)\right)}{\partial v\left(t_{n-1}\right)} \frac{\partial v\left(t_{n-1}\right)}{\partial v_0}\right] = 0
$$

We can rewrite this accordingly:

$$
\left[\frac{\partial i\left(v\left(t_n\right)\right)}{\partial v\left(t_n\right)} + \frac{1}{\Delta t}\frac{\partial q\left(v\left(t_n\right)\right)}{\partial v\left(t_n\right)}\right]\frac{\partial v\left(t_n\right)}{\partial v_0} = \frac{1}{\Delta t}\frac{\partial q\left(v\left(t_{n-1}\right)\right)}{\partial v\left(t_{n-1}\right)}\frac{\partial v\left(t_{n-1}\right)}{\partial v_0}
$$

The expression on the left-hand side is just the Jacobian, $\boldsymbol{J}_f$ for the time domain iterations we had before so

$$
\frac{\partial v\left(t_n\right)}{\partial v_0} = \frac{1}{\Delta t}\boldsymbol{J}_f^{-1}\frac{\partial q\left(v\left(t_{n-1}\right)\right)}{\partial v\left(t_{n-1}\right)}\frac{\partial v\left(t_{n-1}\right)}{\partial v_0}
$$

For a capacitor, we know the charge on the plates relates to the voltage like

$$
C(v) = \frac{\partial q}{\partial v}
$$

And we finally get

$$
\frac{\partial v\left(t_n\right)}{\partial v_0} = \frac{1}{\Delta t}\boldsymbol{J}_f^{-1}C\left(v\left(t_{n-1}\right)\right)\frac{\partial v\left(t_{n-1}\right)}{\partial v_0} \tag{7.48}
$$

The dependence of the state at timestep $t_n$ on the state at time $t = 0$ can so be achieved by multiplying the previous dependent matrix by the effective capacitance matrix and the Jacobian for the circuit equations at the new timestep.

## 7.5   Harmonic Balance Theory

Let us look at the circuit equations in terms of the function $f(v_k)$ similar to the discussion in [6]:

$$f\big(v(t)\big) = i\big(v(t)\big) + \dot{q}\big(v(t)\big) + u \tag{7.49}$$

Let us Fourier transform this equation. We find

$$F(V) = I(V) + \Omega Q(V) + U \tag{7.50}$$

Here the capital $V$ now represents the Fourier coefficients for the voltages at each node. We have used

$$F\big(\dot{q}(v(t))\big) = F\big(\dot{q}_k(v_l(t))\big) = F\left(\frac{dq_k(v_l)}{dv_l}\frac{dv_l(t)}{dt}\right) = F\left(\frac{dq_k(v_l)}{dv_l}\sum_m j\omega_m v_{lm} e^{j\omega_m t}\right)$$

For linear systems, we get

$$F\big(\dot{q}(v(t))\big) = F\big(\dot{q}_k(v_l(t))\big) = F\left(C_{kl}\frac{dv_l(t)}{dt}\right) = C_{kl}\sum_m j\omega_m v_{lm} = \Omega Q(V)$$

$$\Omega = \begin{pmatrix} j\omega & \cdots & 0 \\ \vdots & \ddots & \vdots \\ 0 & \cdots & j\omega \end{pmatrix}, \omega = \big(\omega_1 \cdots \omega_k\big)$$

For nonlinear functions $i, q$, the procedure is to first inverse Fourier transform the voltages in the time domain, solve the time evolution problem there (remember it is harmonic), and then Fourier transform back into the frequency domain. There are many ways to solve Eq. 7.50, but we will use what is often called harmonic Newton which is very similar to Newton-Raphson we used earlier in the time domain. We define the Jacobian

$$J_{ij} = \frac{\partial F_i}{\partial V_j}$$

We get

$$J(V) = \frac{\partial I(V)}{\partial V} + \Omega\frac{\partial Q(V)}{\partial V}$$

Applying the Newton-Raphson algorithm, we find the iteration

$$V^{j+1} = V^{(j)} - J^{-1}\left(V^{(j)}\right)F\left(V^{(j)}\right)$$

The matrix $J$ is referred to as the harmonic Jordan or conversion matrix. It tells you how a Fourier component at some node $j$ couples to another Fourier component at node $i$. Note the matrix is really a matrix within a matrix. Each circuit node contains a set of Fourier vector components that couples to Fourier components at other nodes. So now procedure is very similar to the one we followed in the time domain case. We set up the matrices $F$ and $J$ and iterate along. The boundary conditions set by $U$ has fixed components, and we can just easily iterate our way to the right solution. We do have many more variables; each node has k Fourier components, but once the iterations have converged, we are done.

## 7.6  Matrix Solvers: Brief Theory

Matrix solvers are one of the most important algorithms used in modern society. It is used far beyond mere engineering hallway applications. With the appearance of artificial intelligence on the hardware scene, such systems are also incorporated directly into hardware and are as such very much part of everyday life. From an engineering/scientific viewpoint, we end up with matrix equations simply because of the way we quantize or discretize the problems we are investigating. For the circuit application, we have simply a finite number of unknown voltages and currents, so there a matrix equation is natural. But also in more continuous systems like electromagnetic field solvers, the matrix equation appear because the systems we are investigating are put on a discrete grid within which key properties are constant or varying slowly. It is a very important field of study and notoriously difficult to construct solvers that work all the time for any kind of problem. For circuit systems, we are often helped by the fact the matrices are sparse and then efficient iterative solvers can be used, and we discuss such in Sect. 7.6.3. Sometimes the matrices are dense or small, and then the direct methods are employed. These are discussed in Sects. 7.6.1 and 7.6.2. We follow the presentation in [7, 8] fairly closely.

### 7.6.1  Gauss-Jordan Elimination

The method of Gauss to solve linear systems of equations has a long history [7]. We will here present a condensed version of the mathematical derivations. The Gauss method is simply a set of row and column interchanges with the goal to avoid a small value of the pivot we described in Chap. 2. The method is generally such that the matrix is producing an identity matrix on the left-hand side and so the matrix on the right-hand side is an inverse [7]:

$$Ax = b$$

Multiply by $A^{-1}$

$$A^{-1}Ax = Ix = A^{-1}b$$

where $x$, $b$ can be allowed to be matrices. An appropriate choice of $b$ will now show us the inverse matrix on the right-hand side. Since it requires the right-hand side to be known, it is quite susceptible to round off errors and is rarely the first method one should employ when inverting matrices. The key to the method is the way one adjusts the rows and columns to avoid a small pivot. The best choice is almost always to pick a pivot that is the largest available. Interestingly, the choice of pivot will depend on the original scaling of the problem. It is therefore common to scale the problem such that the largest element is unity; it is often called implicit pivoting. Let us dig into the row and column interchanges some more. One can show pretty easily that the row interchange can be accomplished by multiplying the matrix from the left by a matrix $R$. A Gauss elimination by row is then simply a series of multiplication by some matrix from the left

$$Ax = b$$

$$R_1 R_2 \ldots R_n Ax = Ix = R_1 R_2 \ldots R_n b$$

where we use the set of left-hand operation to define the inverse matrix. Column interchanges correspond to right-hand multiplications of a matrix $C$. We get

$$Ax = b$$

$$AC_1C_1^{-1}x = b$$

$$AC_1C_2C_2^{-1}C_1^{-1}x = b$$

$$AC_1C_2 \ldots C_nC_n^{-1} \ldots C_2^{-1}C_1^{-1}x = b$$

The matrix multiplications $AC_1C_2 \ldots C_n = I$, and we find

$$x = C_1C_2 \ldots C_n b$$

We notice from this equation that we first need to multiply the right-hand side by the *last* matrix we find, so all matrices $C$ need to be stored in memory. This makes the column interchanges very expensive, and one is often left with these operations being very simple ones.

## 7.6.2   LU Decomposition

A much better general method to find the inverse of matrices that we exemplified in Chap. 2 is the LU decomposition method. For this method to work, we do not need the right-hand side of the equation, but we can proceed by rewriting the matrix directly [7]

$$A x = LU x = b$$

where $U$ is an upper-right triangular and $L$ is a lower-left triangular matrix. We then simply have to solve a set of equations $U x = y$ and $L y = b$. Since the $LU$ matrices are triangular, it is a set of straightforward substitutions. This sounds good, but the alert reader no doubt questions the complexity of decomposing a matrix in such a way. There is in fact an extraordinary procedure known as Crout's method that quite easily decomposes a matrix into these two matrices. Let us write

$$L = \begin{pmatrix} l_{11} & \cdots & 0 \\ \vdots & \ddots & \vdots \\ l_{n1} & \cdots & l_{nn} \end{pmatrix} \quad U = \begin{pmatrix} u_{11} & \cdots & u_{1n} \\ \vdots & \ddots & \vdots \\ 0 & \cdots & u_{nn} \end{pmatrix}$$

It turns out the diagonal of $L$ can always be chosen to be unity $l_{ii} = 1$. Let us write out explicitly a row of the matrix equation $LU = A$:

$$l_{i1} u_{1j} + l_{i2} u_{2j} + \ldots + l_{in} u_{nj} = a_{ij}$$

Given the triangular nature of $L, U$, we see the number of terms in the sum on the left depends strongly on whether $i = j$ or $i < j$:

$$i = j : l_{i1} u_{1i} + l_{i2} u_{2i} + \ldots + l_{ii} u_{ii} = a_{ii}$$

$$i < j : l_{i1} u_{1j} + l_{i2} u_{2j} + \ldots + l_{ii} u_{ij} = a_{ij}$$

$$i > j : l_{i1} u_{1j} + l_{i2} u_{2j} + \ldots + l_{ij} u_{jj} = a_{ij}$$

The Crout's algorithm now gives the following steps:

- Set $l_{ii} = 1$
- For each $j = 1, 2, \ldots, N$, do:

  1. For $i = 1, 2, \ldots j$, use (1) and (2) to solve for $u_{ij} : u_{ij} = a_{ij} - \sum_{k=1}^{i-1} l_{ik} u_{kj}$
  2. For $i = j + 1, j + 1, \ldots, N$, use (3) to solve for $a_{ij}$:

$$l_{ij} = \frac{1}{u_{jj}} a_{ij} - \sum_{k=1}^{j-1} a_{ik} u_{kj}$$

- Working through these steps, one finds the procedure can be used to replace the original $A$ with a new

$$
\begin{pmatrix}
u_{11} & u_{12} & u_{1n} \\
l_{21} & u_{22} & u_{2n} \\
l_{n1} & l_{n2} & u_{nn}
\end{pmatrix}
$$

which are the *LU* components! The reader realizes naturally that pivoting is again a critical procedure in completing the calculations. One in fact does not LU-decompose $A$ but rather a row-permutated version of it.

### 7.6.3   *Iterative Matrix Solvers*

The idea of an iterative solver is fairly easy to understand, [7]. Suppose we have an approximate solution to $Ax = b$. Let us call it $x_0$. We then have $x = x_0 + \delta x$. We end up with

$$
Ax = A\left(x_0 + \delta x\right) = b
$$

We can further write $b = b_0 + \delta b$, where $A\delta x = \delta b$ and $Ax_0 = b_0$. We can now put these relations into the matrix equation, and we find

$$
Ax = A\left(x_0 + \delta x\right) = b_0 + \delta b \rightarrow A\delta x = b - b_0
$$

where the right-hand side is known. We now simply need to solve this equation for $\delta x$, and we then simply subtract it from our original guess $x_0$ to end up with a new guess at the solution. Questions regarding convergence, etc. are obviously important, and there is a lot of formal theory regarding this which we will now describe in some cursory fashion.

We saw in Chap. 2 that an iterative method to solve matrix equations belonging to a class of solutions known as Krylov subspace methods can be very efficient in particular when solving large sparse matrices. It involved the idea of using projection to iterate through to the correct solution to solve [8]

$$
Ax = b
$$

Let us attempt a first guess to a solution $x \approx x_0$. We define the residue

$$
r_0 = Ax_0 - b \tag{7.51}
$$

In all likelihood, this residue is not zero. One can now define a string of vectors as

$$r_0, Ar_0, A^2 r_0, \ldots, A^{m-1} r_0$$

This string of vectors make up a subspace, the Krylov subspace, $\mathcal{K}_m$. We will not prove this here but see [8] for many examples and more details.

**General Procedure**

The general iterative procedure is now approximate: we have defined the Krylov subspace $\mathcal{K}_m$. Let us now annotate with $\mathcal{L}_m$ another subspace:

1. We take an approximate solution $x_m$ from the subspace made up of the vectors from the space $x_0 + \mathcal{K}_m$.
2. Require that $b - Ax_m \perp \mathcal{L}_m$. In other words, the resulting residue cannot be part of the subspace $\mathcal{L}_m$. Check error. Go back to 1.

All these techniques create some kind of polynomial approximation of the matrix inverse; the choice of $\mathcal{L}_m$ will be important. Sometimes, $\mathcal{L}_m = \mathcal{K}_m$; at other times $\mathcal{L}_m = A\mathcal{K}_m$. There are many varieties of this method. We will discuss Krylov subspaces and point out some of their properties.

**Krylov Subspaces**

We assume here we have a Krylov subspace made up of

$$\mathcal{K}_m(A, v) = \operatorname{span}\left\{r_0, Ar_0, A^2 r_0, \ldots, A^{m-1} r_0\right\}$$

As we proceed with the iterations, the index, $m$, increases, and so the dimension of the subspace is also increasing. The vector $v$ has associated with it a minimal polynomial that is non-zero of lowest degree such that $p(A)v = 0$. This degree is often called the grade of $v$ with respect to $A$. The grade of $v$ cannot exceed the dimension $n$ of the space $R^n$. There are some interesting properties of this subspace. We have:

1. $\mathcal{K}_m$ is the subspace of all vectors that can be written as a polynomial of $p(A)v$ where $p$ is a polynomial of degree $\leq m - 1$.
2. If we let $\gamma$ denote the grade of $v$, we have $\mathcal{K}_m = \mathcal{K}_\gamma$ for $m > = \gamma$.
3. $\dim\left(\mathcal{K}_m\right) = m \leftrightarrow \operatorname{grade}(v) \geq m$

The number of steps of a Krylov subspace method is limited by the maximal Krylov subspace dimension $d$. A projection process breaks down in step $m$ if no $x_m$ is found or if it is not unique. We are of course interested in methods that ensure existence and uniqueness for each $x_m$ as we step along $m \leq d$. A method is called *well-defined* if it terminates with the exact solution in step $d$. It is called the *finite termination property*. This depends on the properties of the matrix $A$. Let us now assume the matrix $A$ is non-singular. We then also note the conditions

$$x_m \in x_0 + K(A, r_0) \qquad r_m \in r_0 + AK(A, r_0)$$

Imply the error $x - x_m$ and the residual $r_n$ can be written in the polynomial form

$$x - x_m = p_m(A)(x - x_0), \qquad r_m = p_m(A)r_0$$

where $p_m$ is a polynomial of degree at most $m$ and with value one at the origin. We have only discussed the very preliminaries of these powerful methods and for more details see for example the discussions in [7, 8].

**Convergence Properties**
Since these methods have a finite number of steps, the idea of convergence is a little different compared to, for example, Newton-Raphson, where we could get closer and closer to the precise solution by just taking more steps. Here, once you have reached maximal subspace dimension $d$, you are done and should be at the exact solution (if the method is well-defined as we just discussed). We will not go into the details here but suffice it to say the idea of convergence and rate of convergence is quite different and does not have their usual meaning [8].

# References

1. Hachtel, G. D., Brayton, R. K., & Gustavson, F. G. (1971). The sparse tableau approach to network analysis and design. *IEEE Transactions on Circuit Theorym, 18*(1), 101–113.
2. Ho, C.-W., Zein, A., Ruehli, A. E., & Brennan, P. A. (1975). The modified nodal approach to network analysis. *IEEE Transactions on Circuits and Systems, 22*, 504–509.
3. Najm, F. N. (2010). *Circuit simulation.* Hobroken: Wiley.
4. Lambert, J. D. (1991). *Numerical methods for ordinary differential systems.* Chichester: Wiley & Sons.
5. Pedro, J., Root, D., Xu, J., & Nunes, L. (2018). *Nonlinear circuit simulation and modeling: fundamentals for microwave design* (The Cambridge RF and microwave engineering series). Cambridge: Cambridge University Press. https://doi.org/10.1017/9781316492963.
6. Kundert, K., White, J., & Sangiovanni-Vicentelli. (1990). *Steady-state methods for simulating analog and microwave circuits.* Norwell, MA: Kluwer Academic Publications.
7. Press, W. H., Teukolsky, S. A., Vetterling, W. T., & Flannery, B. P. (2007). *Numerical recipes.* Cambridge: Cambridge University Press.
8. Saad, Y. (2003). *Iterative method for sparse linear systems* (2nd ed.). Philadelphia: Society for Industrial and Applied Mathematics.

# Appendix A Complete Python Code Lists for All Examples

This appendix contains the complete codes for all the various simulation examples studied in the book. First, a word of caution:

- The main goal of the code has been to make it readable for people without any code expertise in their background. Therefore, a lot of the cleverness of Python has perhaps not been utilized fully.
- The code segments have not been debugged much beyond the displayed netlists, so if other netlists are attempted, it is likely there will be errors.

## A.1 Introduction

Here we will first define the various variables we are using and show an example of a Python implementation of what we just discussed in the previous sections.

### A.1.1 Variables

The python code will be discussed here with variable definitions, etc.

**DeviceCount** – Counter for number of devices in the netlist

**NDevices** – Max number of devices

**DevType**[] – Array of length NDevices where each element describes the device type for each element

**DevLabel**[] – Array of length NDevices where each element contains a string with the device's label.

**DevNode1**[] – Array of length NDevices where each element contains a string with the device's first node name

**DevNode2**[] – Array of length NDevices where each element contains a string with the device's second node name

© Springer Nature Switzerland AG 2021
M. Sahrling, *Analog Circuit Simulators for Integrated Circuit Designers*,
https://doi.org/10.1007/978-3-030-64206-8

**DevNode3**[] – Array of length NDevices where each element contains a string with the device's third node name

**DevNode4**[] – Array of length NDevices where each element contains a string with the device's fourth node name

**DevValue**[] – Array of length NDevices where each element contains a number with the device's value, for example, resistance

**Nodes**[] – List of string node names. It is dynamically allocated as the netlist is read.

**NumberOfNodes** – Length of Nodes[]

**STA_matrix**[][] – The STA matrix with size [NumberOfNodes+DeviceCount] [NumberOfNodes+DeviceCount]

**STA_rhs**[] – The right hand side of the equation. It has size [NumberOfNodes+DeviceCount].

### A.1.2  Basic Structure

The STA_matrix is defined such that the first NumberOfNodes rows contain the nodes where KCL is described. The next DeviceCount rows contain the branch equation for each device. The STA_rhs follows this structure.

### A.1.3  Netlist Syntax

The netlists follows the basic SPICE netlist format.

The SPICE netlist format was originally conceived by the team behind the SPICE simulator at UC Berkeley. The source file for any version of spice has the following format:

TITLE
ELEMENT DESCRIPTIONS
.MODEL STATEMENTS
OUTPUT COMMANDS
.END

Important points:

- The first line traditionally is a circuit name label preceded by a *. Here we just keep that tradition going and start all netlists with such a row. It is treated as a comment.
- Comment entire lines by beginning them with '*'.
- Only independent sources can contain the ground node "0."

**The Circuit Description**

A circuit description in SPICE, which is frequently called a netlist, consists of a statement defining each circuit element. Connections are described by naming nodes. (The usual names are actually numbers.) One node name has a defined meaning. Node 0 is ground! This node can only be connected to an independent source.

No other elements can have ground explicit in the node list. Instead one defines a node like vss which connects to ground via an independent source.

The format of an element description is

$$\text{<letter> <name> <n1> <n2>} \dots [\text{mname}][\text{parvals}]$$

where <...> must be present and [...] is optional.

- <letter> is a single letter denoting the component type.
- <name> is a unique alpha-num combination describing the particular instance of this component..
- <ni> is the name of a node.
- [mname] is the (optional) modelname.
- [parvals] are (sometimes optional) parametervalues.

**Sign Conventions**

A current is considered positive going into a device port and negative otherwise.

**Passive Elements**

The <letter> that begins an element instance denotes the circuit element. The passive elements are

R or r for resistors, L or l for inductors, and C or c for capacitors.

This <letter> is followed by a unique instancename and then (in order) the nodes associated with + and – voltage and the value of the associated parameter (R, L, or C). The value needs to be in real numbers.

Examples:

- R 1 50 20000
- cloadn IN GND 250e-15
- L41 22 21 4e-9

**Independent Sources**

$$\text{V<name> <n+> <n-> } [\text{type}] \text{<val>}$$

defines an independent voltage source with its + terminal at node n+ and its - node at node n-.

$$\text{I<name> <n+> <n-> } [\text{type}] \text{<val>}$$

defines an independent current source whose current flows through the source from node n+ to node n-.

*Only independent sources can have node 0 defined. All other devices cannot. This is to reduce the number of conditional statements in the code.*

**Examples:**

- Vdd 4 0 5 defines a 5 V source with the + terminal connected at node 4 and the - terminal connected at node 0 (ground).
- Ibias 18 4 15e-3
- Vin vin 0 sin(Offset Amplitude freq phase) defines a sinewave with a certain Offset, Amplitude, freq and phase.
- Vpwl vin 0 pwl( t1 val1 t2 val2) defines a piecewise linear device with time/value points defined by the list of pairs.

**Bipolar Junction Transistors**

The BJT also requires both a netlist statement and a .MODEL. A BJT is included in the netlist with a statement of the form

$$Q<name> <nc> <nb> <ne> <model - name>$$

where the collector is connected at node nc, the base at node nb, and the emitter at node ne.

**Example:** Q3 6 3 0 my-npn corresponds to

The model-name is defined as

$$.MODEL <model - name> Early =, K =$$

where Early means the Early voltage as discussed in Chap. 3 and K is the current $I_0$ as defined in Sect. 3.4.3.

**MOSFETs**

The BJT again requires both a netlist statement and a .MODEL. A MOSFET is included in the netlist with a statement of the form

$$M<name> <nd> <ng> <ns> <model - name>$$

where the drain, gate, source, and body are connected at nodes nd, ng, ns, and nb, respectively. The length L and width W are optional.

The model-name is defined as

$$.MODEL <model - name> K =, VT =, lambdaT =$$

corresponding to the basic parameters in Sect. 3.4.2

**A.1.4 Control Statements**

There are four types of control lines possible: .options, .ic, .write, and .plot. The syntax is trivial, but beware of trailing spaces:

.options parameter1=val1 parameter2=val2
.ic v(out)=1 v(in)=0 * This statement only controls voltage initial conditions at this point.
.write filename v(out) v(in) * This write the solution values for voltage nodes out and in to file 'filename'
.plot v(out) v(in) * This plots the voltages out in to the Python output window.


## A.2 AnalogDef.py

```python
#!/usr/bin/env python3
# -*- coding: utf-8 -*-
"""
Created on Tue Jan 28 12:04:22 2020

@author: mikael
"""
import sys
import re
import math
import matplotlib.pyplot as plt
import numpy
from scipy.fftpack import fft, ifft


Vthermal=1.38e-23*300/1.602e-19
TINY=1e-5

def readnetlist(netlist,modeldict,ICdict,Plotdict,Writedict,Opt
iondict,DevType,DevValue,DevLabel,DevNode1,DevNode2,DevNode3,DevM
odel,Nodes,MaxNDevices):
    try:
        myfile=open(netlist,'r')
    except:
        print('netlist file',netlist,' not found')
        sys.exit()
    DeviceCount=0
    if len(modeldict)==0:
        print('Warning: model dictionary is empty!')
    line=myfile.readline()
    while line !='' :
        DevType[DeviceCount]='empty'
        if line[0]=='*':
            print('comment')
```

```
        if line[0]=='v':
            print('VoltSource')
            DevType[DeviceCount]='VoltSource'
        if line[0]=='i':
            print('CurrSource')
            DevType[DeviceCount]='CurrentSource'
        if line[0]=='r':
            print('resistor')
            DevType[DeviceCount]='resistor'
        if line[0]=='f':
            print('Special Oscillator filter found')
            DevType[DeviceCount]='oscfilter'
        if line[0]=='l':
            print('inductor')
            DevType[DeviceCount]='inductor'
        if line[0]=='c':
            print('capacitor')
            DevType[DeviceCount]='capacitor'
        if line[0]=='m':
            print('transistor')
            DevType[DeviceCount]='transistor'
        if line[0]=='q':
            print('bipolar')
            DevType[DeviceCount]='bipolar'
        if re.split(' ',line)[0]=='.ic':
            print('Initial Condition Statement')
            lineSplit=re.split(' ',line)
            for i in range(len(lineSplit)-1):
                ConditionSplit=re.split('\(|\)|=|\n',re.split('
',line)[i+1])
                if len(ConditionSplit)>2:
                    try:
                        ICdict[i]={}
                        ICdict[i]['NodeName']=ConditionSplit[1]
                        ICdict[i]['Value']=float(ConditionSplit[
3])
                    except:
                        print('Syntax Error in .ic statement')
                        sys.exit()
                else:
                    print('Warning: Odd characters in IC state-
ment \'',ConditionSplit,'\'')
        if re.split(' ',line)[0]=='.plot':
            print('Plot Statement')
            lineSplit=re.split(' ',line)
```

```
                for i in range(len(lineSplit)-1):
                    ConditionSplit=re.split('\(|\)|=|\n',re.split('
',line)[i+1])
                    if len(ConditionSplit)>2:
                        try:
                            Plotdict[i]={}
                            Plotdict[i]['NodeName']=ConditionSplit[
1]
                        except:
                            print('Syntax Error in .plot statement')
                            sys.exit()
                    else:
                        print('Warning: Odd characters in .plot state-
ment \'',ConditionSplit,'\'')
            if re.split(' ',line)[0]=='.write':
                print('Write Statement')
                lineSplit=re.split(' ',line)
                Writedict[0]={}
                Writedict[0]['filename']=lineSplit[1]
                for i in range(len(lineSplit)-2):
                    ConditionSplit=re.split('\(|\)|=|\n',re.split('
',line)[i+2])
                    if len(ConditionSplit)>2:
                        try:
                            Writedict[i+1]={}
                            Writedict[i+1]['NodeName']=ConditionSpl
it[1]
                        except:
                            print('Syntax Error in .write statement
')
                            sys.exit()
                    else:
                        print('Warning: Odd characters in .write
statement \'',ConditionSplit,'\'')
            if re.split(' ',line)[0]=='.options':
                print('Option Statement')
                lineSplit=re.split(' ',line)
                for i in range(len(lineSplit)-1):
                    ConditionSplit=re.split('=|\n',re.split(' ',lin
e)[i+1])
                    if len(ConditionSplit)>=2:
                        try:
                            Optiondict[ConditionSplit[0]]=float(Cond
itionSplit[1])
                        except:
```

```
                    try:
                        Optiondict[ConditionSplit[0]]=Condi
tionSplit[1]
                    except:
                        print('Syntax Error in .options sta
tement')
                        sys.exit()
                else:
                    print('Warning: Odd characters in .options
statement \'',ConditionSplit,'\'')
        if DevType[DeviceCount]!='empty':
            if DevType[DeviceCount] != 'transistor' and
DevType[DeviceCount] != 'bipolar':
                try:
                    DevLabel[DeviceCount]=line.split(' ')[0]
                except:
                    print('Syntax Error in line:',line)
                    sys.exit();
                try:
                    DevNode1[DeviceCount]=line.split(' ')[1]
                except:
                    print('Syntax Error in line:',line)
                    sys.exit()
                if DevType[DeviceCount] != 'VoltSource' and
DevType[DeviceCount]    !=    'CurrentSource'    and    DevNode
1[DeviceCount]=='0':
                    print('Error: Node \'0\' only allowed for i
ndependent sources')
                    print('line',line)
                    sys.exit()
                try:
                    DevNode2[DeviceCount]=line.split(' ')[2]
                except:
                    print('Syntax Error in line:',line)
                    sys.exit()
                if DevType[DeviceCount] != 'VoltSource' and
DevType[DeviceCount]        !=        'CurrentSource'        and
DevNode2[DeviceCount]=='0':
                    print('Error: Node \'0\' only allowed for
independent sources')
                    print('line',line)
                    sys.exit()
                try:
                 DevValue[DeviceCount]=float(line.split(' ')[3])
                except:
```

```
                              print('Value is not a number')
                              if DevType[DeviceCount] != 'VoltSource' and
DevType[DeviceCount] != 'CurrentSource':
                                  sys.exit(0)
                              srcdict={}
                              try:
                                  DevValue[DeviceCount]=re.split((' |\('),
line)[3]
                              except:
                                  print('Syntax Error in line:',line)
                                  sys.exit();
                              srcdict[0]={}
                              srcdict[0]['type']=DevValue[DeviceCount]
                              if DevValue[DeviceCount]=='pwl':
                                  DoneReadingPoints=False
                                  pnt=1
                                  while not DoneReadingPoints:
                                      try:
                                          TimePnt=float(re.split((' |\('),
line)[2+pnt*2])
                                      except:
                                          DoneReadingPoints=True
                                      if not DoneReadingPoints:
                                          srcdict[pnt]={}
                                          srcdict[pnt]['time']=TimePnt
                                          try:
                                              SrcPnt=float(re.split((' |\
(|\)'),line)[3+pnt*2])
                                          except:
                                              print('Syntax Error in lin
e:',line)
                                              sys.exit();
                                          srcdict[pnt]['value']=SrcPnt
                                          pnt=pnt+1
                              if DevValue[DeviceCount]=='sin':
     #                            srcdict[1]={}
                                  try:
                                      srcdict['Offset']=float(re.split((' |\
('),line)[4])
                                  except:
                                      print('Syntax Error in line:',line)
                                      sys.exit();
                                  try:
                                      srcdict['Amplitude']=float(re.split(
(' |\('),line)[5])
```

```
                    except:
                        print('Syntax Error in line:',line)
                        sys.exit();
                    try:
                        srcdict['Freq']=float(re.split((' |\
('),line)[6])

                    except:
                        print('Syntax Error in line:',line)
                        sys.exit();
                    try:
                         srcdict['TDelay']=float(re.split(('
|\('),line)[7])

                    except:
                        print('Syntax Error in line:',line)
                        sys.exit();
                    try:
                        srcdict['Theta']=float(re.split((' |
\(|\)'),line)[8])

                    except:
                        print('Syntax Error in line:',line)
                        sys.exit();
                DevValue[DeviceCount]=srcdict
                if DevNode1[DeviceCount] not in Nodes and
DevNode1[DeviceCount]!='0':
                    Nodes.append(DevNode1[DeviceCount])
                if DevNode2[DeviceCount] not in Nodes and
DevNode2[DeviceCount]!='0':
                    Nodes.append(DevNode2[DeviceCount])
            else:
                try:
                    DevLabel[DeviceCount]=line.split(' ')[0]
                except:
                    print('Syntax Error in line:',line)
                    sys.exit();
                try:
                    DevNode1[DeviceCount]=line.split(' ')[1]
                except:
                    print('Syntax Error in line:',line)
                    sys.exit();
                if DevNode1[DeviceCount]=='0':
                    print('Error: Node \'0\' only allowed for
independent sources')
                    print('line',line)
                    sys.exit()
                try:
```

```
                            DevNode2[DeviceCount]=line.split(' ')[2]
                    except:
                        print('Syntax Error in line:',line)
                        sys.exit();
                    if DevNode2[DeviceCount]=='0':
                        print('Error: Node \'0\' only allowed for i
ndependent sources')
                        print('line',line)
                        sys.exit()
                    try:
                        DevNode3[DeviceCount]=line.split(' ')[3]
                    except:
                        print('Syntax Error in line:',line)
                        sys.exit();
                    if DevNode3[DeviceCount]=='0':
                          print('Error: Node \'0\' only allowed for
independent sources')
                        print('line',line)
                        sys.exit()
                    try:
                        DevModel[DeviceCount]=line.split(' ')[4]
                        DevModel[DeviceCount]=DevModel[DeviceCount].
rstrip('\n')
                        modelIndex=findmodelIndex(modeldict,DevModel
[DeviceCount])
                    except:
                        print('Syntax Error in line4:',line)
                        sys.exit();
                        DevValue[DeviceCount]=-1/modeldict[modelIndex]
['K']#-1/K
                        if DevNode1[DeviceCount] not in Nodes and
DevNode1[DeviceCount]!='0':
                        Nodes.append(DevNode1[DeviceCount])
                        if DevNode2[DeviceCount] not in Nodes and
DevNode2[DeviceCount]!='0':
                        Nodes.append(DevNode2[DeviceCount])
                        if DevNode3[DeviceCount] not in Nodes and
DevNode3[DeviceCount]!='0':
                        Nodes.append(DevNode3[DeviceCount])
                DeviceCount+=1
                if DeviceCount>=MaxNDevices:
                    print('Too many devices in the netlist: Max is
set to ',MaxNDevices)
                    sys.exit()
            line=myfile.readline()
```

```python
        return DeviceCount

    def readmodelfile(filename):
        modeldict={}
        index=0
        modelfile=open(filename,'r')
        line=modelfile.readline()
        while line != '' :
            modeldict[index]={}
            name=line.split(' ')[0]
            print('Reading model ',name)
            modeldict[index]['modelName']=name
            for i in range(3):
                dum=line.split(' ')[i+1]
                try:
                    dum.index("=")
                except:
                    print('Syntax error in model file, line ',line)
                    sys.exit()
                Parname=dum.split('=')[0]
                try:
                    ParValue=float(dum.split('=')[1])
                except:
                    print('Syntax error: Parameter',Parname,' value
is not a number',dum)
                modeldict[index][Parname]=ParValue
            index=index+1
            line=modelfile.readline()
        return modeldict

    def findmodelIndex(modeldict,name):
        for i in range(len(modeldict)):
            if modeldict[i]['modelName'] == name:
                return i
        print('model name ',name,' is not found in modelfile' )
        sys.exit()

    def getSourceValue(DevValue,SimTime):
        if type(DevValue)==float:
            return DevValue
        if type(DevValue)==dict:
            if DevValue[0]['type']=='sin':
                A=DevValue['Amplitude']
                freq=DevValue['Freq']
                Offset=DevValue['Offset']
```

```
                    TDelay=DevValue['TDelay']
   #                Theta=DevValue['Theta']
                            return   Offset+A*math.sin(freq*2*math.
pi*(SimTime-TDelay))
            if DevValue[0]['type']=='pwl':
                TimeIndex=1
                    while SimTime >= DevValue[TimeIndex]['time']  and
TimeIndex<len(DevValue)-1:
                        TimeIndex=TimeIndex+1
                if SimTime>=DevValue[len(DevValue)-1]['time']:
                        return DevValue[len(DevValue)-1]['value']
                else:
                        PrevTime=DevValue[TimeIndex-1]['time']
                        NextTime=DevValue[TimeIndex]['time']
                        PrevValue=DevValue[TimeIndex-1]['value']
                        NextValue=DevValue[TimeIndex]['value']
                    return (NextValue-PrevValue)*(SimTime-PrevTime)/
(NextTime-PrevTime)+PrevValue


   def findParameter(modeldict,modelname,parameterName):
        for i in range(len(modeldict)):
            if modeldict[i]['modelName']==modelname:
                try:
                    return modeldict[i][parameterName]
                except:
                        print('Error: Parameter ',parameterName,' not
found in ',modeldict[i])


   def setupDicts(SimDict,SetupDict,Optdict,DevType,DevValue,DevLa
bel,DevNode1,DevNode2,DevNode3,DevModel,Nodes,MatrixSize,Jacobian
,STA_matrix,STA_rhs,STA_nonlinear,sol,solm1,solm2,f):
        SetupDict['NumberOfNodes']=10
        SetupDict['NumberOfCurrents']=10
        SetupDict['DeviceCount']=10
        SetupDict['Nodes']=Nodes
        SetupDict['DevNode1']=DevNode1
        SetupDict['DevNode2']=DevNode2
        SetupDict['DevNode3']=DevNode3
        SetupDict['DevValue']=DevValue
        SetupDict['DevType']=DevType
        SetupDict['DevModel']=DevModel
        SetupDict['MatrixSize']=MatrixSize
        SetupDict['Jacobian']=Jacobian
        SetupDict['STA_matrix']=STA_matrix
        SetupDict['STA_rhs']=STA_rhs
```

```
        SetupDict['STA_nonlinear']=STA_nonlinear
        SetupDict['Vthermal']=1.38e-23*300/1.602e-19
        Optdict['reltol']=1e-3
        Optdict['iabstol']=1e-7
        Optdict['vabstol']=1e-6
        Optdict['lteratio']=2
        Optdict['MaxTImeStep']=1e-11
        Optdict['FixedTimeStep']='False'
        Optdict['GlobalTruncation']='True'
        Optdict['deltaT']=3e-13
        Optdict['MaxSimulationIterations']=200000
        Optdict['MaxSimTime']=1e-8
        Optdict['MaxNewtonIter']=5
        SimDict['deltaT']=1e-12
        SimDict['sol']=sol
        SimDict['solm1']=solm1
        SimDict['solm2']=solm2
        SimDict['f']=f


    def plotdata(Plotdict,NumberOfNodes,retime,reval,Nodes):
        if len(Plotdict)> 0:
            ax = plt.subplot(111)
            for j in range(NumberOfNodes):
                for i in range(len(Plotdict)):
                    if Plotdict[i]['NodeName']==Nodes[j]:
                        ax.plot(retime, reval[j], label=Nodes[j])
            plt.title('Voltage vs time')
            ax.legend(loc='upper center', bbox_to_anchor=(1.2, 0.97),
shadow=True, ncol=2)
            plt.xlabel('time [s]')
            plt.ylabel('Voltage [V]')
            plt.show()


    def printdata(Printdict,NumberOfNodes,retime,reval,Nodes):
        if len(Printdict)> 0:
            fp=open(Printdict[0]['filename'],"w+")
            fp.write('time ')
            for i in range(len(Printdict)-1):
                fp.write('%s ' % Printdict[i+1]['NodeName'])
            fp.write('\n')
            for i in range(len(retime)):
                fp.write("%g " % retime[i])
                for j in range(NumberOfNodes):
                    for k in range(len(Printdict)-1):
                        if Printdict[k+1]['NodeName']==Nodes[j]:
```

```
                            fp.write("%g " % reval[j][i])
                    fp.write('\n')
            fp.close()

    def build_SysEqns(SetupDict, SimDict, modeldict):
        DeviceCount=SetupDict['DeviceCount']
        DevType=SetupDict['DevType']
        deltaT=SimDict['deltaT']
        for i in range(DeviceCount):
            if DevType[i]=='resistor':
                build_SysEqn_resistor(i, SetupDict)
            if DevType[i]=='capacitor':
                build_SysEqn_capacitor(deltaT, i, SetupDict, SimDict)
            if DevType[i]=='inductor':
                build_SysEqn_inductor(deltaT, i, SetupDict, SimDict)
            if DevType[i]=='VoltSource':
                build_SysEqn_VSource(i, SetupDict)
            if DevType[i]=='CurrentSource':
                build_SysEqn_ISource(i, SetupDict)
            if DevType[i]=='transistor':
                build_SysEqn_MOS(i, SetupDict, SimDict, modeldict)
            if DevType[i]=='bipolar':
                build_SysEqn_bipolar(i, SetupDict, SimDict, modeldict)

    def update_SysEqns(SimTime, SetupDict, SimDict, modeldict):
        DeviceCount=SetupDict['DeviceCount']
        DevType=SetupDict['DevType']
        deltaT=SimDict['deltaT']
        for i in range(DeviceCount):
            if DevType[i]=='capacitor':
                update_SysEqn_capacitor(deltaT, i, SetupDict, SimDict)
            if DevType[i]=='inductor':
                update_SysEqn_inductor(deltaT, i, SetupDict, SimDict)
            if DevType[i]=='VoltSource':
                update_SysEqn_VSource(i, SimTime, SetupDict)
            if DevType[i]=='CurrentSource':
                update_SysEqn_ISource(i, SimTime, SetupDict)
            if DevType[i]=='transistor':
                update_SysEqn_MOS(i, SetupDict, SimDict, modeldict)
            if DevType[i]=='bipolar':
                    update_SysEqn_bipolar(i, SetupDict, SimDict,
modeldict)

    def build_Jacobian(SetupDict, SimDict, modeldict):
        DeviceCount=SetupDict['DeviceCount']
```

```
      DevType=SetupDict['DevType']
      for i in range(DeviceCount):
          if DevType[i]=='transistor':
              build_Jacobian_MOS(i, SetupDict, SimDict, modeldict)
          if DevType[i]=='bipolar':
              build_Jacobian_bipolar(i, SetupDict, SimDict, modeld
ict)

  def build_Jacobian_HB(SetupDict, Simdict, modeldict):
      DeviceCount=SetupDict['DeviceCount']
      DevType=SetupDict['DevType']
      for i in range(DeviceCount):
          if DevType[i]=='transistor':
                  build_Jacobian_MOS_HB(i, SetupDict, Simdict,
modeldict)

  def update_SysEqns_HB(SetupDict, SimDict, modeldict):
      DeviceCount=SetupDict['DeviceCount']
      DevType=SetupDict['DevType']
      TotalHarmonics=SetupDict['TotalHarmonics']
      for i in range(DeviceCount):
          for row in range(TotalHarmonics):
              if DevType[i]=='transistor':
                  update_SysEqn_MOS_HB(i, row, SetupDict, SimDict,
modeldict)
              if DevType[i]=='bipolar':
                  print('Error: Harmonic Balance for Bipolar Tran
sistors not implemented')
                  sys.exit(0)

  def build_SysEqns_HB(SetupDict, SimDict, modeldict):
      DeviceCount=SetupDict['DeviceCount']
      DevType=SetupDict['DevType']
      TotalHarmonics=SetupDict['TotalHarmonics']
      for i in range(DeviceCount):
          for row in range(TotalHarmonics):
              if DevType[i]=='resistor':
                  build_SysEqn_resistor_HB(i, row, SetupDict)
              if DevType[i] == 'oscfilter':
                  build_SysEqn_oscfilter_HB(i, row, SetupDict)
              if DevType[i]=='capacitor':
                  build_SysEqn_capacitor_HB(i, row, SetupDict,
SimDict)
              if DevType[i]=='inductor':
                  build_SysEqn_inductor_HB(i, row, SetupDict,
```

```
SimDict)
                if DevType[i]=='VoltSource':
                    build_SysEqn_VSource_HB(i, row, SetupDict)
                if DevType[i]=='CurrentSource':
                    build_SysEqn_ISource_HB(i, row, SetupDict)
                if DevType[i]=='transistor':
                    build_SysEqn_MOS_HB(i, row, SetupDict, SimDict,
modeldict)
                if DevType[i]=='bipolar':
                        print('Error: Harmonic Balance for Bipolar
Transistors not implemented')
                    sys.exit(0)

    def build_SysEqn_oscfilter_HB(DeviceNr, row, SetupDict):
        NumberOfNodes=SetupDict['NumberOfNodes']
        TotalHarmonics=SetupDict['TotalHarmonics']
        DevValue=SetupDict['DevValue']
        DevNode1=SetupDict['DevNode1']
        DevNode2=SetupDict['DevNode2']
        Nodes=SetupDict['Nodes']
        STA_matrix=SetupDict['STA_matrix']
        Jacobian_Offset=SetupDict['Jacobian_Offset']
        if row==Jacobian_Offset+1 or row==Jacobian_Offset-1:
            OscFilterValue=DevValue[DeviceNr]
        else:
            OscFilterValue=1e18
        STA_matrix[(NumberOfNodes+DeviceNr)*TotalHarmonics+row][(Num
berOfNodes+DeviceNr)*TotalHarmonics+row]=-OscFilterValue
            STA_matrix[(NumberOfNodes+DeviceNr)*TotalHarmonics+row]
[Nodes.index(DevNode1[DeviceNr])*TotalHarmonics+row]=1
            STA_matrix[Nodes.index(DevNode1[DeviceNr])*TotalHarmonics+
row][(NumberOfNodes+DeviceNr)*TotalHarmonics+row]=1
            STA_matrix[(NumberOfNodes+DeviceNr)*TotalHarmonics+row]
[Nodes.index(DevNode2[DeviceNr])*TotalHarmonics+row]=-1
            STA_matrix[Nodes.index(DevNode2[DeviceNr])*TotalHarmonics+
row][(NumberOfNodes+DeviceNr)*TotalHarmonics+row]=-1

    def build_SysEqn_resistor_HB(DeviceNr, row, SetupDict):
        NumberOfNodes=SetupDict['NumberOfNodes']
        DevValue=SetupDict['DevValue']
        TotalHarmonics=SetupDict['TotalHarmonics']
        DevNode1=SetupDict['DevNode1']
        DevNode2=SetupDict['DevNode2']
        Nodes=SetupDict['Nodes']
        STA_matrix=SetupDict['STA_matrix']
```

```
        STA_matrix[(NumberOfNodes+DeviceNr)*TotalHarmonics+row][(Num
berOfNodes+DeviceNr)*TotalHarmonics+row]=-DevValue[DeviceNr]
            STA_matrix[(NumberOfNodes+DeviceNr)*TotalHarmonics+row]
[Nodes.index(DevNode1[DeviceNr])*TotalHarmonics+row]=1
            STA_matrix[Nodes.index(DevNode1[DeviceNr])*TotalHarmonics+
row][(NumberOfNodes+DeviceNr)*TotalHarmonics+row]=1
            STA_matrix[(NumberOfNodes+DeviceNr)*TotalHarmonics+row]
[Nodes.index(DevNode2[DeviceNr])*TotalHarmonics+row]=-1
            STA_matrix[Nodes.index(DevNode2[DeviceNr])*TotalHarmonics+
row][(NumberOfNodes+DeviceNr)*TotalHarmonics+row]=-1


    def build_SysEqn_capacitor_HB(DeviceNr, row, SetupDict, SimDict):
        NumberOfNodes=SetupDict['NumberOfNodes']
        DevValue=SetupDict['DevValue']
        TotalHarmonics=SetupDict['TotalHarmonics']
        DevNode1=SetupDict['DevNode1']
        DevNode2=SetupDict['DevNode2']
        Nodes=SetupDict['Nodes']
        STA_matrix=SetupDict['STA_matrix']
        omegak=SetupDict['omegak']
        STA_matrix[(NumberOfNodes+DeviceNr)*TotalHarmonics+row][(Num
berOfNodes+DeviceNr)*TotalHarmonics+row]=-1
            STA_matrix[(NumberOfNodes+DeviceNr)*TotalHarmonics+row]
[Nodes.index(DevNode1[DeviceNr])*TotalHarmonics+row]=1j*omegak[ro
w]*DevValue[DeviceNr]
            STA_matrix[Nodes.index(DevNode1[DeviceNr])*TotalHarmonics+
row][(NumberOfNodes+DeviceNr)*TotalHarmonics+row]=1
            STA_matrix[(NumberOfNodes+DeviceNr)*TotalHarmonics+row]
[Nodes.index(DevNode2[DeviceNr])*TotalHarmonics+row]=-1j*omegak[r
ow]*DevValue[DeviceNr]
            STA_matrix[Nodes.index(DevNode2[DeviceNr])*TotalHarmonics+
row][(NumberOfNodes+DeviceNr)*TotalHarmonics+row]=-1


    def build_SysEqn_inductor_HB(DeviceNr, row, SetupDict, SimDict):
        NumberOfNodes=SetupDict['NumberOfNodes']
        DevValue=SetupDict['DevValue']
        TotalHarmonics=SetupDict['TotalHarmonics']
        DevNode1=SetupDict['DevNode1']
        DevNode2=SetupDict['DevNode2']
        Nodes=SetupDict['Nodes']
        STA_matrix=SetupDict['STA_matrix']
        omegak=SetupDict['omegak']
        STA_matrix[(NumberOfNodes+DeviceNr)*TotalHarmonics+row][(Num
berOfNodes+DeviceNr)*TotalHarmonics+row]=-1j*omegak[row]*DevValue
[DeviceNr]
```

```
        STA_matrix[(NumberOfNodes+DeviceNr)*TotalHarmonics+row]
[Nodes.index(DevNode1[DeviceNr])*TotalHarmonics+row]=1
        STA_matrix[Nodes.index(DevNode1[DeviceNr])*TotalHarmonics+
row][(NumberOfNodes+DeviceNr)*TotalHarmonics+row]=1
        STA_matrix[(NumberOfNodes+DeviceNr)*TotalHarmonics+row]
[Nodes.index(DevNode2[DeviceNr])*TotalHarmonics+row]=-1
        STA_matrix[Nodes.index(DevNode2[DeviceNr])*TotalHarmonics+
row][(NumberOfNodes+DeviceNr)*TotalHarmonics+row]=-1

    def build_SysEqn_VSource_HB(DeviceNr, row, SetupDict):
        NumberOfNodes=SetupDict['NumberOfNodes']
        DevValue=SetupDict['DevValue']
        TotalHarmonics=SetupDict['TotalHarmonics']
        DevNode1=SetupDict['DevNode1']
        DevNode2=SetupDict['DevNode2']
        DevLabel=SetupDict['DevLabel']
        Nodes=SetupDict['Nodes']
        STA_matrix=SetupDict['STA_matrix']
        STA_rhs=SetupDict['STA_rhs']
        Jacobian_Offset=SetupDict['Jacobian_Offset']
        STA_matrix[(NumberOfNodes+DeviceNr)*TotalHarmonics+row][(Num
berOfNodes+DeviceNr)*TotalHarmonics+row]=0
        if DevNode1[DeviceNr] != '0' :
            STA_matrix[(NumberOfNodes+DeviceNr)*TotalHarmonics+row]
[Nodes.index(DevNode1[DeviceNr])*TotalHarmonics+row]=1
            STA_matrix[Nodes.index(DevNode1[DeviceNr])*TotalHarmoni
cs+row][(NumberOfNodes+DeviceNr)*TotalHarmonics+row]=1
        if DevNode2[DeviceNr] != '0' :
            STA_matrix[(NumberOfNodes+DeviceNr)*TotalHarmonics+row]
[Nodes.index(DevNode2[DeviceNr])*TotalHarmonics+row]=-1
            STA_matrix[Nodes.index(DevNode2[DeviceNr])*TotalHarmoni
cs+row][(NumberOfNodes+DeviceNr)*TotalHarmonics+row]=-1
        if DevLabel[DeviceNr] != 'vinp' and DevLabel[DeviceNr]
!= 'vinn':
            STA_rhs[(NumberOfNodes+DeviceNr)*TotalHarmonics+row]=ge
tSourceValue(DevValue[DeviceNr],0)*(row==Jacobian_Offset)
        if(DevLabel[DeviceNr] == 'vinp'):
            STA_rhs[(NumberOfNodes+DeviceNr)*TotalHarmonics+row]=.5
*((row==Jacobian_Offset+1)+(row==Jacobian_Offset-1))
        if(DevLabel[DeviceNr] == 'vinn'):
            STA_rhs[(NumberOfNodes+DeviceNr)*TotalHarmonics+
row]=-.5*((row==Jacobian_Offset+1)+(row==Jacobian_Offset-1))
        if(DevLabel[DeviceNr] == 'vin'):
            STA_rhs[(NumberOfNodes+DeviceNr)*TotalHarmonics+
row]=-.02*((row==Jacobian_Offset+1)+(row==Jacobian_
```

```
Offset-1))+0.2*(row==Jacobian_Offset)

  def build_SysEqn_ISource_HB(DeviceNr, row, SetupDict):
     NumberOfNodes=SetupDict['NumberOfNodes']
     DevValue=SetupDict['DevValue']
     TotalHarmonics=SetupDict['TotalHarmonics']
     DevNode1=SetupDict['DevNode1']
     DevNode2=SetupDict['DevNode2']
     Nodes=SetupDict['Nodes']
     STA_matrix=SetupDict['STA_matrix']
     STA_rhs=SetupDict['STA_rhs']
     Jacobian_Offset=SetupDict['Jacobian_Offset']
     STA_matrix[(NumberOfNodes+DeviceNr)*TotalHarmonics+row][(Num
berOfNodes+DeviceNr)*TotalHarmonics+row]=1
     STA_rhs[(NumberOfNodes+DeviceNr)*TotalHarmonics+row]=getSou
rceValue(DevValue[DeviceNr],0)*(row==Jacobian_Offset)
     if DevNode1[DeviceNr] != '0' and DevNode2[DeviceNr]!='0':
         STA_matrix[Nodes.index(DevNode1[DeviceNr])*TotalHarmoni
cs+row][(NumberOfNodes+DeviceNr)*TotalHarmonics+row]=1
         STA_matrix[Nodes.index(DevNode2[DeviceNr])*TotalHarmoni
cs+row][(NumberOfNodes+DeviceNr)*TotalHarmonics+row]=-1
     elif DevNode2[DeviceNr] != '0' :
         STA_matrix[Nodes.index(DevNode2[DeviceNr])*TotalHarmoni
cs+row][(NumberOfNodes+DeviceNr)*TotalHarmonics+row]=-1
     elif DevNode1[DeviceNr] != '0' :
         STA_matrix[Nodes.index(DevNode1[DeviceNr])*TotalHarmoni
cs+row][(NumberOfNodes+DeviceNr)*TotalHarmonics+row]=1

    def build_SysEqn_MOS_HB(DeviceNr, row, SetupDict, SimDict,
modeldict):
       NumberOfNodes=SetupDict['NumberOfNodes']
       TotalHarmonics=SetupDict['TotalHarmonics']
       NSamples=SetupDict['NSamples']
       DevNode1=SetupDict['DevNode1']
       DevNode2=SetupDict['DevNode2']
       DevNode3=SetupDict['DevNode3']
       DevModel=SetupDict['DevModel']
       Nodes=SetupDict['Nodes']
       sol=SimDict['sol']
       STA_matrix=SetupDict['STA_matrix']
       STA_nonlinear=SetupDict['STA_nonlinear']
     lambdaT=findParameter(modeldict,DevModel[DeviceNr],'lambdaT')
       VT=findParameter(modeldict,DevModel[DeviceNr],'VT')
       K=findParameter(modeldict,DevModel[DeviceNr],'K')
       Vg=[0 for i in range(TotalHarmonics)]
```

```
        Vs=[0 for i in range(TotalHarmonics)]
        Vd=[0 for i in range(TotalHarmonics)]
        TransistorOutputTime=[0 for i in range(NSamples)]
        TransistorOutputFreq=[0 for i in range(TotalHarmonics)]
    #     if row==0:
        for j in range(TotalHarmonics):
                Vg[j]=sol[Nodes.index(DevNode2[DeviceNr])*TotalHa
rmonics+j]
                Vs[j]=sol[Nodes.index(DevNode3[DeviceNr])*TotalHa
rmonics+j]
                Vd[j]=sol[Nodes.index(DevNode1[DeviceNr])*TotalHa
rmonics+j]
        TransistorOutputTime=TransistorModel(idft(Vg,TotalHarmonics
),idft(Vs,TotalHarmonics),idft(Vd,TotalHarmonics),NSamples,    K,
VT, lambdaT)
        TransistorOutputFreq=dft(TransistorOutputTime,NSamples)
        STA_matrix[(NumberOfNodes+DeviceNr)*TotalHarmonics+row][(Num
berOfNodes+DeviceNr)*TotalHarmonics+row]=-1
         STA_matrix[Nodes.index(DevNode1[DeviceNr])*TotalHarmonics+
row][(NumberOfNodes+DeviceNr)*TotalHarmonics+row]=1
         STA_matrix[Nodes.index(DevNode3[DeviceNr])*TotalHarmonics+
row][(NumberOfNodes+DeviceNr)*TotalHarmonics+row]=-1
        STA_nonlinear[(NumberOfNodes+DeviceNr)*TotalHarmonics+row]=
TransistorOutputFreq[row]

    def update_SysEqn_MOS_HB(DeviceNr, row, SetupDict, SimDict,
modeldict):
        NumberOfNodes=SetupDict['NumberOfNodes']
        TotalHarmonics=SetupDict['TotalHarmonics']
        NSamples=SetupDict['NSamples']
        DevNode1=SetupDict['DevNode1']
        DevNode2=SetupDict['DevNode2']
        DevNode3=SetupDict['DevNode3']
        DevModel=SetupDict['DevModel']
        Nodes=SetupDict['Nodes']
        sol=SimDict['sol']
        STA_nonlinear=SetupDict['STA_nonlinear']
       lambdaT=findParameter(modeldict,DevModel[DeviceNr],'lambdaT')
        VT=findParameter(modeldict,DevModel[DeviceNr],'VT')
        K=findParameter(modeldict,DevModel[DeviceNr],'K')
        STA_nonlinear=SetupDict['STA_nonlinear']
        Vg=[0 for i in range(TotalHarmonics)]
        Vs=[0 for i in range(TotalHarmonics)]
        Vd=[0 for i in range(TotalHarmonics)]
        TransistorOutputTime=[0 for i in range(NSamples)]
```

```
        TransistorOutputFreq=[0 for i in range(TotalHarmonics)]
#       if row==0: This worked on the toplevel but not here ...
        for j in range(TotalHarmonics):
                Vg[j]=sol[Nodes.index(DevNode2[DeviceNr])*TotalHa
rmonics+j]
                Vs[j]=sol[Nodes.index(DevNode3[DeviceNr])*TotalHa
rmonics+j]
                Vd[j]=sol[Nodes.index(DevNode1[DeviceNr])*TotalHa
rmonics+j]
        TransistorOutputTime=TransistorModel(idft(Vg,TotalHarmonics
),idft(Vs,TotalHarmonics),idft(Vd,TotalHarmonics),NSamples,     K,
VT, lambdaT)
        TransistorOutputFreq=dft(TransistorOutputTime,NSamples)
        STA_nonlinear[(NumberOfNodes+DeviceNr)*TotalHarmonics+row]=
TransistorOutputFreq[row]

  def build_SysEqn_resistor(DeviceNr, SetupDict):
      NumberOfNodes=SetupDict['NumberOfNodes']
      DevValue=SetupDict['DevValue']
      DevNode1=SetupDict['DevNode1']
      DevNode2=SetupDict['DevNode2']
      Nodes=SetupDict['Nodes']
      STA_matrix=SetupDict['STA_matrix']
                                STA_matrix[NumberOfNodes+DeviceNr]
[NumberOfNodes+DeviceNr]=-DevValue[DeviceNr]
                        STA_matrix[NumberOfNodes+DeviceNr][Nodes.
index(DevNode1[DeviceNr])]=1
                        STA_matrix[Nodes.index(DevNode1[DeviceNr])]
[NumberOfNodes+DeviceNr]=1
                        STA_matrix[NumberOfNodes+DeviceNr][Nodes.
index(DevNode2[DeviceNr])]=-1
                        STA_matrix[Nodes.index(DevNode2[DeviceNr])]
[NumberOfNodes+DeviceNr]=-1

  def build_SysEqn_capacitor(deltaT, DeviceNr, SetupDict, SimDict):
      NumberOfNodes=SetupDict['NumberOfNodes']
      DevValue=SetupDict['DevValue']
      DevNode1=SetupDict['DevNode1']
      DevNode2=SetupDict['DevNode2']
      Nodes=SetupDict['Nodes']
      STA_matrix=SetupDict['STA_matrix']
      STA_rhs=SetupDict['STA_rhs']
      method=SetupDict['method']
      sol=SimDict['sol']
      solm1=SimDict['solm1']
```

```python
    deltaT=SimDict['deltaT']
    STA_matrix[NumberOfNodes+DeviceNr][NumberOfNodes+DeviceNr]=1
                    STA_matrix[Nodes.index(DevNode1[DeviceNr])]
[NumberOfNodes+DeviceNr]=1
                    STA_matrix[Nodes.index(DevNode2[DeviceNr])]
[NumberOfNodes+DeviceNr]=-1
        if method=='trap':
                        STA_matrix[NumberOfNodes+DeviceNr][Nodes.
index(DevNode1[DeviceNr])]=-2.0*DevValue[DeviceNr]/deltaT
            STA_matrix[NumberOfNodes+DeviceNr][Nodes.index(DevNode2
[DeviceNr])]=2.0*DevValue[DeviceNr]/deltaT
              STA_rhs[NumberOfNodes+DeviceNr]=-2*DevValue[DeviceNr]/
deltaT*(sol[Nodes.index(DevNode1[DeviceNr])]-sol[Nodes.
index(DevNode2[DeviceNr])])-sol[NumberOfNodes+DeviceNr]
        elif method=='gear2':
                        STA_matrix[NumberOfNodes+DeviceNr][Nodes.
index(DevNode1[DeviceNr])]=-3.0/2.0*DevValue[DeviceNr]/deltaT
            STA_matrix[NumberOfNodes+DeviceNr][Nodes.index(DevNode2
[DeviceNr])]=3.0/2.0*DevValue[DeviceNr]/deltaT
                STA_rhs[NumberOfNodes+DeviceNr]=DevValue[DeviceNr]/
deltaT*(-2*(sol[Nodes.index(DevNode1[DeviceNr])]-sol[Nodes.index(
DevNode2[DeviceNr])])+1/2*(solm1[Nodes.
index(DevNode1[DeviceNr])]-solm1[Nodes.
index(DevNode2[DeviceNr])]) )
        elif method=='be':
                        STA_matrix[NumberOfNodes+DeviceNr][Nodes.
index(DevNode1[DeviceNr])]=-1.0*DevValue[DeviceNr]/deltaT
            STA_matrix[NumberOfNodes+DeviceNr][Nodes.index(DevNode2
[DeviceNr])]=1.0*DevValue[DeviceNr]/deltaT
                STA_rhs[NumberOfNodes+DeviceNr]=-DevValue[DeviceNr]/
deltaT*(sol[Nodes.index(DevNode1[DeviceNr])]-sol[Nodes.
index(DevNode2[DeviceNr])])
        else:
            print('Warning: unknown integration method',method)

    def update_SysEqn_capacitor(deltaT, DeviceNr, SetupDict,
SimDict):
        NumberOfNodes=SetupDict['NumberOfNodes']
        DevValue=SetupDict['DevValue']
        DevNode1=SetupDict['DevNode1']
        DevNode2=SetupDict['DevNode2']
        Nodes=SetupDict['Nodes']
        STA_matrix=SetupDict['STA_matrix']
        STA_rhs=SetupDict['STA_rhs']
        method=SetupDict['method']
```

```python
        sol=SimDict['sol']
        solm1=SimDict['solm1']
        if method=='trap':
                        STA_matrix[NumberOfNodes+DeviceNr][Nodes.
index(DevNode1[DeviceNr])]=-2.0*DevValue[DeviceNr]/deltaT
            STA_matrix[NumberOfNodes+DeviceNr][Nodes.index(DevNode2
[DeviceNr])]=2.0*DevValue[DeviceNr]/deltaT
             STA_rhs[NumberOfNodes+DeviceNr]=-2*DevValue[DeviceNr]/
deltaT*(sol[Nodes.index(DevNode1[DeviceNr])]-sol[Nodes.
index(DevNode2[DeviceNr])])-sol[NumberOfNodes+DeviceNr]
        elif method=='gear2':
                        STA_matrix[NumberOfNodes+DeviceNr][Nodes.
index(DevNode1[DeviceNr])]=-3.0/2.0*DevValue[DeviceNr]/deltaT
            STA_matrix[NumberOfNodes+DeviceNr][Nodes.index(DevNode2
[DeviceNr])]=3.0/2.0*DevValue[DeviceNr]/deltaT
             STA_rhs[NumberOfNodes+DeviceNr]=DevValue[DeviceNr]/
deltaT*(-2*(sol[Nodes.index(DevNode1[DeviceNr])]-sol[Nodes.index(
D e v N o d e 2 [ D e v i c e N r ] ) ] ) + 1 / 2 * ( s o l m 1 [ N o d e s .
i n d e x ( D e v N o d e 1 [ D e v i c e N r ] ) ] - s o l m 1 [ N o d e s .
index(DevNode2[DeviceNr])]) )
        elif method=='be':
                        STA_matrix[NumberOfNodes+DeviceNr][Nodes.
index(DevNode1[DeviceNr])]=-1.0*DevValue[DeviceNr]/deltaT
            STA_matrix[NumberOfNodes+DeviceNr][Nodes.index(DevNode2
[DeviceNr])]=1.0*DevValue[DeviceNr]/deltaT
             STA_rhs[NumberOfNodes+DeviceNr]=-DevValue[DeviceNr]/
deltaT*(sol[Nodes.index(DevNode1[DeviceNr])]-sol[Nodes.
index(DevNode2[DeviceNr])])
        else:
            print('Warning: unknown integration method',method)

    def build_SysEqn_inductor(deltaT, DeviceNr, SetupDict, SimDict):
        NumberOfNodes=SetupDict['NumberOfNodes']
        DevValue=SetupDict['DevValue']
        DevNode1=SetupDict['DevNode1']
        DevNode2=SetupDict['DevNode2']
        Nodes=SetupDict['Nodes']
        STA_matrix=SetupDict['STA_matrix']
        STA_rhs=SetupDict['STA_rhs']
        method=SetupDict['method']
        sol=SimDict['sol']
        solm1=SimDict['solm1']
        STA_matrix[NumberOfNodes+DeviceNr][NumberOfNodes+DeviceNr]=1
        STA_matrix[Nodes.index(DevNode1[DeviceNr])][NumberOfNodes+De
viceNr]=1
```

```
                     STA_matrix[Nodes.index(DevNode2[DeviceNr])]
[NumberOfNodes+DeviceNr]=-1
        if method=='trap':
                        STA_matrix[NumberOfNodes+DeviceNr][Nodes.
index(DevNode1[DeviceNr])]=-deltaT/DevValue[DeviceNr]/2
             STA_matrix[NumberOfNodes+DeviceNr][Nodes.index(DevNode2
[DeviceNr])]=deltaT/DevValue[DeviceNr]/2
             STA_rhs[NumberOfNodes+DeviceNr]=sol[NumberOfNodes+Devic
eNr]+deltaT*(sol[Nodes.index(DevNode1[DeviceNr])]-sol[Nodes.
index(DevNode2[DeviceNr])])/(2*DevValue[DeviceNr])
        elif method=='gear2':
                        STA_matrix[NumberOfNodes+DeviceNr][Nodes.
index(DevNode1[DeviceNr])]=-2/3*deltaT/DevValue[DeviceNr]
             STA_matrix[NumberOfNodes+DeviceNr][Nodes.index(DevNode2
[DeviceNr])]=2/3*deltaT/DevValue[DeviceNr]
             STA_rhs[NumberOfNodes+DeviceNr]=4/3*sol[NumberOfNodes+D
eviceNr]-1/3*solm1[NumberOfNodes+DeviceNr]
        elif method=='be':
                        STA_matrix[NumberOfNodes+DeviceNr][Nodes.
index(DevNode1[DeviceNr])]=-deltaT/DevValue[DeviceNr]
             STA_matrix[NumberOfNodes+DeviceNr][Nodes.index(DevNode2
[DeviceNr])]=deltaT/DevValue[DeviceNr]
                  STA_rhs[NumberOfNodes+DeviceNr]=sol[NumberOfNodes
+DeviceNr]
        else:
            print('Warning: unknown integration method',method)

  def update_SysEqn_inductor(deltaT, DeviceNr, SetupDict, SimDict):
        NumberOfNodes=SetupDict['NumberOfNodes']
        DevValue=SetupDict['DevValue']
        DevNode1=SetupDict['DevNode1']
        DevNode2=SetupDict['DevNode2']
        Nodes=SetupDict['Nodes']
        STA_matrix=SetupDict['STA_matrix']
        STA_rhs=SetupDict['STA_rhs']
        method=SetupDict['method']
        sol=SimDict['sol']
        solm1=SimDict['solm1']
        deltaT=SimDict['deltaT']
        if method=='trap':
                        STA_matrix[NumberOfNodes+DeviceNr][Nodes.
index(DevNode1[DeviceNr])]=-deltaT/DevValue[DeviceNr]/2
             STA_matrix[NumberOfNodes+DeviceNr][Nodes.index(DevNode2
[DeviceNr])]=deltaT/DevValue[DeviceNr]/2
             STA_rhs[NumberOfNodes+DeviceNr]=sol[NumberOfNodes+Devic
```

```
eNr]+deltaT*(sol[Nodes.index(DevNode1[DeviceNr])]-sol[Nodes.
index(DevNode2[DeviceNr])])/(2*DevValue[DeviceNr])
      elif method=='gear2':
                          STA_matrix[NumberOfNodes+DeviceNr][Nodes.
index(DevNode1[DeviceNr])]=-2/3*deltaT/DevValue[DeviceNr]
          STA_matrix[NumberOfNodes+DeviceNr][Nodes.index(DevNode2
[DeviceNr])]=2/3*deltaT/DevValue[DeviceNr]
          STA_rhs[NumberOfNodes+DeviceNr]=4/3*sol[NumberOfNodes+D
eviceNr]-1/3*solm1[NumberOfNodes+DeviceNr]
      elif method=='be':
                          STA_matrix[NumberOfNodes+DeviceNr][Nodes.
index(DevNode1[DeviceNr])]=-deltaT/DevValue[DeviceNr]
          STA_matrix[NumberOfNodes+DeviceNr][Nodes.index(DevNode2
[DeviceNr])]=deltaT/DevValue[DeviceNr]
              STA_rhs[NumberOfNodes+DeviceNr]=sol[NumberOfNodes
+DeviceNr]
      else:
          print('Warning: unknown integration method',method)

  def build_SysEqn_VSource(DeviceNr, SetupDict):
      NumberOfNodes=SetupDict['NumberOfNodes']
      DevValue=SetupDict['DevValue']
      DevNode1=SetupDict['DevNode1']
      DevNode2=SetupDict['DevNode2']
      Nodes=SetupDict['Nodes']
      STA_matrix=SetupDict['STA_matrix']
      STA_rhs=SetupDict['STA_rhs']
      if DevNode1[DeviceNr] != '0' :
                          STA_matrix[NumberOfNodes+DeviceNr][Nodes.
index(DevNode1[DeviceNr])]=1
                      STA_matrix[Nodes.index(DevNode1[DeviceNr])]
[NumberOfNodes+DeviceNr]=1
      if DevNode2[DeviceNr] != '0' :
                          STA_matrix[NumberOfNodes+DeviceNr][Nodes.
index(DevNode2[DeviceNr])]=-1
                      STA_matrix[Nodes.index(DevNode2[DeviceNr])]
[NumberOfNodes+DeviceNr]=-1
      STA_matrix[NumberOfNodes+DeviceNr][NumberOfNodes+DeviceNr]=0
      STA_rhs[NumberOfNodes+DeviceNr]=getSourceValue(DevValue[Devi
ceNr],0)

  def update_SysEqn_VSource(DeviceNr, SimTime, SetupDict):
      NumberOfNodes=SetupDict['NumberOfNodes']
      DevValue=SetupDict['DevValue']
      STA_rhs=SetupDict['STA_rhs']
```

```python
        STA_rhs[NumberOfNodes+DeviceNr]=getSourceValue(DevValue[Dev
iceNr],SimTime)

    def build_SysEqn_ISource(DeviceNr, SetupDict):
        NumberOfNodes=SetupDict['NumberOfNodes']
        DevValue=SetupDict['DevValue']
        DevNode1=SetupDict['DevNode1']
        DevNode2=SetupDict['DevNode2']
        Nodes=SetupDict['Nodes']
        STA_matrix=SetupDict['STA_matrix']
        STA_rhs=SetupDict['STA_rhs']
        if DevNode1[DeviceNr] != '0' :
                        STA_matrix[NumberOfNodes+DeviceNr][Nodes.
index(DevNode1[DeviceNr])]=0
                        STA_matrix[Nodes.index(DevNode1[DeviceNr])]
[NumberOfNodes+DeviceNr]=0
        if DevNode2[DeviceNr] != '0' :
                        STA_matrix[NumberOfNodes+DeviceNr][Nodes.
index(DevNode2[DeviceNr])]=0
                        STA_matrix[Nodes.index(DevNode2[DeviceNr])]
[NumberOfNodes+DeviceNr]=0
                        STA_matrix[NumberOfNodes+DeviceNr][NumberOf
Nodes+DeviceNr]=1
                        STA_rhs[NumberOfNodes+DeviceNr]=getSourceVa
lue(DevValue[DeviceNr],0)
        if DevNode1[DeviceNr] != '0' and DevNode2[DeviceNr]!='0':
                        STA_matrix[Nodes.index(DevNode1[DeviceNr])]
[NumberOfNodes+DeviceNr]=1
                        STA_matrix[Nodes.index(DevNode2[DeviceNr])]
[NumberOfNodes+DeviceNr]=-1
        elif DevNode2[DeviceNr] != '0' :
                        STA_matrix[Nodes.index(DevNode2[DeviceNr])]
[NumberOfNodes+DeviceNr]=-1
        elif DevNode1[DeviceNr] != '0' :
                        STA_matrix[Nodes.index(DevNode1[DeviceNr])]
[NumberOfNodes+DeviceNr]=1

    def update_SysEqn_ISource(DeviceNr, SimTime, SetupDict):
        NumberOfNodes=SetupDict['NumberOfNodes']
        DevValue=SetupDict['DevValue']
        STA_rhs=SetupDict['STA_rhs']
        STA_rhs[NumberOfNodes+DeviceNr]=getSourceValue(DevValue[Dev
iceNr],SimTime)

    def build_SysEqn_MOS(DeviceNr, SetupDict, SimDict, modeldict):
```

```
        NumberOfNodes=SetupDict['NumberOfNodes']
        DevValue=SetupDict['DevValue']
        DevNode1=SetupDict['DevNode1']
        DevNode2=SetupDict['DevNode2']
        DevNode3=SetupDict['DevNode3']
        DevModel=SetupDict['DevModel']
        Nodes=SetupDict['Nodes']
        STA_matrix=SetupDict['STA_matrix']
        STA_nonlinear=SetupDict['STA_nonlinear']
        sol=SimDict['sol']
      lambdaT=findParameter(modeldict,DevModel[DeviceNr],'lambdaT')
        VT=findParameter(modeldict,DevModel[DeviceNr],'VT')
        STA_matrix[NumberOfNodes+DeviceNr][NumberOfNodes+DeviceNr]=
DevValue[DeviceNr]
        STA_matrix[NumberOfNodes+DeviceNr][Nodes.index(DevNode1[Dev
iceNr])]=0
        STA_matrix[Nodes.index(DevNode1[DeviceNr])][NumberOfNodes+D
eviceNr]=1
        STA_matrix[NumberOfNodes+DeviceNr][Nodes.index(DevNode3[Dev
iceNr])]=0
        STA_matrix[Nodes.index(DevNode3[DeviceNr])][NumberOfNodes+D
eviceNr]=-1
        VD=sol[Nodes.index(DevNode1[DeviceNr])]
        VG=sol[Nodes.index(DevNode2[DeviceNr])]
        VS=sol[Nodes.index(DevNode3[DeviceNr])]
        Vgs=VG-VS
        Vds=VD-VS
        if DevModel[DeviceNr][0]=='p':
            Vds=-Vds
            Vgs=-Vgs
        if Vds < Vgs-VT :
                        STA_nonlinear[NumberOfNodes+DeviceNr]=2*((
Vgs-VT)*Vds-0.5*Vds**2)
        else :
                        STA_nonlinear[NumberOfNodes+DeviceNr]=(
Vgs-VT)**2*(1+lambdaT*Vds)

   def update_SysEqn_MOS(DeviceNr, SetupDict, SimDict, modeldict):
        NumberOfNodes=SetupDict['NumberOfNodes']
        DevNode1=SetupDict['DevNode1']
        DevNode2=SetupDict['DevNode2']
        DevNode3=SetupDict['DevNode3']
        DevModel=SetupDict['DevModel']
        Nodes=SetupDict['Nodes']
        STA_nonlinear=SetupDict['STA_nonlinear']
```

```
    soltemp=SimDict['soltemp']
  lambdaT=findParameter(modeldict,DevModel[DeviceNr],'lambdaT')
  VT=findParameter(modeldict,DevModel[DeviceNr],'VT')
  VD=soltemp[Nodes.index(DevNode1[DeviceNr])]
  VG=soltemp[Nodes.index(DevNode2[DeviceNr])]
  VS=soltemp[Nodes.index(DevNode3[DeviceNr])]
  Vgs=VG-VS
  Vds=VD-VS
  if DevModel[DeviceNr][0]=='p':
      Vds=-Vds
      Vgs=-Vgs
  if Vgs<VT:
      STA_nonlinear[NumberOfNodes+DeviceNr]=1e-5
  elif Vds < Vgs-VT:
      STA_nonlinear[NumberOfNodes+DeviceNr]=2*((Vgs-VT)*Vds-
0.5*Vds**2)
  else :
      STA_nonlinear[NumberOfNodes+DeviceNr]=(Vgs-VT)**2*(1+
lambdaT*Vds)


def build_Jacobian_MOS(DeviceNr, SetupDict, SimDict, modeldict):
    NumberOfNodes=SetupDict['NumberOfNodes']
    DevNode1=SetupDict['DevNode1']
    DevNode2=SetupDict['DevNode2']
    DevNode3=SetupDict['DevNode3']
    DevModel=SetupDict['DevModel']
    DevValue=SetupDict['DevValue']
    Nodes=SetupDict['Nodes']
    Jacobian=SetupDict['Jacobian']
    soltemp=SimDict['soltemp']
  lambdaT=findParameter(modeldict,DevModel[DeviceNr],'lambdaT')
  VT=findParameter(modeldict,DevModel[DeviceNr],'VT')
  Jacobian[NumberOfNodes+DeviceNr][NumberOfNodes+DeviceNr]=De
vValue[DeviceNr] # due to derfivative leading to double gain
  VD=soltemp[Nodes.index(DevNode1[DeviceNr])]
  VG=soltemp[Nodes.index(DevNode2[DeviceNr])]
  VS=soltemp[Nodes.index(DevNode3[DeviceNr])]
  Vgs=VG-VS
  Vds=VD-VS
  Vgd=VG-VD
  if DevModel[DeviceNr][0]=='p':
      PFET=-1
      Vgs=-Vgs
      Vds=-Vds
      Vgd=-Vgd
```

```
        else:
            PFET=1
        if Vgs<VT :
            Jacobian[NumberOfNodes+DeviceNr][Nodes.index(DevNode1[D
eviceNr])]=TINY
            Jacobian[NumberOfNodes+DeviceNr][Nodes.index(DevNode2[D
eviceNr])]=TINY
            Jacobian[NumberOfNodes+DeviceNr][Nodes.index(DevNode3[D
eviceNr])]=-2*TINY
            Jacobian[Nodes.index(DevNode1[DeviceNr])][NumberOfNodes
+DeviceNr]=1
            Jacobian[Nodes.index(DevNode3[DeviceNr])][NumberOfNodes
+DeviceNr]=-1
        elif Vds <= Vgs-VT:
            Jacobian[NumberOfNodes+DeviceNr][Nodes.index(DevNode1[D
eviceNr])]=PFET*2*(Vgd-VT)
            Jacobian[NumberOfNodes+DeviceNr][Nodes.index(DevNode2[D
eviceNr])]=PFET*2*Vds
            Jacobian[NumberOfNodes+DeviceNr][Nodes.index(DevNode3[D
eviceNr])]=-PFET*2*(Vgs-VT)
            Jacobian[Nodes.index(DevNode1[DeviceNr])][NumberOfNodes
+DeviceNr]=1
            Jacobian[Nodes.index(DevNode3[DeviceNr])][NumberOfNodes
+DeviceNr]=-1
        else :
            Jacobian[NumberOfNodes+DeviceNr][Nodes.index(DevNode1[D
eviceNr])]=PFET*lambdaT*(Vgs-VT)**2
            Jacobian[NumberOfNodes+DeviceNr][Nodes.index(DevNode2[D
eviceNr])]=PFET*2*(Vgs-VT)*(1+lambdaT*Vds)
            Jacobian[NumberOfNodes+DeviceNr][Nodes.index(DevNode3[D
eviceNr])]=PFET*(-2*(Vgs-VT)*(1+lambdaT*Vds)-lambdaT*(Vgs-VT)**2)
            Jacobian[Nodes.index(DevNode1[DeviceNr])][NumberOfNodes
+DeviceNr]=1
            Jacobian[Nodes.index(DevNode3[DeviceNr])][NumberOfNodes
+DeviceNr]=-1

  def build_SysEqn_bipolar(DeviceNr, SetupDict, SimDict, modeldict):
      NumberOfNodes=SetupDict['NumberOfNodes']
      DevValue=SetupDict['DevValue']
      DevNode1=SetupDict['DevNode1']
      DevNode2=SetupDict['DevNode2']
      DevNode3=SetupDict['DevNode3']
      DevModel=SetupDict['DevModel']
      Nodes=SetupDict['Nodes']
      STA_matrix=SetupDict['STA_matrix']
```

```
        STA_nonlinear=SetupDict['STA_nonlinear']
        soltemp=SimDict['soltemp']
        VEarly=findParameter(modeldict,DevModel[DeviceNr],'Early')
        STA_matrix[NumberOfNodes+DeviceNr][NumberOfNodes+DeviceNr]=
DevValue[DeviceNr]
        STA_matrix[NumberOfNodes+DeviceNr][Nodes.index(DevNode1[Dev
iceNr])]=0
        STA_matrix[Nodes.index(DevNode1[DeviceNr])][NumberOfNodes+D
eviceNr]=1
        STA_matrix[NumberOfNodes+DeviceNr][Nodes.index(DevNode3[Dev
iceNr])]=0
        STA_matrix[Nodes.index(DevNode3[DeviceNr])][NumberOfNodes+D
eviceNr]=-1
        VC=soltemp[Nodes.index(DevNode1[DeviceNr])]
        VB=soltemp[Nodes.index(DevNode2[DeviceNr])]
        VE=soltemp[Nodes.index(DevNode3[DeviceNr])]
        Vbe=VB-VE
        Vce=VC-VE
        if Vbe < 0 :
            STA_nonlinear[NumberOfNodes+DeviceNr]=0
        else :
                STA_nonlinear[NumberOfNodes+DeviceNr]=math.exp(Vbe/
Vthermal)*(1+Vce/VEarly)


    def   build_Jacobian_MOS_HB(DeviceNr,   SetupDict,   SimDict,
modeldict):
        NumberOfNodes=SetupDict['NumberOfNodes']
        TotalHarmonics=SetupDict['TotalHarmonics']
        Jacobian_Offset=SetupDict['Jacobian_Offset']
        NSamples=SetupDict['NSamples']
        DevNode1=SetupDict['DevNode1']
        DevNode2=SetupDict['DevNode2']
        DevNode3=SetupDict['DevNode3']
        DevModel=SetupDict['DevModel']
        Nodes=SetupDict['Nodes']
        Jacobian=SetupDict['Jacobian']
        sol=SimDict['sol']
        lambdaT=findParameter(modeldict,DevModel[DeviceNr],'lambdaT')
        VT=findParameter(modeldict,DevModel[DeviceNr],'VT')
        K=findParameter(modeldict,DevModel[DeviceNr],'K')
        Vg=[0 for i in range(TotalHarmonics)]
        Vs=[0 for i in range(TotalHarmonics)]
        Vd=[0 for i in range(TotalHarmonics)]
        gm=[0 for i in range(TotalHarmonics)]
        for j in range(TotalHarmonics):
```

```
                Vg[j]=sol[Nodes.index(DevNode2[DeviceNr])*TotalHa
rmonics+j]
                Vs[j]=sol[Nodes.index(DevNode3[DeviceNr])*TotalHa
rmonics+j]
                Vd[j]=sol[Nodes.index(DevNode1[DeviceNr])*TotalHa
rmonics+j]
        gm=TransistorModel_dIdVg(idft(Vg,TotalHarmonics),idft(Vs,To
talHarmonics),idft(Vd,TotalHarmonics),NSamples,K, VT, lambdaT)
        go=TransistorModel_dIdVd(idft(Vg,TotalHarmonics),idft(Vs,To
talHarmonics),idft(Vd,TotalHarmonics),NSamples,K, VT, lambdaT)
        Jlkm=dft(gm,NSamples)
        Jlko=dft(go,NSamples)
        for j in range(TotalHarmonics):
    #                    Jlk[j]=2*K*(j==Jacobian_Offset)+0j
            Jlkm[j]=Jlkm[j]+TINY*(j==Jacobian_Offset)
            Jlko[j]=Jlko[j]+TINY*(j==Jacobian_Offset)
        for row in range(TotalHarmonics):
            Jacobian[(NumberOfNodes+DeviceNr)*TotalHarmonics+row]
[(NumberOfNodes+DeviceNr)*TotalHarmonics+row]=-1
            Jacobian[Nodes.index(DevNode1[DeviceNr])*TotalHarmonics
+row][(NumberOfNodes+DeviceNr)*TotalHarmonics+row]=1
            Jacobian[Nodes.index(DevNode3[DeviceNr])*TotalHarmonics
+row][(NumberOfNodes+DeviceNr)*TotalHarmonics+row]=-1
            for col in range(TotalHarmonics):
                            if(col-row+Jacobian_Offset>=0   and
col-row+Jacobian_Offset<TotalHarmonics):
                    Jacobian[(NumberOfNodes+DeviceNr)*TotalHarmonic
s+row][Nodes.index(DevNode1[DeviceNr])*TotalHarmonics+col]=Jlko[
col-row+Jacobian_Offset]
                    Jacobian[(NumberOfNodes+DeviceNr)*TotalHarmonic
s+row][Nodes.index(DevNode2[DeviceNr])*TotalHarmonics+col]=Jlkm[
col-row+Jacobian_Offset]
                    Jacobian[(NumberOfNodes+DeviceNr)*TotalHarmonic
s+row][Nodes.index(DevNode3[DeviceNr])*TotalHarmonics+
col]=-Jlkm[col-row+Jacobian_Offset]-Jlko[col-row+Jacobian_Offset]


    def   update_SysEqn_bipolar(DeviceNr,  SetupDict,  SimDict,
modeldict):
        NumberOfNodes=SetupDict['NumberOfNodes']
        DevNode1=SetupDict['DevNode1']
        DevNode2=SetupDict['DevNode2']
        DevNode3=SetupDict['DevNode3']
        DevModel=SetupDict['DevModel']
        Nodes=SetupDict['Nodes']
        STA_nonlinear=SetupDict['STA_nonlinear']
```

```
        soltemp=SimDict['sol']
        VEarly=findParameter(modeldict,DevModel[DeviceNr],'Early')
        VC=soltemp[Nodes.index(DevNode1[DeviceNr])]
        VB=soltemp[Nodes.index(DevNode2[DeviceNr])]
        VE=soltemp[Nodes.index(DevNode3[DeviceNr])]
        Vbe=VB-VE
        Vce=VC-VE
        if Vbe<0:
            STA_nonlinear[NumberOfNodes+DeviceNr]=0
        else :
                STA_nonlinear[NumberOfNodes+DeviceNr]=math.exp(Vbe/
Vthermal)*(1+Vce/VEarly)


    def   build_Jacobian_bipolar(DeviceNr,   SetupDict,   SimDict,
modeldict):
        NumberOfNodes=SetupDict['NumberOfNodes']
        DevValue=SetupDict['DevValue']
        DevNode1=SetupDict['DevNode1']
        DevNode2=SetupDict['DevNode2']
        DevNode3=SetupDict['DevNode3']
        DevModel=SetupDict['DevModel']
        Nodes=SetupDict['Nodes']
        Jacobian=SetupDict['Jacobian']
        soltemp=SimDict['soltemp']
        VEarly=findParameter(modeldict,DevModel[DeviceNr],'Early')
        Jacobian[NumberOfNodes+DeviceNr][NumberOfNodes+DeviceNr]=De
vValue[DeviceNr] # due to derfivative leading to double gain
        VC=soltemp[Nodes.index(DevNode1[DeviceNr])]
        VB=soltemp[Nodes.index(DevNode2[DeviceNr])]
        VE=soltemp[Nodes.index(DevNode3[DeviceNr])]
        Vbe=VB-VE
        Vce=VC-VE
        Vbc=VB-VC
        if Vbe<=0 :
            Jacobian[NumberOfNodes+DeviceNr][Nodes.index(DevNode1[D
eviceNr])]=1e-5
            Jacobian[NumberOfNodes+DeviceNr][Nodes.index(DevNode2[D
eviceNr])]=1e-5
            Jacobian[NumberOfNodes+DeviceNr][Nodes.index(DevNode3[D
eviceNr])]=-1e-5
                        Jacobian[Nodes.index(DevNode1[DeviceNr])]
[NumberOfNodes+DeviceNr]=1
                        Jacobian[Nodes.index(DevNode3[DeviceNr])]
[NumberOfNodes+DeviceNr]=-1
        else :
```

```
                        Jacobian[NumberOfNodes+DeviceNr][Nodes.
index(DevNode1[DeviceNr])]=math.exp(Vbe/Vthermal)/VEarly
                        Jacobian[NumberOfNodes+DeviceNr][Nodes.
index(DevNode2[DeviceNr])]=math.exp(Vbe/Vthermal)*(1+Vce/VEarly)/
Vthermal
                        Jacobian[NumberOfNodes+DeviceNr][Nodes.
index(DevNode3[DeviceNr])]=(-math.exp(Vbe/Vthermal)/VEarly-math.
exp(Vbe/Vthermal)*(1+Vce/VEarly)/Vthermal)
                        Jacobian[Nodes.index(DevNode1[DeviceNr])]
[NumberOfNodes+DeviceNr]=1
                        Jacobian[Nodes.index(DevNode3[DeviceNr])]
[NumberOfNodes+DeviceNr]=-1

    def DidResidueConverge(SetupDict, SimDict ):#NumberOfNodes,
NumberOfCurrents, STA_matrix, sol, f, reltol, iabstol):
        NumberOfNodes=SetupDict['NumberOfNodes']
        NumberOfCurrents=SetupDict['NumberOfCurrents']
        STA_matrix=SetupDict['STA_matrix']
        soltemp=SimDict['soltemp']
        f=SimDict['f']
        reltol=SetupDict['reltol']
        iabstol=SetupDict['iabstol']
        ResidueConverged=True
        node=0
        while ResidueConverged and node<NumberOfNodes:
  # Let us find the maximum current going into node, Nodes[node]
            MaxCurrent=0
            for current in range(NumberOfCurrents):
                MaxCurrent=max(MaxCurrent,abs(STA_matrix[node][Numb
erOfNodes+current]*(soltemp[NumberOfNodes+current])))
            if f[node] > reltol*MaxCurrent+iabstol:
                ResidueConverged=False
            node=node+1
        return ResidueConverged

    def DidUpdateConverge(SetupDict, SimDict ):#NumberOfNodes,
NumberOfCurrents, Jacobian, sol, f, reltol, vabstol, PointLocal,
GlobalTruncation):
        NumberOfNodes=SetupDict['NumberOfNodes']
        Jacobian=SetupDict['Jacobian']
        soltemp=SimDict['soltemp']
        f=SimDict['f']
        reltol=SetupDict['reltol']
        vabstol=SetupDict['vabstol']
        PointLocal=SetupDict['PointLocal']
```

```python
        GlobalTruncation=SetupDict['GlobalTruncation']
        vkmax=SetupDict['vkmax']
    SolutionCorrection=numpy.matmul(numpy.linalg.inv(Jacobian),f)
        UpdateConverged=True
        if PointLocal:
            for node in range(NumberOfNodes):
                vkmax=max(abs(soltemp[node]),abs(soltemp[node]-Solutio
nCorrection[node]))
                if abs(SolutionCorrection[node])>vkmax*reltol+vabstol:
                    UpdateConverged=False
        elif GlobalTruncation:
            for node in range(NumberOfNodes):
                if abs(SolutionCorrection[node])>vkmax*reltol+vabstol:
                    UpdateConverged=False
        else:
            print('Error: Unknown truncation error')
            sys.exit()
        return UpdateConverged


    def DidLTEConverge(SetupDict, SimDict, iteration, LTEIter,
NewtonConverged, timeVector, SimTime, SolutionCorrection):
        NumberOfNodes=SetupDict['NumberOfNodes']
        sol=SimDict['sol']
        soltemp=SimDict['soltemp']
        solm1=SimDict['solm1']
        solm2=SimDict['solm2']
        PointLocal=SetupDict['PointLocal']
        GlobalTruncation=SetupDict['GlobalTruncation']
        lteratio=SetupDict['lteratio']
        vkmax=SetupDict['vkmax']
        reltol=SetupDict['reltol']
        vabstol=SetupDict['vabstol']
        LTEConverged=True
        MaxLTERatio=0
        PredMatrix=[[0 for i in range(2)] for j in range(2)]
        Predrhs=[0 for i in range(2)]
        if iteration>200 and NewtonConverged:
            LTEIter=LTEIter+1
            for i in range(NumberOfNodes):

tau1=(timeVector[iteration-2]-timeVector[iteration-3])

tau2=(timeVector[iteration-1]-timeVector[iteration-3])
                PredMatrix[0][0]=tau2
                PredMatrix[0][1]=tau2*tau2
```

```
                PredMatrix[1][0]=tau1
                PredMatrix[1][1]=tau1*tau1
                Predrhs[0]=sol[i]-solm2[i]
                Predrhs[1]=solm1[i]-solm2[i]
                 Predsol=numpy.matmul(numpy.linalg.inv(PredMatrix),
Predrhs)
                vpred=solm2[i]+Predsol[0]*(SimTime-timeVector[itera
tion-3])+Predsol[1]*(SimTime-timeVector[iteration-3])*(SimTime-
timeVector[iteration-3])
                if PointLocal:
                    for node in range(NumberOfNodes):
                        vkmax=max(abs(soltemp[node]),abs(soltemp[n
ode]-SolutionCorrection[node]))

                        if abs(vpred-soltemp[i])> lteratio*(vkmax*r
eltol+vabstol):
                            LTEConverged=False
                        else:
                            MaxLTERatio=max(abs(vpred-soltemp[i])/
(vkmax*reltol+vabstol),MaxLTERatio)
                elif GlobalTruncation:
                    for node in range(NumberOfNodes):
                        if abs(vpred-soltemp[i])> lteratio*(vkmax*r
eltol+vabstol):
                            LTEConverged=False

                else:
                    print('Error: Unknown truncation error')
                    sys.exit()
        return LTEConverged, MaxLTERatio


    def  UpdateTimeStep(SetupDict,   SimDict,   LTEConverged,
NewtonConverged, val, iteration, NewtonIter, MaxLTERatio, timeVec-
tor, SimTime):
        MatrixSize=SetupDict['MatrixSize']
        FixedTimeStep=SetupDict['FixedTimeStep']
        MaxTimeStep=SetupDict['MaxTimeStep']
        soltemp=SimDict['soltemp']
        sol=SimDict['sol']
        solm1=SimDict['solm1']
        solm2=SimDict['solm2']
        deltaT=SimDict['deltaT']
        ThreeLevelStep=SimDict['ThreeLevelStep']
        Converged=NewtonConverged and LTEConverged
        if Converged:
```

```
            if not NewtonConverged:# We can have a trap for just
skipping the time step reduction or if we do fixed time step just
skip the point
                print('Some trouble converging, skipping')
                NewtonConverged=True
        for i in range(MatrixSize):
            solm2[i]=solm1[i]
            solm1[i]=sol[i]
        if iteration > -1:
            for j in range(MatrixSize):
                val[j][iteration]=soltemp[j]
            timeVector[iteration]=SimTime
        iteration=iteration+1
        if not FixedTimeStep:
            if ThreeLevelStep:
                if 0.9<MaxLTERatio<1.0:
                    deltaT=deltaT/1.1
                else:
                    if MaxLTERatio<0.1:
                        deltaT=1.01*deltaT
            else:
                deltaT=1.001*deltaT
            deltaT=min(deltaT,MaxTimeStep)
#               else:
#                   print('Unchanging')
    else:
        if FixedTimeStep:
            if iteration>100:
                if not NewtonConverged:
                    print('Newton failed to converge',NewtonIter)
                if not LTEConverged:
                    print('LTE failed to converge',NewtonIter)
                sys.exit()
        else:
            SimTime=max(SimTime-deltaT,0)
            deltaT=deltaT/1.1
    return deltaT, iteration, SimTime, Converged

def dft(Samples,N):
    sol=[0+0j for i in range(N+1) ]
    y=fft(Samples)
    for i in range(int(N/2)):
        sol[i]=y[int(N/2)+i]/N
    for i in range(int(N/2)+1):
        sol[int(N/2)+i]=y[i]/N
```

```
        # DC??
    return sol

def idft(Vk, N):
    y=[0 for i in range(N-1)]
    for i in range(int(N/2)+1):
        y[i]=Vk[int(N/2)+i]*(N-1)
    for i in range(int(N/2)-1):
        y[int(N/2)+i+1]=Vk[i+1]*(N-1)
    return ifft(y,N-1)


def TransistorModel(Vg, Vs, Vd, N, K, VT, lambdaT):
    Id=[0 for i in range(N)]
    for i in range(N):
        Vgs=Vg[i]-Vs[i]
        Vds=Vd[i]-Vs[i]
        if Vgs<VT:
            Id[i]=K*0
        elif Vds < Vgs-VT :
            Id[i]=2*K*((Vgs-VT)*Vds-0.5*Vds**2)
        else:
            Id[i]=K*(Vgs-VT)**2*(1+lambdaT*Vds)
    return Id
def TransistorModel_dIdVg(Vg, Vs, Vd, N, K, VT, lambdaT):
    gmm=[0 for i in range(N)]
    for i in range(N):
        Vgs=Vg[i]-Vs[i]
        Vds=Vd[i]-Vs[i]
        if Vgs<VT:
            gmm[i]=K*0
        elif Vds < Vgs-VT :
            gmm[i]=K*2*Vds
        else:
            gmm[i]=2*K*(Vgs-VT)*(1+lambdaT*Vds)
    return gmm
def TransistorModel_dIdVd(Vg, Vs, Vd, N, K, VT, lambdaT):
    goo=[0 for i in range(N)]
    for i in range(N):
        Vgs=Vg[i]-Vs[i]
        Vds=Vd[i]-Vs[i]
        Vgd=Vg[i]-Vd[i]
        if Vgs<VT:
            goo[i]=K*0
        elif Vds < Vgs-VT :
            goo[i]=2*K*(Vgd-VT)
```

```
        else:
            goo[i]=K*lambdaT*(Vgs-VT)**2
      return goo
  def TimeDerivative(inp,wk,N):
      deriv=[0+0j for i in range(N)]
      for i in range(N):
          deriv[i]=inp[i]*wk[i]*1j
      return deriv
```

## A.3   Models.txt

The transistors in the main text have models defined by the following file:

```
nch  K=1e-2 VT=400e-3 lambdaT=1e-5
pch  K=-1e-2 VT=400e-3 lambdaT=1e-5
nch1 K=1e-3 VT=400e-6 lambdaT=1e-5
pch1 K=-1e-3 VT=400e-6 lambdaT=1e-5
nch2 K=1e-3 VT=400e-3 lambdaT=1e-2
pch2 K=-1e-3 VT=400e-3 lambdaT=1e-2
nchp1 K=1e-3 VT=3 lambdaT=1e-2
pchp1 K=-1e-3 VT=3 lambdaT=1e-2
nch3 K=3e-3 VT=400e-6 lambdaT=1e-5
pch3 K=-3e-3 VT=400e-6 lambdaT=1e-5
npn  K=1e-15 Early=100 dum=1e-5
```
**models.txt**

# Index

© Springer Nature Switzerland AG 2021            401
M. Sahrling, *Analog Circuit Simulators for Integrated Circuit Designers*,
https://doi.org/10.1007/978-3-030-64206-8

Printed in the United States
by Baker & Taylor Publisher Services